The Foundations of Geometry

Gerard A. Venema

Department of Mathematics and Statistics
Calvin College

PEARSON

Prentice
Hall

Upper Saddle River, New Jersey 07458

Library of Congress Cataloging-in-Publication Data
Venema, Gerard A.
The foundations of geometry.
1st ed./Gerard A. Venema
    p. cm.
Includes bibliographical references index.
ISBN 0-13-143700-3
1. Geometry
CIP data available

Executive Acquisitions Editor: *George Lobell*
Editor-in-Chief: *Sally Yagan*
Production Editor: *Debbie Ryan*
Senior Managing Editor: *Linda Mihatov Behrens*
Assistant Managing Editor: *Bayani Mendoza de Leon*
Executive Managing Editor: *Kathleen Schiaparelli*
Assistant Manufacturing Manager/Buyer: *Michael Bell*
Marketing Manager: *Halee Dinsey*
Marketing Assistant: *Joon Moon*
Cover Designer: *Bruce Kenselaar*
Art Director: *Jayne Conte*
Director of Creative Services: *Paul Belfanti*
Editorial Assistant: *Jennifer Urban*
Cover Image:
Vasarely, Victor (1908–1997) Gestalt-Zoeld, 1976. Acrylic on Canvas, 240 × 225 cm. Private Collection, Paris, France. © 2004 Artists Rights Society (ARS), New York / ADAGP, Paris Erich Lessing / Art Resource, NY.

Text Credits:
Postulates (pp 868-869), The University of Chicago School Mathematics Project Geometry © 2002 by Pearson Education, Inc., publishing as Pearson Prentice Hall. Used by permission.
Statements of the Postulates, GEOMETRY, STUDENT TEXT PART 1, School Mathematics Study Group, © 1960,1961 Yale University Press.
Statement of Hilberts Axioms. Foundations of Geometry by David Hilbert. © 1971 by Open Court Publishing Company, a division of Carus Publishing.

 © 2006 Pearson Education, Inc.
Pearson Prentice Hall
Pearson Education, Inc.
Upper Saddle River, New Jersey 07458

Pearson Prentice Hall^TM is a trademark of Pearson Education, Inc.
Printed in the United States of America
10 9 8 7 6 5 4 3 2
ISBN 0-13-143700-3

Pearson Education LTD., *London*
Pearson Education Australia PTY, Limited, *Sydney*
Pearson Education Singapore, Pte. Ltd
Pearson Education North Asia Ltd, *Hong Kong*
Pearson Education Canada, Ltd., *Toronto*
Pearson Educacion de Mexico, S.A. de C.V.
Pearson Education - Japan, *Tokyo*
Pearson Education Malaysia, Pte. Ltd

To Pat, Sara, Emily, and Dan.

Contents

Preface

This is a textbook for an undergraduate course in axiomatic geometry. The course is aimed at mathematics majors who have completed the calculus sequence and perhaps a first course in linear algebra, but who have not yet encountered such upper-level mathematics courses as real analysis and abstract algebra. The course will normally be taken by students who are at the junior or senior level, but well-prepared sophomores also benefit from the course.

THE FOUNDATIONS OF GEOMETRY

The primary goal of the course is to study the foundations of geometry. That means returning to the beginnings of geometry, exposing exactly what is assumed there, and building the entire subject on those foundations. Such careful attention to the foundations has a long tradition in geometry, going back more than two thousand years to Euclid and the ancient Greeks. Over the years since Euclid wrote his famous *Elements*, there have been profound changes in the way in which the foundations have been understood. Most of those changes have been byproducts of efforts to understand the true place of Euclid's parallel postulate in the foundations, so the parallel postulate is one of the primary emphases of this book.

ORGANIZATION OF THE BOOK

The course begins with a quick look at Euclid's *Elements*, and Euclid's system of organization is used as motivation for the concept of an axiomatic system. A system of axioms for geometry is then carefully laid out. The axioms used here are based on the real numbers, in the spirit of Birkhoff, and their statements have been kept as close to those in contemporary high school textbooks as is possible.

After the axioms have been stated and certain foundational issues faced, neutral geometry, in which no parallel postulate is assumed, is extensively explored. Next both Euclidean and hyperbolic geometries are investigated from an axiomatic point of view. In order to get as quickly as possible to some of the interesting results of non-Euclidean geometry, the first part of the book focuses exclusively on results regarding lines, parallelism, and triangles. Only after those subjects have been treated separately in neutral, Euclidean, and hyperbolic geometries are results on area, circles, and construction introduced. While the treatment of these subjects does not exactly follow Euclid, it roughly parallels Euclid in the sense that Euclid collected most of his propositions about area in Book II and most of his propositions about circles in Books III and IV. The three chapters covering area, circles, and construction complete the coverage of the major theorems of Books I through VI of the *Elements*.

Next, the more modern notion of a transformation is introduced and some of the standard results regarding transformations of the plane are explored. A complete proof of the classification of the rigid motions of both the Euclidean and hyperbolic planes is included. There is a discussion of how the foundations of geometry can be reorganized to reflect the transformational point of view (as is common practice in contemporary high school geometry textbooks). Specifically, it is possible to replace the Side-Angle-Side Postulate with a postulate that asserts the existence of certain reflections.

The standard models for hyperbolic geometry are carefully constructed and the results of the chapter on transformations are used to verify their properties. The chapter on models can be relatively short because all the hard technical work involved in the constructions is done in the preceding chapter. The final chapter includes a study of some of the polygonal models that have recently been developed to help students understand what it means to say that hyperbolic space is negatively curved. The book ends with a discussion of the practical significance of non-Euclidean geometry and a brief look at the geometry of the real world.

PROOFS

An important secondary goal of the course is to teach the art of writing proofs. There is a growing recognition of the need for a course in which mathematics students learn how to write good proofs. Such a course should serve as a bridge between the lower-level mathematics courses, which are largely technique oriented, and the upper-level courses, which tend to be much more conceptual. This book uses geometry as the vehicle for helping students to write and appreciate proofs. The ability to write proofs is a skill that can only be acquired by actually practicing it, so most of the material on writing proofs is integrated into the course and the attention to proof permeates the entire text. This means that the book can also be used in classes where the students already have experience writing proofs; despite the emphasis on writing proofs, the book is still primarily a geometry text.

Having the geometry course serve as the introduction to proof represents a return to tradition in that the course in Euclidean geometry has for thousands of years been seen as the standard introduction to logic, rigor, and proof in mathematics. Using the geometry course this way makes historical sense because the axiomatic method was first introduced in geometry and geometry remains the branch of mathematics in which that method has had its greatest success. While proof and logical deduction are still emphasized in the standards for high school mathematics, most high school students no longer take a full-year course devoted exclusively to geometry in which there is a sustained introduction to proof. This makes it more important than ever that we teach a good college-level geometry course to all mathematics students. By doing so we can return geometry to its place as the subject in which students first learn to appreciate the importance of clearly spelling out assumptions and deducing results from those assumptions via careful logical reasoning.

The emphasis on proofs makes the course into a do-it-yourself course in that the reader will be asked to supply proofs for many of the key theorems. Students who diligently work the exercises come away from the course with a sense that they have an unusually deep understanding of the material. In this way the student should not only learn the mechanics of good proof writing style but should also come to more fully appreciate the important role proof plays in an understanding of mathematics.

NATIONAL STANDARDS

A third major goal of the course is to implement the recommendations in the recent report on "The Mathematical Education of Teachers" (MET) [12]. Those recommendations, in turn, are based on the "Principles and Standards for School Mathematics" of the National Council of Teachers of Mathematics [54]. The basic recommendation is that "Prospective teachers need mathematics courses that develop a deep understanding of the mathematics they will teach" [12, Part I, page 7]. This course is designed to do precisely that in the area of geometry. A second basic recommendation in MET is that courses for prospective mathematics teachers should make explicit connections with high school mathematics. Again, this book attempts to implement that recommendation in geometry. The goal is to implement the basic recommendations of MET, not necessarily to cover every geometric topic that future teachers need to see; some geometric topics will be included in other courses, such as linear algebra.

An example of the way in which connections with high school geometry have influenced the design of the course is the choice of the axioms that are used as the starting point. The axioms on which the development of the geometry in the text is based are almost exactly those that are used in high school textbooks. While most high school textbooks still include an axiomatic treatment of geometry, there is no standard set of axioms that is common to all high school geometry courses. Therefore, various axiom systems are considered in the text and the merits and advantages of each are discussed. The axioms on which the course is ultimately based are as close as possible to those in the Geometry textbook in the University of Chicago School Mathematics Project (UCSMP) series [71], a textbook that is in wide use at the present time. One of the main goals of the course is to help preservice teachers understand the logical foundations of the geometry course they will teach and that can best be accomplished in the context of axioms that are like the ones they will encounter later in the classroom. There are many other connections with high school geometry that are brought in as the course progresses.

Some of the newer high school mathematics curricula present mathematics in an integrated way that emphasizes connections between the branches of mathematics. There is no separate course in geometry, but rather a geometry thread is woven into all the high school mathematics courses. In order to teach such a course well, the teacher herself needs to have an understanding of the structure of geometry as a coherent subject. This book is intended to provide such an understanding.

One of the recurring themes in "The Mathematical Education of Teachers" [12] is the recommendation that prospective teachers must acquire an understanding of high school mathematics that goes well beyond that of a typical high school graduate. One way in which such understanding of geometry is often measured is in terms of the van Hiele model of geometric thought. This model is described in

Appendix D. The goal of most high school courses is to develop student thinking to Level 3. A goal of this course is to bring students to Level 4 (or to Level 5, depending on whether the first level is numbered 0 or 1). It is recognized, however, that not all students entering the course are already at Level 3 and so the early part of the course is designed to ensure that students are brought to that level first.

HISTORICAL AND PHILOSOPHICAL PERSPECTIVE

A final goal of the course is to present a historical perspective on geometry. Geometry is a dynamic subject that has changed over time. It is a part of human culture that was created and developed by people who were very much products of their time and place. The foundations of geometry have been challenged and reformulated over the years, and beliefs about the relationship between geometry and the real world have been challenged as well.

The material in the book is presented in a way that is sensitive to such historical and philosophical issues. This does not mean that the material is presented in a strictly historical order or that there are lengthy historical discussions but rather that geometry is presented in such a way that the reader can understand and appreciate the historical development of the subject and so that it would be natural to investigate the history of the subject while learning it. Many chapters include suggested readings on the history of geometry that can be used to enrich the course.

Throughout the book there are references to philosophical issues that arise in geometry. For example, one question that naturally occurs to anyone studying non-Euclidean geometry is this: What is the connection between the abstract entities that are studied in a course on the foundations of geometry and properties of physical space? The book does not present dogmatic answers to such questions, but instead simply raises them in an effort to promote student thinking. The hope is that this will serve to counter the common perception that mathematics is a subject in which every question has a single correct answer and in which there is no room for creative ideas or opinions.

TECHNOLOGY

In recent years powerful computer software has been developed that can be used to explore geometry. The study of geometry from this book can be greatly enhanced by such dynamic software and the reader is encouraged to find appropriate ways in which to incorporate this technology into the geometry course. While software can enrich the experience of learning geometry from this book, its use is not required. The book can be read and studied quite profitably without it.

There are several commercially available pieces of software designed for use in a course such as this, and any one of them will serve the purpose. *Geometer's Sketchpad*™ (Key Curriculum Press) is widely used and readily available; it is probably the natural choice if you are just starting out. *Cabri Geometry*™ (Texas Instruments) is less commonly used in college-level courses, but it is also completely adequate. It has some predefined tools, such as an inversion tool and a test for collinearity, that are not included in Sketchpad. *Cinderella*™ (Springer-Verlag) is a newer piece of software that is also very good. It has the advantage that it allows diagrams to be drawn in all three two-dimensional geometries: Euclidean, hyperbolic, and spherical. Another advantage is that it allows diagrams to be easily exported as Java applets. A program called NonEuclid is freely available on the World Wide Web and it can be used to enhance the non-Euclidean geometry in the course. New software is being produced all the time, so you may find that other products are available to you.

This is a course in the foundations of axiomatic geometry, and software will necessarily play a more limited role in such a course than it might in other kinds of geometry courses. Nonetheless, there is an appropriate role for software in a course such as this and the author hopes that the book will demonstrate that. There is no reason for those who love the proofs of Euclid to resist the use of technology. After all, Euclid himself made use of the limited technology that was available to him, namely the compass and straightedge. In the same way we can make good use of modern technology in our study of geometry. It is especially important that future high school teachers learn to understand and appreciate the appropriate use of technology.

In the first part of the course (Chapters 1 through 6), the objective is to carefully expose all the assumptions that form the foundations of geometry and to understand for ourselves how the basic results of geometry are built on those foundations. For most users the software is a black box in the sense that we either don't know what assumptions are built into it or we have only the authors' description of what went into the software. As a result, software is of limited use in this part of the course and it will not be mentioned explicitly in the first six chapters of the book. But you should be using it to draw diagrams and to experiment with what happens when you vary the data in the theorems. During that phase of the course the main function of the software is to illustrate the relationships being studied.

It is in the second half of the course that the software comes into its own. Computer software is ideal for experimenting, exploring, and discovering new relationships. In order to illustrate that, several of the later chapters include sections in which the software is used to explore ideas that go beyond those that are presented in detail and to discover new relationships. In particular, there are such exploration sections in the chapters on Euclidean geometry and circles. The entire chapter on constructions is written as an exploration with only a limited number of proofs or hints provided in the text.

This book is not tied to any one piece of software. It is assumed that you will learn to perform the basic operations from the documentation that comes with your software. There are relatively few operations that are required and the necessary operations are listed at the beginning of each exploration section. An instructor who wishes to teach a course in which the software plays a larger role should probably supplement this text with a lab manual such as *Geometry in Action* by Clark Kimberling [39].

DESIGNING A COURSE

A full-year course will cover essentially all the material in the text. There can be some variation based on instructor and student interest, but most or all of every chapter should be included.

An instructor teaching a one-semester or one-quarter course will be forced to pick and choose. It is important that this be done carefully so that the course reaches some of the interesting and useful material that is to be found in the second half of the book.

Chapters 1 and 2 set the stage for what comes later, so they should definitely be covered in some way. But they can be treated lightly in order to free up time for other things. Topics from Chapters 3 and 4 should be covered as needed, depending on student background. The basic coverage of geometry begins with Chapter 5. Chapters 5 and 6 form the heart of a one-semester course. Those chapters should be included in any course taught from the book. At least some of Chapter 7 should also be included in any course. Starting with Chapter 8, the chapters are largely independent of each other and an instructor can select material from them based on the interests and needs of the class.

Several sample course outlines are included below. Many other variations are possible. It should be noted that the suggested outlines are ambitious and many instructors will choose to cover less.

The suggested course for future high school teachers includes just a brief introduction to each of the topics in later chapters. The idea is that the course should provide enough background so that students can study these topics in more depth later if they need to. It is hoped that this book can later serve as a valuable reference for those who go on to teach geometry courses. The book could be a resource that provides information about rigorous treatments of such topics as parallel lines, area, circles, constructions, transformations, and so on, that are part of the high school curriculum.

POSSIBLE ONE-SEMESTER COURSE OUTLINES

A course emphasizing Euclidean Geometry

Chapter	Topic	Number of weeks
1–4	Preliminaries	≤ 2
5	Axioms	2
6	Neutral geometry	3
7	Euclidean geometry	2
9	Area	1–2
10	Circles	1–2
12	Transformations	2

A course emphasizing non-Euclidean Geometry

Chapter	Topic	Number of weeks
1–4	Preliminaries	$\leqslant 2$
5	Axioms	2
6	Neutral geometry	3
7	Euclidean geometry	1
8	Hyperbolic geometry	2
9	Area	1–2
13	Models	1–2
14	Geometry of space	1

A course for future high school teachers

Chapter	Topic	Number of weeks
1–4	Preliminaries	$\leqslant 2$
5	Axioms	2
6	Neutral geometry	3
7	Euclidean geometry	1
8	Hyperbolic geometry	1
9	Area	1
10	Circles	1
12	Transformations	1
13	Models	1
14	Geometry of space	1

SUPPLEMENTS

There is an *Instructors' Manual* that contains more information about how to teach from the book. In particular, it includes suggestions about what a one-semester course should cover from each chapter and what can be omitted. There is also a table that shows dependencies among the various sections. The *Instructors' Manual* contains solutions to all of the exercises. Instructors should contact their local Prentice Hall sales representative or the Prentice Hall offices in Upper Saddle River, New Jersey, to obtain a copy.

The author maintains a website at

    http://calvin.edu/~venema/geometrybook.html.

The website contains supplementary material, more information about the book, and errata.

ACKNOWLEDGMENTS

I wish to thank all those who helped me put this book together. There are too many such people for me to name them all, but there are a few whose contributions should definitely be acknowledged.

Calvin College students Ginni Baker and Tom Clark both made special contributions to the project. During the summer of 2002 Ginni made many of the diagrams and constructed the first index; the following summer she helped to construct many of the solutions that are included in the *Instructors' Manual*. Tom did an honors project in mathematics education in which he experimented with the polygonal models of the hyperbolic plane that are described in Chapter 14 and helped to devise ways in which they can be brought into the classroom. Many of the diagrams at the end of Chapter 14 are based on the diagrams in his senior thesis [14].

The members of the Calvin College Foundations of Geometry classes in the spring terms of 2001, 2002, 2003, and 2004 showed great patience with early versions of these notes and encouraged me to continue to develop them. Each student contributed in some way to the project, in many cases by asking pertinent questions. My colleague Mike Stob shared his expertise in mathematical logic and his input informed the discussion of consistency in Chapter 13.

Thanks to Mario Lopez of the University of Denver and Ric Ancel of the University of Wisconsin Milwaukee. During the summer of 2003 Mario Lopez and his students read an early version of the manuscript. They offered many helpful suggestions for improving the presentation of the material. In the fall of 2003 Mario Lopez and Ric Ancel both field tested a preliminary version of the book, and their reactions have been most useful.

I wish to thank reviewers Lawrence Cannon, Manfred Dugas, Patricia Hale, Todd Moyer, F. Alexander Norman, David Royster, Franz Rothe, Peter F. Stiller, and Anke Walz for their thoughtful and helpful comments on the manuscript.

I would like to thank George Lobell for his help and encouragement during the preparation of this manuscript, and I thank the entire Prentice Hall production staff for their assistance in getting the manuscript into final form. I especially want to thank Adam Lewenberg for all his help with the LATEX formatting.

Finally, I want to thank Pat and Dan for their loving support throughout this entire process.

Gerard A. Venema
venema@calvin.edu

CHAPTER 1

Euclid's *Elements*

1.1 GEOMETRY BEFORE EUCLID

Geometry is an ancient subject. Its roots go back thousands of years and geometric thinking of one kind or another is found in nearly every human culture. The subject as we know it emerged more than 4000 years ago in Mesopotamia, Egypt, India, and China.

Because the Nile River annually flooded vast areas of land and obliterated property lines, surveying and measuring were important to the ancient Egyptians. This practical interest probably motivated their study of geometry. Egyptian geometry was mostly an empirical science, consisting of many rule-of-thumb procedures that were arrived at through experimentation, observation, and trial and error. Most formulas were approximate ones that appeared to work, or at least gave answers that were close enough for practical purposes. But the ancient Egyptians were also aware of more general principles, such as special cases of the Pythagorean Theorem and formulas for volumes.

The ancient Mesopotamians, or Babylonians, apparently had an even more advanced understanding of geometry. They knew the Pythagorean Theorem long before Pythagoras. They discovered some of the area-based proofs of the theorem that will be discussed in Chapter 9, and apparently knew a general method that generates all triples of integers that are lengths of sides of right triangles. In India, ancient texts apply the Pythagorean Theorem to geometric problems associated with the design of structures. It appears that the Pythagorean Theorem was also discovered in China at roughly the same time.

About 2500 years ago there was a profound change in the way geometry was practiced: Greek mathematicians introduced abstraction, logical deduction, and proof into geometry. They insisted that geometric results be based on logical reasoning from first principles. In theory this made the results of geometry exact, certain, and undeniable, rather than just likely or approximate. It also took geometry

1

out of the realm of everyday experience and made it a subject that studies abstract entities. Since the purpose of this course is to study the logical foundations of geometry, it is natural that we should start with the geometry of the Greeks.

The process of introducing logic into geometry apparently began with Thales of Miletus around 600 B.C. and culminated in the work of Euclid of Alexandria in approximately 300 B.C. Euclid is the most famous of the Greek geometers and his name is still universally associated with the geometry that is studied in schools today. Most of the ideas that are included in what we call "Euclidean Geometry" probably did not originate with Euclid himself; rather, Euclid's contribution was to organize and present the results of Greek geometry in a logical and coherent way. He published his results in a series of thirteen books known as the *Elements*. We begin our study of geometry by examining those *Elements* because they set the agenda for geometry for the next two millennia and more.

1.2 THE LOGICAL STRUCTURE OF EUCLID'S *ELEMENTS*

Euclid's *Elements* are organized according to strict logical rules. Euclid begins each book with a list of definitions of the technical terms he will use in that book. In Book I he next states five "postulates" and five "common notions." These are assumptions that are meant to be accepted without proof. Both the postulates and common notions are basic statements whose truth should be apparent to any reasonable person. They are the starting point for what follows. Euclid recognized that it is not possible to prove everything, that he had to start somewhere, but he attempted to be clear about exactly what his assumptions were.

Most of Euclid's postulates are simple statements of intuitively obvious and undeniable facts about space. For example, Postulate I asserts that it is possible to draw a straight line through any two given points. Postulate II says that a straight line segment can be extended to a longer segment. Postulate III says that it is possible to construct a circle with any given center and radius. Traditionally these first three postulates have been associated with the tools that are used to implement them on a piece of paper. The first two postulates say that two different uses of a straight edge are allowed: A straight edge can be used to draw a line segment through any two points or to extend a given line segment to a longer one. The third postulate says that a compass can be used to construct a circle with a given center and radius. Thus the first three postulates simply permit the familiar straightedge and compass constructions of high school geometry.

The fourth postulate asserts that all right angles are congruent ("equal" in Euclid's terminology). The fifth postulate makes a more subtle and complicated assertion about two lines that are cut by a transversal. These last two postulates are the two technical facts about geometry that Euclid needs in his proofs.

The common notions are also intuitively obvious facts that Euclid plans to use in his development of geometry. The difference between the common notions and the postulates is that the common notions are not peculiar to geometry but are common to all branches of mathematics. They are everyday, common-sense assumptions. Most spell out properties of equality, at least as Euclid used the term *equal*.

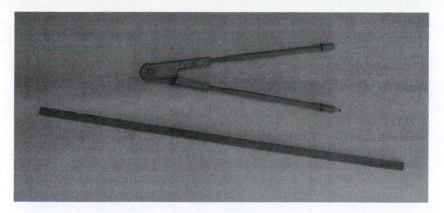

FIGURE 1.1: Euclid's tools: a compass and a straightedge

The largest part of each Book of the *Elements* consists of propositions and proofs. These too are organized in a strict, logical progression. The first proposition is proved using only the postulates, Proposition 2 is proved using only the postulates and Proposition 1, and so on. Thus the entire edifice is built on just the postulates and common notions; once these are granted, everything else follows logically and inevitably from them. What is astonishing is the number and variety of propositions that can be deduced from so few assumptions.

1.3   THE HISTORICAL IMPORTANCE OF EUCLID'S *ELEMENTS*

It is nearly impossible to overstate the importance of Euclid's *Elements* in the development of mathematics and human culture generally. From the time they were written, the *Elements* have been held up as the standard of the way in which careful thought ought to be organized. They became the model for the development of all scientific and philosophical theories. What was especially admired about Euclid's work was the way in which he clearly laid out his assumptions and then used pure logic to deduce an incredibly varied and extensive set of conclusions from them.

Up until the twentieth century, Euclid's *Elements* were the textbook from which all students learned both geometry and logic. Even today the geometry in school textbooks is presented in a way that is remarkably close to that of Euclid. Furthermore, much of what mathematicians did during the next two thousand years centered around tying up loose ends left by Euclid. Countless mathematicians spent their careers trying to solve problems that were raised by Greek geometers of antiquity and trying to improve on Euclid's treatment of the foundations.

Most of the efforts at improvement focused on Euclid's fifth postulate. Even though the statement does not explicitly mention parallel lines, this postulate is usually referred to as "Euclid's parallel postulate." It asserts that two lines that are cut by a transversal must intersect on one side of the transversal if the interior angles on that side of the transversal sum to less than two right angles. In particular, it asserts that the condition on the angles formed by a transversal implies that the two given lines are not parallel. Thus it is really a statement about nonparallel lines. As we shall see later, the postulate can be reformulated in ways that make it more obviously and directly a statement about parallel lines.

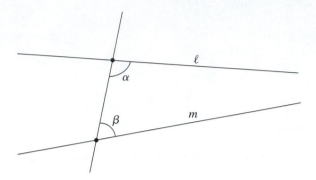

FIGURE 1.2: Euclid's Postulate V: If the sum of α and β is less than two right angles, then ℓ and m must eventually intersect

A quick reading of the postulates (see the next section) reveals that Postulate V is noticeably different from the others. For one thing, its statement is much longer than those of the other postulates. A more significant difference is the fact that it involves a fairly complicated arrangement of lines and also a certain amount of ambiguity in that the lines must be "produced indefinitely." It is not as intuitively obvious or self-evident as the other postulates; it has the look and feel of a proposition rather than a postulate. For these reasons generations of mathematicians tried to improve on Euclid by attempting to prove that Postulate V is a logical consequence of the other postulates or, failing that, they tried at least to replace Postulate V with a simpler, more intuitively obvious postulate from which Postulate V could then be deduced as a consequence. No one ever succeeded in proving the fifth postulate using just the first four postulates, but it was not until the nineteenth century that mathematicians finally understood why that was the case.

It should be recognized that these efforts at improvement were not motivated by a perception that there was anything wrong with Euclid's work. Quite the opposite: Thousands of mathematicians spent enormous amounts of time trying to improve on Euclid precisely because they thought so highly of Euclid's accomplishments. They wanted to take what was universally regarded as the crown of theoretical thought and make it even more wonderful than it already was!

Another important point is that efforts to rework Euclid's treatment of geometry led indirectly to progress in mathematics that went far beyond mere improvements in the *Elements* themselves. Attempts to prove Euclid's Fifth Postulate eventually resulted in the realization that, in some kind of stroke of genius, Euclid had somehow had the great insight to pinpoint one of the deepest properties that a geometry may have. Not only that, but it was discovered that there are alternative geometries in which Euclid's Fifth Postulate fails to hold. These discoveries were made in the early nineteenth century and had far-reaching implications for all of mathematics. They opened up whole new fields of mathematical study; they also produced a revolution in the conventional view of how mathematics relates to the real world and forced a new understanding of the nature of mathematical truth.

The story of how Euclid's Parallel Postulate inspired all these developments is one of the most interesting in the history of mathematics. That story will unfold in the course of our study of geometry in this book. It is only in the light of that story that the current organization of the foundations of geometry can be properly understood.

1.4   A LOOK AT BOOK I OF THE *ELEMENTS*

In order to give more substance to our discussion, we now take a direct look at parts of Book I of the *Elements*. All Euclid's postulates are stated below as well as selected definitions and propositions. The excerpts included here are chosen to illustrate the points that will be made in the following section. The translation into English is by Sir Thomas Little Heath (1861–1940).

Some of Euclid's Definitions

Definition 1. A *point* is that which has no part.

Definition 2. A *line* is breadthless length.

Definition 4. A *straight line* is a line which lies evenly with the points on itself.

Definition 10. When a straight line set up on a straight line makes the adjacent angles equal to one another, each of the equal angles is *right*, and the straight line standing on the other is called a *perpendicular* to that on which it stands.

Definition 11. An *obtuse* angle is an angle greater than a right angle.

Definition 12. An *acute* angle is an angle less than a right angle.

Euclid's Postulates

Postulate I. *To draw a straight line from any point to any point.*

Postulate II. *To produce a finite straight line continuously in a straight line.*

Postulate III. *To describe a circle with any center and distance.*

Postulate IV. *That all right angles are equal to one another.*

Postulate V. *That, if a straight line falling on two straight lines makes the interior angles on the same side less than two right angles, the two straight lines, if produced indefinitely, meet on that side on which are the angles less than the two right angles.*

Euclid's Common Notions

Common Notion I. *Things which equal the same thing are also equal to one another.*

Common Notion II. *If equals be added to equals, the wholes are equal.*

Common Notion III. *If equals be subtracted from equals, the remainders are equal.*

Common Notion IV. *Things which coincide with one another are equal to one another.*

Common Notion V. *The whole is greater than the part.*

Three of Euclid's Propositions and their proofs

Proposition 1. *On a given finite straight line to construct an equilateral triangle.*

Let AB be the given finite straight line. Thus it is required to construct an equilateral triangle on the straight line AB. With center A and distance AB let the circle BCD be described [Post. III]; again, with center B and distance BA let the circle ACE be described [Post. III]; and from the point C, in which the circles cut one another, to the points A, B let the straight lines CA, CB be joined [Post. I].

Now, since the point A is the center of the circle CDB, AC is equal to AB [Def. 15]. Again, since the point B is the center of the circle CAE, BC is equal to BA [Def. 15]. But CA was also proved equal to AB; therefore each of the straight lines CA, CB is equal to AB. And things which are equal to the same thing also equal one another [C.N. I]; therefore CA is also equal to CB. Therefore the three straight lines CA, AB, BC are equal to one another. Therefore the triangle ABC is equilateral; and it has been constructed on the given finite straight line AB.

Being what it was required to do.

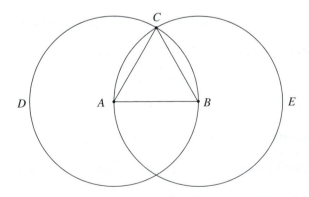

FIGURE 1.3: Euclid's diagram for Proposition 1

Proposition 4. *If two triangles have the two sides equal to two sides respectively, and have the angles contained by the equal straight lines equal, they will also have the base equal to the base, the triangle will be equal to the triangle, and the remaining angles will be equal to the remaining angles respectively, namely those which the equal sides subtend.*

Let ABC, DEF be two triangles having the two sides AB, AC equal to the two sides DE, DF respectively, namely AB to DE and AC to DF, and the angle BAC equal to the angle EDF. I say that the base BC is also equal to the base EF, the triangle ABC will be equal to the triangle DEF, and the remaining angles will be equal to the remaining angles respectively, namely those which the equal sides subtend, that is, the angle ABC to the angle DEF, and the angle ACB to the angle DFE.

For, if the triangle ABC be applied to the triangle DEF, and if the point A be placed on the point D and the straight line AB on DE, then the point B will also coincide with E, because AB is equal to DE. Again, AB coinciding with DE, the straight line AC will also coincide with DF, because the angle BAC is equal to the angle EDF; hence the point C will also coincide with the point F, because AC is again equal to DF.

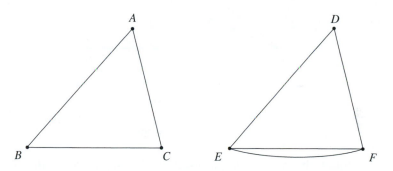

FIGURE 1.4: Euclid's diagram for Proposition 4

But B also coincided with E; hence the base BC will coincide with the base EF. [For if, when B coincides with E and C with F, the base BC does not coincide with the base EF, two straight lines will enclose a space: which is impossible. Therefore the base BC will coincide with EF] and will be equal to it [C.N. 4]. Thus the whole triangle ABC will coincide with the whole triangle DEF, and will be equal to it. And the remaining angles also coincide with the remaining angles and will be equal to them, the angle ABC to the angle DEF, and the angle ACB to the angle DFE.

Therefore etc. Being what it was required to prove.

Proposition 16. *In any triangle, if one of the sides be produced, the exterior angle is greater than either of the interior and opposite angles.*

Let ABC be a triangle, and let one side of it BC be produced to D; I say that the exterior angle ACD is greater than either of the interior and opposite angles CBA, BAC.

Let AC be bisected at E [Prop. 10] and let BE be joined and produced in a straight line to F; let EF be made equal to BE [Prop. 3], let FC be joined [Post. I], and let AC be drawn through to G [Post. II]. Then since AE is equal to EC, and BE to EF, the two sides AE, EB are equal to the two sides CE, EF respectively; and the angle AEB is equal to the angle FEC, for they are vertical angles [Prop. 15]. Therefore the base AB is equal to the base FC, and the triangle ABE is equal to the triangle CFE, and the remaining angles are equal to the remaining angles respectively, namely those which the equal sides subtend [Prop. 4]; therefore the angle BAE is equal to the angle ECF. But the angle ECD is greater than the angle ECF [C.N. 5]; therefore the angle ACD is greater than the angle BAE.

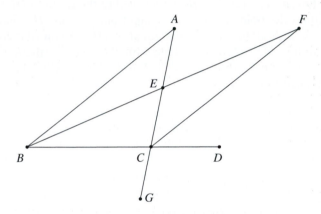

FIGURE 1.5: Euclid's diagram for Proposition 16

Similarly, also if BC is bisected, the angle BCG, that is, the angle ACD [Prop. 15], can be proved greater than the angle ABC as well.

Therefore etc.   Q.E.D.

1.5 A CRITIQUE OF EUCLID'S *ELEMENTS*

As indicated earlier, the *Elements* have been the subject of a great deal of interest over the thousands of years since Euclid wrote them and the study of Euclid's method of organizing his material has inspired mathematicians to even greater levels of logical rigor. Originally attention was focused on Euclid's postulates, especially his fifth postulate, but efforts to clarify the role of the postulates eventually led to the realization that there are difficulties with other parts of Euclid's *Elements* as well. When the definitions and propositions are examined in the light of modern standards of rigor it becomes apparent that Euclid did not achieve all the goals he set for himself—or at least that he did not accomplish everything he was traditionally credited with having done.

Euclid purports to define all the technical terms he will use.[1] However, an examination of his definitions shows that he did not really accomplish this. The first few definitions are somewhat vague, but suggestive of intuitive concepts. An example is the very first definition, in which *point* is defined as "that which has no part." This does indeed suggest something to most people, but it is not really a rigorous definition in that it does not stipulate what sorts of objects are being considered. It is somehow understood from the context that it is only geometric objects which cannot be subdivided that are to be called points. Even then it is not completely clear what a point is: Apparently a point is pure location and has no size whatsoever. But there is nothing in the physical world of our experience that has those properties exactly. Thus we must take point to be some kind of idealized abstract entity and admit that its exact nature is not adequately explained by the definition. Similar comments could be made about Euclid's definitions of *line* and *straight line*.

By contrast, later definitions are more complete in that they define one technical word in terms of others that have been defined previously. Examples are Definitions 11 and 12 in which *obtuse angle* and *acute angle* are defined in terms of the previously defined *right angle*. From the point of view of modern rigor there is still a gap in these definitions because Euclid does not specify what it means for one angle to be greater than another. The difference is that these definitions could be made rigorous. If Euclid were to spell out first what it means for one angle to be greater than another and also defined what a right angle is, then these definitions would be complete and usable.

Such observations have led to the realization that there are actually two kinds of technical terms. It is not really possible to define all terms; just as some statements must be accepted without proof and the other propositions proved as consequences, so some terms must be left undefined. Other technical terms can then be defined using the undefined terms and previously defined terms. This distinction will be made precise in the next chapter.

A careful reading of Euclid's proofs reveals some gaps there as well. The proof of Proposition 1 is a good example. In one sense, the proposition and its proof are simple and easy to understand. In modern terminology the proposition asserts the following: *Given two points A and B, it is possible to construct a third point C such that △ABC is an equilateral triangle.* Euclid begins with the segment from A to B. He then uses Postulate III to draw two circles of radius *AB*, one centered at *A* and the other centered at *B*. He takes *C* to be one of the two points at which the circles intersect and uses Postulate I to fill in the sides of a triangle. Euclid completes the proof by using the common notions to explain why the triangle he has constructed must be equilateral. The written proof is supplemented by a diagram that makes the construction clear and convincing.

---

[1] Some scholars suggest that the definitions included in the *Elements* were not in Euclid's original, but were added later. Even if that is the case, the observations made here about the definitions are still valid.

Closer examination shows, however, that Euclid assumed more than just what he stated in the postulates. In particular there is nothing explicitly stated in the postulates that would guarantee the existence of a point C at which the two circles intersect. The existence of C is taken for granted because the diagram clearly shows the circles intersecting in two points. There are, however, situations in which no point of intersection exists; one such example will be studied in Chapter 5. So Euclid is using "facts" about his points and lines that are undoubted and intuitively obvious to most readers, but which have not been explicitly stated in the postulates.

Euclid's Proposition 4 is the familiar Side-Angle-Side Congruence Condition from high school geometry. This proposition is not just a construction like Proposition 1 but asserts a logical implication: If two sides and the included angle of one triangle are congruent to the corresponding parts of a second triangle, then the remaining parts of the two triangles must also be congruent. Euclid's method of proof is interesting. He takes one triangle and "applies" it to the other triangle. By this we understand that he means to pick up the first triangle, move it, and carefully place one vertex at a time on the corresponding vertices of the second triangle. This is often called Euclid's *method of superposition*. It is quite clear from an intuitive point of view that this operation should be possible, but again the objection can be raised that Euclid is using unstated assumptions about triangles. Over the years geometers have come to realize that the ability to move geometric objects around without distorting their shapes cannot be taken for granted. The need to include an explicit assumption about motions of triangles will be discussed further in Chapter 5.

Another interesting aspect of the proof is the fact that part of it is enclosed in square brackets. (See the words starting with, "For if ..." in the third paragraph of the proof.) These words are in brackets because it is believed that they are not part of Euclid's original proof, but were inserted later.[2] They were added to justify Euclid's obvious assumption that there is only one straight line segment joining two points. Postulate I states that there exists a straight line joining two points, but here Euclid needs the stronger statement that there is only one such line. The fact that these words were added in antiquity is an indication that already then some readers of the *Elements* recognized that Euclid was using unstated assumptions.

Euclid's Proposition 16 is the result we now know as the Exterior Angle Theorem. This theorem and its proof will be discussed in Chapter 6. For now we merely point out that Euclid's proof depends on a relationship that appears to be obvious from the diagram provided, but which Euclid does not actually prove. Euclid wants to show that the interior angle $\angle BAC$ is smaller than the exterior angle $\angle ACD$. He first constructs the points E and F, and then uses the Vertical Angles Theorem (Proposition 15) and Side-Angle-Side to conclude that $\angle BAC$ is congruent to $\angle ACF$. Euclid assumes that F is in the interior of $\angle ACD$ and uses Common Notion 5 to conclude that $\angle BAC$ is smaller than $\angle ACD$. However he provides no justification for the assertion that F is in the interior of angle $\angle ACD$. In Chapter 5 we carefully state postulates that will allow us to fill in this gap when we prove the Exterior Angle Theorem in Chapter 6.

---

[2]See [30, page 249].

The preceding discussion is not meant to suggest that Euclid is wrong in his conclusions or that his work is in any way flawed. Rather, the point is that standards of mathematical rigor have changed since Euclid's day. Euclid thought of his postulates as statements of self-evident truths about the real world. He stated the key geometric facts as postulates, but felt free to bring in other spatial relationships when they were needed and were obvious from the diagrams. As we will see in the next chapter, we no longer see postulates in the same way Euclid did. Instead we try to use the postulates to state *all* the assumptions that are needed in order to prove our theorems. If that is our goal, we will need to assume much more than is stated in Euclid's postulates.

1.6   FINAL OBSERVATIONS ABOUT THE *ELEMENTS*

One aspect of Euclid's proofs that should be noted is the fact that each statement in the proof is justified by appeal to one of the postulates, common notions, definitions, or previous propositions. These references are placed immediately after the corresponding statements. They were probably not written explicitly in Euclid's original and therefore Heath encloses them in square brackets. This aspect of Euclid's proofs serves as an important model for the proofs we will write later in this course.

The words "Therefore etc." found near the end of the proofs are also not in Euclid's original. In the Greek view, the proof should culminate in a full statement of what had been proved. Thus Euclid's proof would have ended with a complete restatement of the conclusion of the proposition. Heath omits this reiteration of the conclusion and simply replaces it with "etc." Notice that the proof of Proposition 1 ends with the phrase "Being what it was required to do," while the proof of Proposition 4 ends with "Being what it was required to prove." The difference is that Proposition 1 is a construction while Proposition 4 is a logical implication. Later Heath uses the Latin abbreviations Q.E.F. and Q.E.D. for these phrases.

There are many features of Euclid's work that strike the modern reader as strange. One is the spare purity of Euclid's geometry. The points and lines are pure geometric forms that float in the plane with no fixed location. All of us have been trained since childhood to identify points on a line with numbers and points in the plane with pairs of numbers. That concept would have been foreign to Euclid; he did not mix the notions of number and point the way we do. The identification of number and point did not occur until the time of Descartes in the seventeenth century and it was not until the twentieth century that the real numbers were incorporated into the statements of the postulates of geometry. It is important to recognize this if we are to understand Euclid.

Euclid (really Heath) also uses language in a way that is different from contemporary usage. For example, what Euclid calls a line we would call a curve. We reserve the term *line* for what Euclid calls a "straight line." More precisely, what Euclid calls a straight line we would call a line segment (finitely long, with two endpoints). This distinction is more than just a matter of definitions; it indicates a philosophical difference. In Euclid, straight lines are potentially infinite in that they can always be extended to be as long as is needed for whatever construction is being considered, but he never considers the entire infinite line all at once. Since the time of Georg Cantor in the nineteenth century, mathematicians have been comfortable with sets that are actually infinite, so we usually think of the line as already being infinitely long and do not worry about the need to extend it.

Euclid chose to state his postulates in terms of straightedge and compass constructions. His propositions then often deal with the question of what can be constructed using those two instruments. For example, Proposition 1 really asserts the following: *Given a line segment, it is possible to construct, using only straightedge and compass, an equilateral triangle having the given segment as base.* In some ways Euclid identifies constructibility with existence. One of the major problems that the ancient Greeks never solved is the question of whether or not a general angle can be trisected. From a modern point of view the answer is obvious: any angle has a measure (in degrees, for example) which is a real number; simply dividing that real number by 3 gives us an angle that is one-third the original. But the question the Greeks were asking was whether or not the smaller angle can always be constructed from the original using only straightedge and compass. Such constructibility questions will be discussed in Chapter 11.

In this connection it is worthwhile to observe that the tools Euclid chose to use reflect the same pure simplicity that is evident throughout his work. His straightedge has no marks on it whatsoever. He did not allow a mark to be made on it that could be preserved when the straightedge is moved to some other location. In modern treatments of geometry we freely allow the use of a ruler, but we should be sure to note that a ruler is much more than a straightedge: It not only allows straight lines to be drawn, but it also measures distances at the same time. Euclid's compass, in the same way, is what we would now call a "collapsing" compass. It can be used to draw a circle with a given center and radius (where "radius" means a line segment with the center as one endpoint), but it cannot be moved to some other location and used to draw a different circle of the same radius. When the compass is picked up to be moved, it collapses and does not remember the radius of the previous circle. In contemporary treatments of geometry the compass has been supplemented by a protractor, which is a device for measuring angles. Euclid did not rely on numerical measurements of angles and he did not identify angles with the numbers that measure them the way we do.

SUGGESTED READING

1. Chapters 1 and 2 of *Journey Through Genius*, [23].
2. Part I (pages 1–49) of *Euclid's Window*, [50].
3. Chapters 1–4 of *Geometry: Our Cultural Heritage*, [35].
4. Chapters I–IV of *Mathematics in Western Culture*, [40].
5. Chapters 1 and 2 of *The Non-Euclidean Revolution*, [70].
6. Chapters 1 and 2 of *A History of Mathematics*, [37].

EXERCISES

1.1. A *quadrilateral* is a four-sided figure in the plane. Consider a quadrilateral whose successive sides have lengths a, b, c, and d. Ancient Egyptian geometers used the formula

$$A = \frac{1}{4}(a + c)(b + d)$$

to calculate the area of a quadrilateral. Check that this formula gives the correct answer for rectangles but not for parallelograms.

1.2. An ancient Egyptian document, known as the *Moscow papyrus*, suggests that the area of a circle can be determined by finding the area of a square whose side has length $\frac{8}{9}$ the diameter of the circle. What value of π is implied by this formula? How close is it to the correct value?

1.3. The familiar Pythagorean Theorem states that if $\triangle ABC$ is a right triangle with right angle at vertex C and a, b, and c, are the lengths of the sides opposite vertices A, B, and C, respectively, then $a^2 + b^2 = c^2$. Ancient proofs of the theorem were based on diagrams like those in Fig. 1.6. Explain how the two diagrams together can be used to provide a proof for the theorem.

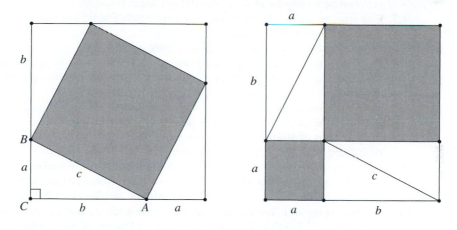

FIGURE 1.6: Proof of The Pythagorean Theorem

1.4. A *Pythagorean triple* is a triple (a, b, c) of positive integers such that $a^2 + b^2 = c^2$. A Pythagorean triple (a, b, c) is *primitive* if a, b, and c have no common factor. The tablet *Plimpton 322* indicates that the ancient Babylonians discovered the following method for generating all primitive Pythagorean triples. Start with relatively prime (i.e., no common factors) positive integers u and v, $u > v$, and then define $a = u^2 - v^2, b = 2uv$, and $c = u^2 + v^2$.
   (a) Verify that (a, b, c) is a Pythagorean triple.
   (b) Verify that a, b, and c are all even if u and v are both odd.
   (c) Verify that (a, b, c) is a primitive Pythagorean triple in case one of u and v is even and the other is odd.

Every Pythagorean triple (a, b, c) with b even is generated by this Babylonian process. The proof of that fact is significantly more difficult than the exercises above but can be found in most modern number theory books.

1.5. The ancient Egyptians had a well-known interest in pyramids. According to the *Moscow papyrus*, they developed the following formula for the volume of a truncated pyramid with square base:

$$V = \frac{h}{3}(a^2 + ab + b^2).$$

In this formula, the base of the pyramid is an $a \times a$ square, the top is a $b \times b$ square and the height of the truncated pyramid (measured perpendicular to the base) is h. One fact you learned in high school geometry is that that volume of a pyramid is one-third the area of the base times the height. Use that fact along with some high school geometry and algebra to verify that the Egyptian formula is exactly correct.

1.6. Explain how to complete the following constructions using only compass and straightedge. (You probably learned to do this in high school.)
   (a) Given a line segment $\overline{AB}$, construct the perpendicular bisector of $\overline{AB}$.
   (b) Given a line ℓ and a point P not on ℓ, construct a line through P that is perpendicular to ℓ.
   (c) Given an angle $\angle BAC$, construct the angle bisector.

1.7. Can you prove the following assertions using only Euclid's postulates and common notions? Explain your answer.
   (a) Every line has at least two points lying on it.
   (b) For every line there is at least one point that does not lie on the line.
   (c) For every pair of points $A \neq B$, there is only one line that passes through A and B.

1.8. A *rhombus* is a quadrilateral in which all four sides have equal lengths. The *diagonals* are the line segments joining opposite corners. Use the first five Propositions of Book I of the *Elements* to show that the diagonals of a rhombus divide the rhombus into four congruent triangles.

1.9. A *rectangle* is a quadrilateral in which all four angles have equal measures. (Hence they are all right angles.) Use the propositions in Book I of the *Elements* to show that the diagonals of a rectangle are congruent and bisect each other.

1.10. The following well-known argument illustrates the danger in relying too heavily on diagrams.[3] Find the flaw in the "Proof." (The proof uses familiar high school notation that will be explained later in this textbook. For example, $\overline{AB}$ denotes the segment from A to B and $\overleftrightarrow{AB}$ denotes the line through points A and B.)

False Proposition. *If $\triangle ABC$ is any triangle, then side $\overline{AB}$ is congruent to side $\overline{AC}$.*

Proof. Let ℓ be the bisector of $\angle BAC$ and let G be the point at which ℓ intersects $\overline{BC}$. Either ℓ is perpendicular to $\overline{BC}$ or it is not. We give a different argument for each case.

Assume, first, that ℓ is perpendicular to $\overline{BC}$. Then $\triangle AGB \cong \triangle AGC$ by Angle-Side-Angle and therefore $\overline{AB} \cong \overline{AC}$.

---

[3]This fallacy is apparently due to W. W. Rouse Ball (1850–1925) and first appeared in the original 1892 edition of [4].

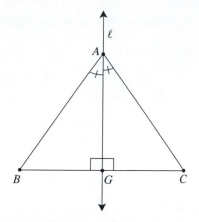

FIGURE 1.7: One possibility: the angle bisector is perpendicular to the base

Now suppose that ℓ is not perpendicular to $\overline{BC}$. Then ℓ intersects the perpendicular bisector of $\overline{BC}$ in a point D. There are three possibilities: either D is inside $\triangle ABC$, D is on $\triangle ABC$, or D is outside $\triangle ABC$. The three possibilities are illustrated in Fig. 1.8.

The point M is the midpoint of the segment $\overline{BC}$. In every case we drop perpendiculars from D to the lines $\overleftrightarrow{AB}$ and $\overleftrightarrow{AC}$ and call the feet of those perpendiculars E and F, respectively.

Consider first the case in which D is on $\triangle ABC$. Then $\triangle ADE \cong \triangle ADF$ by Angle-Angle-Side, so $\overline{AE} \cong \overline{AF}$ and $\overline{DE} \cong \overline{DF}$. Also $\overline{BD} \cong \overline{CD}$ since $D = M$ is the midpoint of $\overline{BC}$. It follows from the Hypotenuse-Leg Theorem that $\triangle BDE \cong \triangle CDF$ and therefore $\overline{BE} \cong \overline{CF}$. Hence $\overline{AB} \cong \overline{AC}$ by addition.

Next consider the case in which D is inside $\triangle ABC$. We have $\triangle ADE \cong \triangle ADF$ just as before, so again $\overline{AE} \cong \overline{AF}$ and $\overline{DE} \cong \overline{DF}$. Also $\triangle BMD \cong \triangle CMD$ by Side-Angle-Side and hence $\overline{BD} \cong \overline{CD}$. Applying the Hypotenuse-Leg Theorem gives $\triangle BDE \cong \triangle CDF$ and therefore $\overline{BE} \cong \overline{CF}$ as before. It follows that $\overline{AB} \cong \overline{AC}$ by addition.

Finally consider the case in which D is outside $\triangle ABC$. Once again we have $\triangle ADE \cong \triangle ADF$ by Angle-Angle-Side, so again $\overline{AE} \cong \overline{AF}$ and $\overline{DE} \cong \overline{DF}$. Just as before, $\triangle BMD \cong \triangle CMD$ by Side-Angle-Side and hence $\overline{BD} \cong \overline{CD}$. Applying the Hypotenuse-Leg Theorem gives $\triangle BDE \cong \triangle CDF$ and therefore $\overline{BE} \cong \overline{CF}$. It then follows that $\overline{AB} \cong \overline{AC}$, this time by subtraction.   □

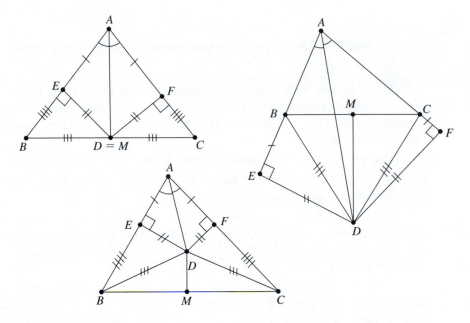

FIGURE 1.8: Three possible locations for D in case the angle bisector is not perpendicular to the base

CHAPTER 2

Axiomatic Systems and Incidence Geometry

The last chapter examined Euclid's scheme for organizing geometry. Over the years that scheme has been refined into what is known as an *axiomatic system*. The basic structure of an axiomatic system was inspired by Euclid's system of organization, but there are also several significant ways in which a modern axiomatic system differs from Euclid's system.

In this book the foundations of geometry will be described as an axiomatic system. It is often thought that every branch of mathematics should be formalized as an axiomatic system, but the goal of axiomatizing all of mathematics has never been fully achieved. Geometry remains the branch of elementary mathematics in which the axiomatic method has been the most successful and has been used most extensively.

This chapter presents the basic definition of an axiomatic system, examines the various parts of an axiomatic system, and explains their relationships. In addition, the chapter contains an example of a simple axiomatic system for geometry. This lays the groundwork for our presentation of plane geometry as an axiomatic system in Chapter 5 and also allows us to understand more clearly what it means to say that Euclid's fifth postulate is independent of his other postulates.

2.1   UNDEFINED AND DEFINED TERMS

The first part of an axiomatic system is a list of *undefined terms*. These are the technical words that will be used in the subject. As was pointed out in the previous chapter, Euclid attempted to define all his terms. But we now recognize that that is not an achievable goal. A standard dictionary appears to contain a definition of every word in a language, but there will inevitably be some circularity in the definitions. So, rather than attempting to define every term that we will use, we simply take certain key words to be undefined and work from there.

In geometry, we usually take such words as *point* and *line* to be undefined. In other parts of mathematics, the words *set* and *element of* are undefined.

There is still a place for definitions and defined terms in an axiomatic system. The aim is to start with a minimal number of undefined terms and then to define other technical words using the original undefined terms and previously defined terms. Thus the role of definitions is just to allow statements to be made concisely. For example, we will define three points to be collinear if there is one line such that all three points lie on that line. It is much more clear and concise to say that three points are noncollinear than it is to say that there does not exist a single line such that all three points lie on that line.

2.2 AXIOMS

The second part of an axiomatic system is a list of *axioms*. The words *axiom* and *postulate* are used interchangeably and mean exactly the same thing in this book.[1] The axioms are statements that are accepted without proof. They are where the subject begins. Everything else in the system should be logically deduced from them.

The axioms are what give meaning to the undefined terms. Thus, for example, we do not define exactly what a point or a line is, but in the axioms we spell out completely what it is about points and lines that will be used in our development of geometry. In that limited sense the axioms act as definitions for the undefined terms.

All relevant assumptions are to be stated in the axioms and the only properties of the undefined terms that may be used in the subsequent development of the subject are those that are explicitly spelled out in the axioms. Hence we will allow ourselves to use those and only those properties of points and lines that have been stated in our axioms—any other properties or facts about points and lines that we know from our intuition or previous experience are not to be used until and unless they have been proven from the axioms.

One of the goals of this course is to present geometry as an axiomatic system. This will require a much more extensive list of axioms than Euclid used. The reason for this is that we must include all the assumptions that will be needed in the proofs and not allow ourselves to rely on diagrams or any intuitive but unstated properties of points and lines the way Euclid did.

2.3 THEOREMS

The final part, usually by far the largest part, of an axiomatic system consists of the theorems and their proofs. Again there are two different words that are used synonymously: the words *theorem* and *proposition* will mean the same thing in

---

[1]Generally *postulate* will be used when a particular postulate is being named, while the word *axiom* will be used in a more generic sense to refer to unproven assumptions.

this course.[2]   In this third part of an axiomatic system we work out the logical consequences of the axioms.

Just as in Euclid's *Elements*, there is a strict logical organization that applies. The first theorem is proved using only the axioms. The second theorem is proved using the first theorem together with the axioms, and so on.

In the next chapter we will have more to say about theorems and proofs as well as the rules of logic that are to be used in proofs.

2.4   MODELS

In an axiomatic system the undefined terms do not in themselves have any definite meaning (other than that explicitly stated in the axioms).   The terms may be interpreted in any way that is consistent with the axioms.   An *interpretation* of an axiomatic system is a particular way of giving meaning to the undefined terms in that system. An interpretation is called a *model* for the axiomatic system if the axioms are correct (true) statements in that interpretation. Since the theorems in the system were all logically deduced from the axioms and nothing else, we know that all the theorems will automatically be correct and true statements in any model.

We say that a statement in our axiomatic system is *independent* of the axioms if it is impossible to either prove or disprove the statement as a logical consequence of the axioms. A good way to show that a statement is independent of the axioms is to exhibit one model for the system in which the statement is true and another model in which it is false. As we shall see, that is exactly the way in which it was eventually shown that Euclid's fifth postulate is independent of Euclid's other postulates.

The axioms in an axiomatic system are said to be *consistent* if no logical contradiction can be derived from them. This is obviously a property we would want our axioms to have. Again it is a property that can be verified using models. If there exists a model for an axiomatic system, then the system must be consistent. The existence of a model for Euclidean geometry and thus the consistency of Euclid's postulates was taken for granted until relatively recently. Our study of geometry will repeat the historical pattern: We will first study various geometries as axiomatic systems and only address the questions of consistency and existence of models in Chapter 13.   Again the use of models will allow us to answer a basic question regarding the foundations of geometry.

2.5   AN EXAMPLE OF AN AXIOMATIC SYSTEM

In order to clarify what an axiomatic system is, we study the important example of *incidence geometry*. For now we simply look at the axioms and various models for this system; a more extensive discussion of theorems and proofs in incidence geometry is delayed until the next chapter.

---

[2]There are also two other words that are used for theorem. A *lemma* is a theorem that is stated as a step toward some more important result. Usually a lemma is not an end in itself but is used as a way to organize a complicated proof by breaking it down into steps of manageable size. A *corollary* is a theorem that can be quickly and easily deduced from a previously stated theorem.

Let us take the three words *point, line,* and *lie on* (as in "point P lies on line ℓ") to be our undefined terms. The word *incident* is also used in place of *lie on,* so the two statements "P lies on ℓ" and "P is incident with ℓ" mean the same thing. For that reason the axioms for this relationship are called *incidence axioms.* One advantage of the word incident is that it can be used symmetrically: We can say that P is incident with ℓ or that ℓ is incident with P; both statements mean exactly the same thing.

There are three incidence axioms. The word *distinct* that is used in their statements simply means "not equal."

Incidence Axiom 1. *For every pair of distinct points P and Q there exists exactly one line ℓ such that both P and Q lie on ℓ.*

Incidence Axiom 2. *For every line ℓ there exist at least two distinct points P and Q such that both P and Q lie on ℓ.*

Incidence Axiom 3. *There exist three points that do not all lie on any one line.*

The axiomatic system with the three undefined terms and the three axioms listed above is called *incidence geometry.* We usually also call a model for the axiomatic system *an incidence geometry.* Before giving examples of incidence geometries, it is convenient to introduce a defined term.

Definition 2.5.1. Three points A, B, and C are *collinear* if there exists one line ℓ such that all three of the points A, B, and C all lie on ℓ. The points are *noncollinear* if there is no such line ℓ.

Using this definition we can give a more succinct statement of Incidence Axiom 3: *There exist three noncollinear points.*

EXAMPLE 2.5.2   The three-point plane

Interpret *point* to mean one of the three symbols A, B, C; interpret *line* to mean a pair of points;[3] and interpret *lie on* to mean "is an element of." In this interpretation there are three lines, namely $\{A, B\}$, $\{A, C\}$, and $\{B, C\}$. Since any pair of distinct points determines exactly one line and no one line contains all three points, this is a model for incidence geometry. ∎

Be sure to notice that this "geometry" contains only three points. It is an example of a *finite geometry*, which is a geometry that contains only a finite number of points. It is customary to picture such geometries by drawing a diagram in which the points are represented by dots and the lines by segments joining them. So the diagram for the three-point plane looks like a triangle (see Fig. 2.1). Don't be misled by the diagram: the "points" on the sides of the triangle are *not* points in the geometry. The diagram is strictly schematic, meant to illustrate relationships and is not to be taken as a literal picture of the geometry.

---

[3] A *pair of points* is an unordered pair, or simply a set of two points.

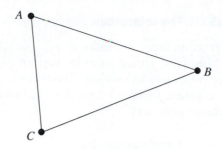

FIGURE 2.1: The three-point plane

EXAMPLE 2.5.3   The three-point line

Interpret *point* to mean one of the three symbols A, B, C, but this time interpret *line* to mean the set of all points. This geometry contains only one line, namely $\{A, B, C\}$. In this interpretation Incidence Axioms 1 and 2 are satisfied, but Incidence Axiom 3 is not satisfied. Hence the three-point line is not a model for incidence geometry. ■

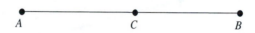

FIGURE 2.2: The three-point Line

EXAMPLE 2.5.4   Four-point geometry

Interpret *point* to mean one of the four symbols A, B, C, D; interpret *line* to mean a pair of points and interpret *lie on* to mean "is an element of." In this interpretation there are six lines, namely $\{A, B\}, \{A, C\}, \{A, D\}, \{B, C\}, \{B, D\}$, and $\{C, D\}$. Since any pair of distinct points determines exactly one line and no one line contains three distinct points, this is a model for incidence geometry. ■

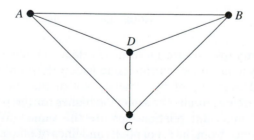

FIGURE 2.3: Four-point geometry

EXAMPLE 2.5.5 The interurban

In this interpretation there are three *points*, namely the cities of Grand Rapids, Holland, and Muskegon (three cities in western Michigan). A *line* consists of a railroad line from one city to another. There is one railroad line joining each pair of distinct cities. Hence there are three lines in this interpretation. Again, this is a model for incidence geometry. ∎

EXAMPLE 2.5.6 Fano's geometry

Interpret *point* to mean one of the seven symbols A, B, C, D, E, F, G; interpret *line* to mean one of the seven three-point sets listed below and interpret *lie on* to mean "is an element of." The seven lines are

$$\{A, B, C\}, \{C, D, E\}, \{E, F, A\}, \{A, G, D\}, \{C, G, F\}, \{E, G, B\}, \{B, D, F\}.$$

It is easy to check that Fano's geometry[4] is also is a model for incidence geometry. ∎

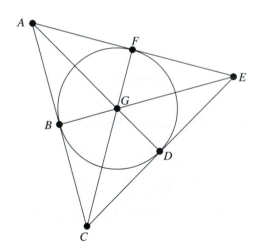

FIGURE 2.4: Fano's geometry

The examples above illustrate the fact that the undefined terms in a given axiomatic system can be interpreted in widely different ways. No one of the models is preferred over any of the others. Notice that three-point geometry and the interurban are essentially the same; the names for the points and lines are different, but all the important relationships are the same. We could easily construct a correspondence from the set of points and lines of one model to the set of points and lines of the other model. The correspondence would preserve all the relationships

---

[4]Named for Gino Fano, 1871–1952.

that are important in the geometry (such as incidence). Models that are related in this way are called *isomorphic* models and a function relating them is an *isomorphism*.

All the models described so far have been finite geometries. Of course the geometries with which we are most familiar are not finite. We next describe three infinite geometries.

EXAMPLE 2.5.7   The Cartesian plane

In this model a *point* is any ordered pair (x, y) of real numbers. A *line* is the collection of points whose coordinates satisfy a linear equation of the form $ax + by + c = 0$, where a, b, and c are real numbers and a and b are not both 0. More specifically, three real numbers a, b, and c, with a and b not both 0, determine the line ℓ consisting of all pairs (x, y) such that $ax + by + c = 0$. A point (x, y) is said to *lie on* the line if the coordinates of the point satisfy the equation. This is just the coordinate (or Cartesian) plane model for high school Euclidean geometry. It is also a model for incidence geometry. We will use the symbol $\mathbb{R}^2$ to denote the set of points in this model.   ∎

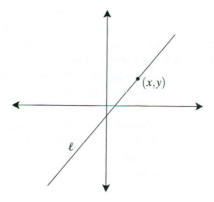

FIGURE 2.5: The Cartesian plane

EXAMPLE 2.5.8   The sphere

Interpret *point* to mean a point on the surface of a round 2-sphere in three-dimensional space. Specifically, a point is an ordered triple (x, y, z) of real numbers such that $x^2 + y^2 + z^2 = 1$. A *line* is interpreted to mean a great circle on the sphere and *lie on* is again interpreted to mean "is an element of." We will use the symbol $\mathbb{S}^2$ to denote the set of points in this model.

Be sure to notice that we are only considering points on the surface of the sphere. In other contexts the term *sphere* is sometimes used to refer to a solid three-dimensional object; that is not what is meant here. The sphere is two-dimensional in the sense that a very small piece of it would be hard to distinguish from an equally small piece of the plane or the disk.

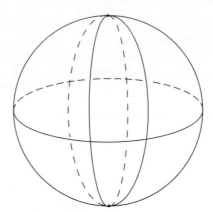

FIGURE 2.6: The sphere

A *great circle* is a circle on the sphere whose radius is equal to that of the sphere. A great circle is the intersection of a plane through the origin in 3-space with the sphere. Two points on the sphere are *antipodal* (or opposite) if they are the two points at which a line through the origin intersects the sphere. Two given antipodal points on the sphere lie on an infinite number of different great circles and so this geometry does not satisfy Incidence Axiom 1. If A and B are two points on the sphere that are not antipodal, then A and B determine a unique plane through the origin in 3-space and thus lie on a unique great circle. Hence "most" pairs of points determine a unique line in this geometry.

Since this interpretation does not satisfy Incidence Axiom 1, it is not a model for incidence geometry. Note that Incidence Axioms 2 and 3 are correct statements in this interpretation. Another important observation about the sphere is that there are no parallel lines: Any two distinct great circles on the sphere intersect in a pair of points. ■

EXAMPLE 2.5.9   The Klein disk

Interpret *point* to mean a point in the Cartesian plane such that the point lies inside the unit circle. In other words, a point is an ordered pair (x, y) of real numbers such that $x^2 + y^2 < 1$. A *line* is the part of a Euclidean line that lies inside the circle and *lie on* has its usual Euclidean meaning. This is a model for incidence geometry. ■

The Klein disk is an infinite model for incidence geometry, just like the familiar Cartesian plane is. (In this context *infinite* means that the number of points is unlimited, not that distances are unbounded.) The two models are obviously different in superficial ways. But they are also quite different with respect to some of the deeper relationships that are important in geometry. We illustrate this in the next section by studying parallel lines in each of the various geometries we have described.

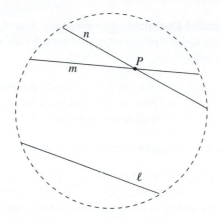

FIGURE 2.7: The Klein disk

2.6 THE PARALLEL POSTULATES

In this section we investigate parallelism in incidence geometry. The purpose of the investigation is to clarify what it means to say that the Euclidean Parallel Postulate is independent of the other axioms of geometry.

We begin with a definition of parallel, which becomes our second defined term in incidence geometry. In high school geometry, parallel lines could be characterized in many different ways and so there were several different, but equivalent, definitions of parallel. We must choose one of those definitions and make it the official definition of parallel. We choose the simplest characterization: the lines do not intersect. That definition fits best because it can be made using only the undefined terms of incidence geometry. It is also Euclid's definition of what it means for two lines in a plane to be parallel (Definition 23, Appendix A). Obviously this is not the right definition to use for lines in 3-space, but the geometry studied in this book is restricted to the geometry of the two-dimensional plane. Note that, according to this definition of parallel, a line is not parallel to itself.

Definition 2.6.1. Two lines ℓ and m are said to be *parallel* if there is no point P such that P lies on both ℓ and m. The notation for parallelism is $\ell \parallel m$.

There are three different parallel postulates that will be useful in this course. The first is called the Euclidean Parallel Postulate, even though it is not actually one of Euclid's postulates. We will see later that (in the right context) it is logically equivalent to Euclid's fifth postulate. This formulation of the Euclidean Parallel Postulate is often called *Playfair's Postulate* for reasons that will be explained later.

Euclidean Parallel Postulate. *For every line ℓ and for every point P that does not lie on ℓ, there is exactly one line m such that P lies on m and $m \parallel \ell$.*

There are other possibilities besides the Euclidean one. We state two of them.

Elliptic Parallel Postulate. *For every line ℓ and for every point P that does not lie on ℓ, there is no line m such that P lies on m and $m \parallel \ell$.*

Hyperbolic Parallel Postulate. *For every line ℓ and for every point P that does not lie on ℓ, there are at least two lines m and n such that P lies on both m and n and both m and n are parallel to ℓ.*

These are not new axioms for incidence geometry. Rather they are additional statements that may or may not be satisfied by a particular model for incidence geometry. We illustrate with some examples.

EXAMPLE 2.6.2   Parallelism in three- and four-point geometries

In the three-point plane any two lines intersect. Therefore there are no parallel lines and this model satisfies the Elliptic Parallel Postulate. On the other hand, each line in four-point geometry is disjoint from exactly one other. Thus, for example, the line $\{A, B\}$ is parallel to the line $\{C, D\}$ and no others. There is exactly one parallel line that is incident with each point that does not lie on $\{A, B\}$. Since the analogous statement is true for every line, four-point geometry satisfies the Euclidean Parallel Postulate. ∎

EXAMPLE 2.6.3   Parallelism in five-point geometry

Five-point geometry is the model for incidence geometry in which *point* is interpreted to be one of the letters A, B, C, D, and E and a *line* is a pair of points. Consider the line $\{A, B\}$ and the point C that does not lie on $\{A, B\}$. Observe that C lies on two different lines, namely $\{C, D\}$ and $\{C, E\}$, that are parallel to $\{A, B\}$. Since this happens for each line and for each point that does not lie on that line, five-point geometry satisfies the Hyperbolic Parallel Postulate. ∎

EXAMPLE 2.6.4   Parallelism in the Cartesian plane

In the Cartesian plane $\mathbb{R}^2$ the Euclidean Parallel Postulate holds and neither of the other two parallel postulates does. The fact that the Cartesian plane satisfies the Euclidean Parallel Postulate is probably familiar to you from your high school geometry course. ∎

EXAMPLE 2.6.5   Parallelism on the sphere

The sphere $\mathbb{S}^2$ satisfies the Elliptic Parallel Postulate and neither of the other two parallel postulates. The reason for this is simply that there are no parallel lines on the sphere. ∎

EXAMPLE 2.6.6   Parallelism in the Klein disk

The Klein disk satisfies the Hyperbolic Parallel Postulate and neither of the other parallel postulates. The fact that the Klein disk satisfies the Hyperbolic Parallel Postulate is illustrated in Fig. 2.7. ∎

Conclusion

We can conclude from the preceding examples that each of the parallel postulates is independent of the axioms of incidence geometry. For example, the fact that there are some models for incidence geometry that satisfy the Euclidean Parallel Postulate and there are other models that do not shows that neither the Euclidean Parallel Postulate nor its negation can be proved as a theorem in incidence geometry. The examples make it clear that it would be fruitless to try to prove any of the three parallel postulates in incidence geometry.

This is exactly how it was eventually shown that Euclid's fifth postulate is independent of his other postulates. Understanding that proof is one of the major goals of this course. Later in the book we will construct two models for geometry, both of which satisfy all of Euclid's assumptions other than his fifth postulate. One of the models satisfies Euclid's fifth postulate while the other does not. (It satisfies the Hyperbolic Parallel Postulate.) This shows that it is impossible to prove Euclid's fifth postulate using only Euclid's other postulates. It will take us most of the course to develop those models. One of the reasons it is such a difficult task is that we have to dig out *all* of Euclid's assumptions—not just the assumptions stated in the postulates, but all the unstated ones as well.

2.7 AXIOMATIC SYSTEMS AND THE REAL WORLD

An axiomatic system, as defined in this chapter, is obviously just a refinement of Euclid's system for organizing geometry. It should be recognized, however, that these refinements have profound implications for our understanding of the place of mathematics in the world.

The ancient Greeks revolutionized geometry by making it into an abstract discipline. Before that time, mathematics and geometry had been closely tied to the physical world. Geometry was the study of one aspect of the real world, just like physics or astronomy. In fact, the word *geometry* literally means "to measure the earth."

Later Greek geometry, on the other hand, is about relationships between ideal, abstract objects. In this view, geometry is not just about the physical world in which we live our everyday lives, but it also gives us information about an ideal world of pure forms. In the view of Greek philosophers such as Plato, this ideal world was, if anything, more real than the physical world of our existence. The relationships in the ideal world are eternal and pure. The Greeks presumably thought of a postulate as a statement about relationships that really pertain in that ideal world. The postulates are true statements that can be accepted without proof because they are self-evident truths about the way things really are in the ideal world. So the Greeks distanced geometry from the physical world by making it abstract, but at the same time they kept it firmly rooted in the real world where it could give them true and reliable information about actual spatial relationships.

We have no direct knowledge of how Euclid himself understood the significance of his geometry. All we know about his thinking is what we find written in the *Elements* and those books are remarkably terse by modern standards. But

Euclid lived approximately 100 years after Plato, so it seems reasonable to assume that he was influenced by Plato's ideas. In any event, it is quite clear from reading the *Elements* that Euclid thought of geometry as being about real things and this is precisely why he felt free to use intuitively obvious facts about points and lines in his proofs.

The view of geometry as an axiomatic system (as described in this chapter) moves us well beyond the Platonic view. In our effort to spell out completely what our assumptions are, we have been led to make geometry much more relative and detached from reality. We do not apply the terms true or false to the axioms in any absolute sense. An axiom is simply a statement that may be true or may be false in any particular situation, it just depends on how we choose to interpret the undefined terms. Thus our efforts to introduce abstraction and rigor into geometry have led us to drain the meaning out of such everyday terms as point and line. Since the words can now mean just about anything we want them to, we must wonder whether they any longer have any real content.

The naïve view is that geometry is the study of space and spatial relationships. We usually think of geometry as a science that gives us true and reliable information about the world in which we live. The view of geometry as an axiomatic system detached from the real world is a bit disturbing to most of us.

Some mathematicians have promoted the view that mathematics is just a logical game in which we choose an arbitrary set of axioms and then see what we can deduce using the rules of logic. Most professional mathematicians, however, have a profound sense that the mathematics they study is about real things. The fact that mathematics has such incredibly powerful and practical applications is evidence that it is much more than a game.

It is surprisingly difficult to resolve the kinds of philosophical issues that are raised by these observations. The mathematical community's thinking on these matters has evolved over time and there have been several amazing revolutions in the conventional understanding of what the correct views should be. Those views will be explored in this book as the historical development of geometry unfolds. We do not attempt to give definitive answers; instead we simply raise the questions and encourage the reader to think about them. In particular, the following questions should be recognized and should be kept in mind as the development of geometry is worked out in this book.

1. Are the theorems of geometry true statements about the world in which we live?
2. What physical interpretation should we attach to the terms *point* and *line*?
3. What are the axioms that describe the geometry of the space in which we live?
4. Is it worthwhile to study arbitrary axiom systems or should we restrict our attention to just those axiom systems that appear to describe the real world?

At this point you might be asking yourself why it would be thought desirable to make mathematics so abstract and therefore to get into the kind of difficult issues that have been raised here. That is one question we can answer. The answer is that abstraction is precisely what gives mathematics its power. By identifying certain key features in a given situation, listing exactly what it is about those features that is to be studied, and then studying them in an abstract setting detached from the original context, we are able to see that the same kinds of relationships hold in many apparently different contexts. We are able to study the important relationships in the abstract without a lot of other irrelevant information cluttering up the picture and obscuring the underlying structure. Once things have been clarified in this way, the kind of logical reasoning that characterizes mathematics becomes an incredibly powerful and effective tool. The history of mathematics is full of examples of surprising practical applications of mathematical ideas that were originally discovered and developed by people who were completely unaware of the eventual applications.

EXERCISES

2.1. It is said that Hilbert once illustrated his contention that the undefined terms in a geometry should not have any inherent meaning by claiming that is should be possible to replace *point* by *beer mug* and *line* by *table* in the statements of the axioms. Consider three friends sitting around one table. Each person has one beer mug. At the moment all the beer mugs are resting on the table. Suppose we interpret point to mean beer mug, line to mean the table, and lie on to mean resting on. Is this a model for incidence geometry? Explain. Is this interpretation isomorphic to any of the examples in the text?

2.2. One-point geometry contains just one point and no line. Which incidence axioms does one-point geometry satisfy? Explain. Which parallel postulates does one-point geometry satisfy? Explain.

2.3. Consider a small mathematics department consisting of Professors Alexander, Bailey, Curtis, and Dudley with three committees: curriculum committee, personnel committee, and social committee. Interpret *point* to mean a member of the department, interpret *line* to be a departmental committee, and interpret *lie on* to mean that the faculty member is a member of the specified committee.
   (a) Suppose the committee memberships are as follows: Alexander, Bailey, and Curtis are on the curriculum committee; Alexander and Dudley are on the personnel committee; and Bailey and Curtis are on the social committee. Is this a model for Incidence Geometry? Explain.
   (b) Suppose the committee memberships are as follows: Alexander, Bailey and Curtis are on the curriculum committee; Alexander and Dudley are on the personnel committee; and Bailey and Dudley are on the social committee. Is this a model for incidence geometry? Explain.
   (c) Suppose the committee memberships are as follows: Alexander and Bailey are on the curriculum committee, Alexander and Curtis are on the personnel committee, and Dudley and Curtis are on the social committee. Is this a model for incidence geometry? Explain.

2.4. A three-point geometry is an incidence geometry that satisfies the following additional axiom: *There exist exactly three points.*
  (a) Find a model for three-point geometry.
  (b) How many lines does any model for three-point geometry contain? Explain.
  (c) Explain why any two models for three-point geometry must be isomorphic. (An axiomatic system with this property is said to be *categorical.*)

2.5. Interpret *point* to mean one of the four vertices of a square, *line* to mean one of the sides of the square, and *lie on* to mean that the vertex is an endpoint of the side. Which incidence axioms hold in this interpretation? Which parallel postulates hold in this interpretation?

2.6. Draw a schematic diagram of five-point geometry (see Example 2.6).

2.7. Which parallel postulate does Fano's geometry satisfy? Explain.

2.8. Which parallel postulate does the three-point line satisfy? Explain.

2.9. Under what conditions could a geometry satisfy more than one of the parallel postulates? Explain. Could an incidence geometry satisfy more than one of the parallel postulates? Explain.

2.10. Consider a finite model for incidence geometry that satisfies the following additional axiom: *Every line has exactly three points lying on it.* What is the minimum number of points in such a geometry? Explain your reasoning.

2.11. Find a finite model for Incidence Geometry in which there is one line that has exactly three points lying on it and there are other lines that have exactly two points lying on them.

2.12. Find interpretations for the words *point, line,* and *lie on* that satisfy the following conditions.
  (a) Incidence Axioms 1 and 2 hold, but Incidence Axiom 3 does not.
  (b) Incidence Axioms 2 and 3 hold, but Incidence Axiom 1 does not.
  (c) Incidence Axioms 1 and 3 hold, but Incidence Axiom 2 does not.

2.13. For any interpretation of incidence geometry there is a *dual* interpretation. For each point in the original interpretation there is a line in the dual and for each line in the original there is point in the dual. A point and line in the dual are considered to be incident if the corresponding line and point are incident in the original interpretation.
  (a) What is the dual of the three-point plane? Is it a model for incidence geometry?
  (b) What is the dual of the three-point line? Is it a model for incidence geometry?
  (c) What is the dual of four-point geometry? Is it a model for incidence geometry?
  (d) What is the dual of Fano's geometry?

CHAPTER 3

Theorems, Proofs, and Logic

3.1 THE PLACE OF PROOF IN MATHEMATICS
3.2 MATHEMATICAL LANGUAGE
3.3 STATING THEOREMS
3.4 WRITING PROOFS
3.5 INDIRECT PROOF
3.6 THE THEOREMS OF INCIDENCE GEOMETRY

In this chapter we take a more careful look at the third part of an axiomatic system: the theorems and proofs. Both require extra care. Most of us have enough experience with mathematics to know that the ability to write good proofs is a skill that must be learned, but we often overlook the fact that a necessary prerequisite to good proof writing is good statements of theorems. The chapter includes both an emphasis on the careful use of language in mathematical statements and an introduction to good proof-writing practices. It also provides practice at writing proofs in the relatively simple context of incidence geometry.

A major goal of this course is to teach the art of writing proofs and it is not expected that the reader is already proficient at it. The main way in which one learns to write proofs is by actually writing them, so the the remainder of the book will provide lots of opportunities for practice. This chapter will simply lay out a few basic principles and then those principles will be put to work in the rest of the course. The brief introduction provided in this chapter will not make you an instant expert at writing proofs, but it will equip you with the basic tools you need to get started. You can refer back to this chapter as necessary in the remainder of the course. If you want to go beyond the bare-bones treatment here, the book by Donald Knuth on *Mathematical Writing* [42] is recommended as a good source of additional ideas.

3.1 THE PLACE OF PROOF IN MATHEMATICS

An emphasis on proof is one of the characteristics of mathematics that distinguishes it from other disciplines. In mathematics a result is not considered to be complete until a proof has been found. It is not enough to accumulate overwhelming amounts of experimental evidence or to point out that the result is consistent with all known evidence; a result is only accepted when it can be proved. You have already seen that the origins of this aspect of mathematics are ancient and that geometry is the branch of mathematics in which the concept of proof first emerged. Despite all the changes that have taken place in mathematics over the years and despite the fact that the goal of organizing the various branches of mathematics as axiomatic systems has never been fully achieved, the practice of presenting mathematical results in the theorem-proof format remains the standard in the discipline. So anyone who hopes to become a professional in mathematics must master the form.

The goal of this course is to teach more than just the mechanics of good proof-writing style. The real goal is to help you learn to *appreciate* proofs. By having the experience of working out many of the proofs for yourself you will come to a much deeper understanding of the material and should experience the profound sense of satisfaction that comes from truly understanding mathematical arguments. There is no better branch of mathematics in which to experience that feeling of satisfaction than geometry. When you experience it for yourself you will begin to understand why proof and the Euclidean axiomatic approach have such a powerful hold over mathematics.

It must be acknowledged that an emphasis on proof and rigor is not the only thing that distinguishes mathematics as a discipline. Mathematics can mean many different things in different situations. It is often called a language or the study of patterns. Sometimes the discipline of mathematics is described in terms of problem solving. Other times mathematics is said to be the subject in which we investigate numerical and spatial relationships and try to come to understand them in any way we can. Reasoning by analogy can be a powerful tool in mathematics (see Chapter 14, for example). In many circumstances it is just as important to develop an intuitive understanding of a topic as it is to provide a rigorous proof. Thus it would be incorrect to claim that proof is all there is to mathematics. Nonetheless, proof remains one of the hallmarks of mathematics.

Occasionally there is a tension in mathematics between the need to present intuitive explanations of ideas and the desire to present strict logical proofs of theorems. Sometimes we are forced to choose between presenting mathematical ideas in a way that motivates them intuitively and shows how they were discovered and presenting them in the strict logical deductive order that is demanded in an axiomatic system. While that tension is real, a goal of this course is to help you understand that the two ways of presenting mathematics are not necessarily incompatible but can coexist quite comfortably.

Even though there are many other completely legitimate ways of coming to understand geometry, this book makes the deliberate choice to use geometry as a vehicle to teach an appreciation of proofs.

3.2 MATHEMATICAL LANGUAGE

The most important step toward proving a theorem is to state it clearly and carefully in precise, unambiguous language. To illustrate this point, consider the following proposition in incidence geometry.

Proposition 3.2.1. *Lines that are not parallel intersect in one point.*

Compare that statement with the following.

Proposition 3.2.2. *If ℓ and m are two distinct lines that are not parallel, then there exists exactly one point P such that P lies on both ℓ and m.*

Both are correct statements of the same theorem. But the second statement is much better, at least as far as we are concerned, because it clearly states where the proof should begin (with two lines ℓ and m such that $\ell \neq m$ and $\ell \nparallel m$) and where it should end (with a point P that lies on both ℓ and m). This provides the framework within which we can build a proof. Contrast that with the first statement. In the first statement it is much less clear how to begin a proof. In fact it is not possible to begin the proof until we have at least mentally translated the first statement into language that is closer to that in the second statement. When we start to do that we realize that the first statement is not precise enough. For example, it does not clearly say whether it is an assertion about two (or more?) particular lines or whether it is making an assertion that applies to all lines. Writing good proofs requires clear thinking and clear thinking begins with clear statements.

In mathematics the word *statement* refers to any assertion that can be classified as either true or false (but not both). Here is an example: Dan is tiny. The statements of geometry often involve assertions that objects (such as points or lines) satisfy certain conditions (such as parallelism). Such statements must be preceded by definitions of the terms used. For example, it is not possible to determine if the assertion "Dan is tiny" is true or false unless we have a precise definition of what *tiny* means in this context. It is obvious that the word *tiny* might mean one thing in one situation (for example, in microbiology) and something completely different in another context (such as astronomy). So we would need a definition of the form, "A person *x* is said to be *tiny* if ...." Once you have that definition, you can check to see whether or not a particular person named Dan satisfies the conditions in the definition.

One of the distinctions that must be made clear is whether you are asserting that *every* object of a certain type satisfies the condition or whether you are simply asserting that there is one that does. This is specified through the use of *quantifiers*. There are two quantifiers: the existential quantifier (written ∃) and the universal quantifier (written ∀).

As the name implies, the existential quantifier asserts that something exists. Example: There exists a point P such that P does not lie on ℓ. If this statement occurs as part of the conclusion of a theorem then your proof need only exhibit one point P that does not lie on ℓ. It is important to be clear about this so that you don't find yourself trying to prove, for example, that every point does not lie on ℓ. The universal quantifier, on the other hand, is used to say that some property holds for all objects in a certain class. Example: For every point P not on ℓ, the distance from P to ℓ is positive. In this case the strategy of the proof would be completely different; instead of exhibiting one particular P for which the distance is positive we should give a universal proof that works for every P not on ℓ. Such a proof would begin with a statement such as this: "Let P be a point that does not lie on ℓ." The proof should then go on to demonstrate that the distance from P to ℓ is positive, using only the fact that P does not lie on ℓ.

We will often want to negate statements. Specifically, given one statement we will want to write down a second statement that asserts the opposite of the first. There is a sense in which it is easy to negate a statement: simply say, "It is not true that ...." But this is not helpful. In order to extract some useful information from the fact that a statement is false, it is necessary to understand that negation interchanges the two quantifiers. For example, consider this statement: Every angle is acute. Stated more precisely, it says, For every angle α, α is acute. The negation is, There exists an angle α such that α is not acute. Here is another example: There exists a point that does not lie on ℓ. The negation is, Every point lies on ℓ. Better yet is this statement: For every point P, P lies on ℓ.

The word *unique* is often used in connection with the existential quantifier. For example, here is a statement that is important in geometry: For every line ℓ and for every point P there is a unique line m such that P lies on m and m is perpendicular to ℓ. The word *unique* in this statement indicates that there is exactly one line m satisfying the stated conditions. A proof of this assertion should have two parts. First, there should be a proof that there is a line m satisfying the conditions and, second, a proof that there cannot be two different lines m and n satisfying the conditions. The usual strategy for the second half of the proof is to start with the assumption that m and n are lines that satisfy the property and then to prove that m and n must, in fact, be equal to each other. The symbol ! is used to indicate uniqueness; so the notation "$\exists ! ...$" should be read "there exists a unique ...."

Simple statements can be combined into compound statements using the words *and* and *or*. The use of *and* is easy to understand; it means that both the statements are true. The use of *or* in mathematics differs from the way the word is used in ordinary language. In mathematics *or* is always used in a nonexclusive way; it means that one or the other of the statements is true and allows the possibility that both are true. For example, $x > 0$ or $x < 1$. Every real number x satisfies this condition. Contrast that with the following statement: You are either for me or against me. We understand from the tone of the statement that it means you are one or the other but not both. Mathematical language eliminates such ambiguities and the word *or* always has the nonexclusive meaning when it is used in a mathematical statement.

Negation interchanges *and* and *or*. In other words, we have the following laws (for any statements S and T).

not (S and T) = (not S) or (not T)

not (S or T) = (not S) and (not T)

For example, if it is not true that $x > 0$ and $x < 1$, then either $x \leq 0$ or $x \geq 1$. The two rules stated above are known as *DeMorgan's laws*.

3.3 STATING THEOREMS

A *conditional statement* is a compound statement of the form "If ..., then ..." in which the first set of dots represents a statement called the *hypothesis* (or *antecedent*) and the second set of dots represents a statement called the *conclusion* (or *consequent*). A *theorem* is a conditional statement that has been proved true. Note that a conditional statement may be either true or false, but it is not called a theorem unless it is true and has been (or can be) proved. This means that there are no false theorems, just statements that have the form of theorems but turn out not to be theorems. A theorem *does not* assert that the conclusion is true without the hypothesis, only that the conclusion is true whenever the hypothesis is. *Every theorem should be stated in if-then form.*[1] Another way in which a conditional statement can be written is this: hypothesis implies conclusion. Notation: Hypothesis $\Rightarrow$ Conclusion.

The statement $H \Rightarrow C$ means that C is true in every case in which H is true. In other words, the statement rules out the possibility that H is true while C is false. For example, here is an easy theorem from high school algebra: If x is a real number such that $x > 1$, then $2x > 2$. The conclusion $2x > 2$ is not true (or false) by itself, but it is true in case x is a real number for which the statement $x > 1$ is true. The theorem does not give any information about the conclusion in case the hypothesis is false. Here is another example from high school algebra: If x is a real number and $x^2 < 0$, then $x = 3$. This is considered to be a logically correct statement because the conclusion is true of every x for which the hypothesis is true. (There are no x for which the hypothesis is true.) In this case we usually say that the theorem is *vacuously true* since the theorem is true only because there is no way the hypotheses can be satisfied.

For every conditional statement there are two related statements called the *converse* and the *contrapositive*. The converse of $P \Rightarrow Q$ is $Q \Rightarrow P$ and the contrapositive is *not $Q \Rightarrow$ not P*. The converse to a conditional statement is an entirely different statement; the fact that $P \Rightarrow Q$ is a theorem tells us nothing about whether or not the converse is a theorem. On the other hand, the contrapositive is logically equivalent to the original statement. Here is a simple example: If $x = 2$, then $x^2 = 4$. This is a correct theorem. Its converse, however, is not correct. ($x^2 = 4 \nRightarrow x = 2$.) The contrapositive is this: If $x^2 \neq 4$, then $x \neq 2$. The contrapositive is a correct statement and is just a negative way of restating the original theorem.

Consider another simple example: If $x = 0$, then $x^2 = 0$. This time both the statement and its converse are true. The phrase "if and only if" (abbreviated iff) is used to indicate that the implication goes both ways. In other words, P iff Q (or $P \Leftrightarrow Q$) means $P \Rightarrow Q$ and $Q \Rightarrow P$. Thus we could say that $x = 0$ iff $x^2 = 0$. An if-and-only-if statement is really two theorems in one and the proof should reflect this; there should normally be separate proofs of each of the two implications.

There is a sense in which the fact that the theorem and its contrapositive are equivalent is obvious: the contrapositive is just a negative way of saying the same thing. On the other hand, it can be confusing to explain the equivalence because negations are piled on negations. One simple way to explain the equivalence is

---

[1] As is the case with most rules, this one allows some exceptions. Here is a well-known theorem from calculus: *π is irrational.* In this case the hypotheses are hidden in the definition of π. While such a statement does qualify as a theorem, it is not a model we should adopt for this course. The practice of stating the hypotheses explicitly will serve us well as we learn to write proofs. The example does illustrate the fact that theorem statements are context dependent and there are often unstated hypotheses that are assumed in a given setting.

through a *truth table*. This is a good way to think of it because the truth table also sheds light on the meaning of the theorem itself.

Consider the statement $H \Rightarrow C$. The hypothesis and the conclusion can each be either true or false. Thus there are four possibilities for H and C and the statement $H \Rightarrow C$ is true in three of the four cases. The various possibilities are shown in the following truth table.

H	C	$H \Rightarrow C$
True	True	True
True	False	False
False	True	True
False	False	True

It is the second half of the table that often confuses beginners; these are the cases in which the theorem is vacuously true. Since the conditional statement is true in three of the four cases, a proof is simply an argument that rules out the fourth possibility. If we now expand the table to include the negations of H and C as well as the contrapositive of the theorem we see that the contrapositive is true exactly when the theorem is. (The third and sixth columns are identical.)

H	C	$H \Rightarrow C$	*not H*	*not C*	*not C* $\Rightarrow$ *not H*
True	True	True	False	False	True
True	False	False	False	True	False
False	True	True	True	False	True
False	False	True	True	True	True

Sometimes it is more convenient to prove the contrapositive of a theorem than it is to prove the theorem itself. This is perfectly legitimate because the original statement is logically equivalent to the contrapositive.

Negating a conditional statement can be tricky. The statement $P \Rightarrow Q$ means that Q is true whenever P is. The negation of $P \Rightarrow Q$ is the assertion that that it is possible for P to be true while Q is false. Note that the negation of a conditional statement is not another conditional statement. Consider the following example of a proposed theorem: *If x is irrational, then x^2 is irrational.* This statement is false[2] because there are some irrational numbers whose square is rational [e.g., $\sqrt{2}$ is irrational while $(\sqrt{2})^2$ is rational]. It is true that there are some irrational numbers whose squares are irrational, but it takes only one example to show that the conditional statement is false. For this reason we normally demonstrate that a conditional statement is false by producing a counterexample.

In practice many theorem statements are actually somewhat more complicated than the simple ideal we have been discussing in this section. Here is a common form: *For every x in some class, if the hypothesis P is true of x, then the conclusion*

---

[2]The real theorem is this: *If x is rational, then x^2 is rational.*

Q is also true of x. This can be written in symbols as $(\forall x)(P(x) \Rightarrow Q(x))$. Most of the example theorems mentioned in this section have precisely that form. In those examples the variable x has been a number, but in the remainder of the course the variables will be geometric objects such as points or lines. Since many theorems have this form, our theorems will often not be stated in the strict "if...then..." form. Look ahead at the statement of Theorem 6.4.6 to see a typical theorem statement that takes a more complicated form. Stating a theorem in the form $(\forall x)(P(x) \Rightarrow Q(x))$ allows us to see that the comments about negating quantifiers that appeared earlier in the section are closely related to the comments about counterexamples in the preceding paragraph; specifically, the negation of $(\forall x)(P(x) \Rightarrow Q(x))$ is $(\exists x)(P(x)$ and not $Q(x))$.

3.4   WRITING PROOFS

A proof consists of a sequence of steps that lead us logically from the hypothesis to the conclusion. Each step should be justified by a reason. There are six kinds of reasons that can be given:

- by hypothesis
- by axiom
- by previous theorem
- by definition
- by an earlier step in this proof
- by one of the rules of logic

At the beginning of the course we will follow Euclid's practice of writing the reasons in parentheses after the statements. We will eventually drop that style as we develop more proficiency at writing and reading proofs. But for now, it is very important to spell out all your justifications.

In high school you may have learned to write your proofs in two columns, with the statements in one column and the reasons in another. We will not do that in this course, not even at the outset, because we are aiming to write proofs that can be read by fellow humans; in order to facilitate this, the proofs should be written in ordinary paragraph form. For the same reason we will not follow the high school custom of numbering the statements in a proof.

It is helpful to distinguish between the proof itself and the written argument that is used to communicate the proof to other people. The proof is a sequence of logical steps that lead from the hypothesis to the conclusion. The written argument lays out those steps for the reader, and the writer has an obligation to write them in a way that the reader can understand without undue effort. So the written proof is both a listing of the logical steps and an explanation of the reasoning that went into them.

Obviously you need to know who your audience is. You should assume that the reader is someone, like a fellow student, who has approximately the same background that you have. It is important to remember that written proofs have a subjective aspect to them. They are written for a particular audience and how many details you include will depend on who is to read the proof. As you and the rest of the class learn more geometry together, you will share a larger and larger set of common experiences. You can draw on those experiences and assume that your readers will know many of the justifications for steps in the proofs. Later in the course you will be able to leave many of the reasons unstated; this will allow you to concentrate on the essential new ideas in a proof and not obscure them with a lot of detailed information that is already well known to your readers. But don't be too quick to jump ahead to that level. Our aim in this course is to lay out all our assumptions in the axioms and then to base our proofs on those assumptions and no others. Only by being explicit about our reasons for each step can we discipline ourselves to use only those assumptions and not bring in any hidden assumptions that are based on previous experience or on diagrams.

You are encouraged to include in your written proofs more than just a list of the logical steps in the proof. In order to make the proofs more readable, you should also include information for the reader about the structure of the logical argument you are using, what you are assuming, and where the proof is going. Such statements are not strictly necessary from a logical point of view, but they make an enormous difference in the readability of a proof. For now our goal is not so much to prove the theorems as it is to learn to write good proofs, so we will write more than is strictly necessary and not worry about the risk of being pedantic. A good analogy: When a child is learning to walk, she has to think consciously about each individual step; later the steps happen without any conscious thought.

The beginning of a proof is marked by the word ***Proof***. It is also a good idea to include an indication that you have reached the end of a proof. Traditionally the end of a proof was indicated with the abbreviation QED, which stands for the Latin phrase *quod erat demonstrandum* (which was to be demonstrated). In this book we mark the end of our proofs with the symbol □.

Like each individual proof, the overall structure of the collection of theorems and proofs in an axiomatic system should be logical and sequential. Within any given proof, it is legitimate to appeal only to the axioms, a theorem that has been previously stated and proved, a definition that has been stated earlier, or to an earlier step in the same proof. The rules of logic that are listed as one possible type of justification for a step in a proof are the rules that are explained in this chapter. They include such rules as the rules for negating compound statements that were described above and the rules for indirect proofs that will be described below.

There is one last point related to the justification of the steps in a proof that is specific to this course in the foundations of geometry. We intend to build geometry on the real number system. Hence we will base many steps in our proofs on known facts about the real numbers. For example, if we have proved that $x + z = y + z$, we will want to conclude that $x = y$. Technically, this falls under the heading "by previous theorem," but we will usually say something like "by algebra" when we bring in some fact about the algebra of real numbers. The next chapter will list many of the important properties of the real numbers that are assumed in this course. A few of them have names (such as Trichotomy and the Archimedean Property) and those names should be mentioned when the properties are used.

3.5 INDIRECT PROOF

Indirect proof is one proof form that should be singled out for special consideration because you will find that it is one you will often want to use. The straightforward strategy for proving $P \Rightarrow Q$, called *direct proof*, is to start by assuming that P is true and then to use a series of logical deductions to conclude Q. But the statement $P \Rightarrow Q$ means that Q is true if P is, so the the real purpose of the proof is to rule out the possibility that P is true while Q is false. The indirect strategy is to begin by assuming that P is true *and* that Q is false, and then to show that this leads to a logical contradiction. If it does, then we know that it is impossible for both P and the negation of Q to hold simultaneously and therefore Q must be true whenever P is.[3] This indirect form of proof is called *proof by contradiction*. It goes by the official name *reductio ad absurdum*, which we will abbreviate as RAA.

The reason this proof form often works so well is that you have more information with which to work. In a direct argument you assume only the hypothesis P and work from there. In an indirect argument you begin by assuming both the original hypothesis P and also the additional hypothesis *not Q*. You can make use of both assumptions in your proof. In order to help clarify what is going on in an indirect proof, we will give a special name to the additional hypothesis *not Q*; we will call it the *RAA hypothesis* to distinguish it from the standard hypothesis P.

Indirect proofs are often confused with direct proofs of the contrapositive. They are not the same, however, since in a direct proof of the contrapositive we assume only the negation of the conclusion while in an indirect proof we assume and use both the hypothesis and the negation of the conclusion. Suppose we want to prove the theorem P implies Q. A direct proof of the contrapositive would start with *not Q* and conclude *not P*. One way to formulate the argument would be to start by assuming P (the hypothesis) and *not Q* (the RAA hypothesis), then to use the same proof as before to conclude *not P* and finally to end by saying that we must reject the RAA hypothesis because we now have both P and *not P*, an obvious contradiction. While this is logically correct, it is considered to be bad form and sloppy thinking; this way of organizing a proof should, therefore, usually be avoided.[4]

An even worse misuse of proof by contradiction is the following. Suppose again that we want to prove the theorem P implies Q. We assume the hypothesis P. Then we also suppose *not Q* (RAA hypothesis). After that we proceed to prove that P implies Q. At that point in the proof we have both Q and *not Q*. That is a contradiction, so we reject the RAA hypothesis and conclude Q. In this case the structure of an indirect argument has been erected around a direct proof, thus obscuring the real proof. Again this is sloppy thinking. It is an abuse of indirect proof and should *always* be avoided.

---

[3]Indirect proof is based on two laws of logic: the Law of the Excluded Middle, which asserts that any statement must be either true or false, and the Law of Non-Contradiction, which asserts that no statement can be both true and false. These laws are accepted by essentially all practicing mathematicians.

[4]We will occasionally find good reason to formulate our proofs this way even though it is generally not a good idea.

Use RAA proofs when they are helpful, but don't misuse them. A proof is often discovered as an indirect proof because we can suppose the conclusion is false and explore the consequences. Once you have found a proof, you should reexamine it to see if the logic can be simplified and the essence of the proof presented more directly.

3.6 THE THEOREMS OF INCIDENCE GEOMETRY

We illustrate the lessons of this chapter with some theorems and proofs from incidence geometry. Proofs of the theorems below are to be based on the three incidence axioms that were stated in §2.5. One of the hardest lessons to be learned in writing the proofs is that we may use only what it actually says in the axioms, nothing more.

Theorem 3.6.1. *If ℓ and m are distinct, nonparallel lines, then there exists a unique point P such that P lies on both ℓ and m.*

Proof. Let ℓ and m be two lines such that $\ell \neq m$ and $\ell \nparallel m$ (hypothesis). We must prove two things: first, that there is a point that lies on both ℓ and m and, second, that there is only one such point.

There is a point P such that P lies on both ℓ and m (negation of the definition of parallel). Suppose there exists a second point Q, different from P, such that Q also lies on both ℓ and m (RAA hypothesis). Then ℓ is the unique line such that P and Q lie on ℓ and m is the unique line such that P and Q lie on m (Incidence Axiom 1). Hence $\ell = m$ (property of equality). But this contradicts the hypothesis that ℓ and m are distinct. Hence we must reject the RAA hypothesis and conclude that no such point Q exists. □

Commentary on the Proof. It is a good idea to begin each proof by restating exactly what the hypotheses give you. (In high school these were called the "givens.") It is also a good idea to state next exactly what it is that you need to prove. This will often amount to nothing more than a restatement of the conclusions of the theorem, but stating them again, in terms of the notation you have used for the hypotheses, helps to focus your thinking and explains to the reader where the proof is going. Once those two preliminaries are out of the way, the real proof begins. In this case there are two things to prove since the conclusion "there exists a unique" means both that there exists one and that there exists only one. The first proof is direct (just a simple invocation of a definition) while the second proof is indirect.

The form of the first proof should be noted. All that we need to do is to look up the definition of parallel and then negate it. When we do that we see that the first conclusion is immediate. This is a useful hint about how to get a proof off the ground: it is often just a matter of going to the definitions of the terms used in the hypotheses and making use of them.

If we were to completely separate the two proofs, we could formulate the second one as a direct proof of the contrapositive (if the point of intersection is not unique, then the lines are not distinct). It is better to formulate the theorem the way we have, however, so that existence and uniqueness can be asserted together.

The reason for each statement is given in parentheses at the end of the sentence. Of course this is only true of those sentences that correspond to logical steps in the proof. The extra, explanatory sentences do not need any justification; their purpose is simply to make the written proof more readable and understandable.

Here are several other theorems from incidence geometry. You can practice what you have learned in this chapter by writing proofs for them.

Theorem 3.6.2. *If ℓ is any line, then there exists at least one point P such that P does not lie on ℓ.*

Proof. Exercise 3.8. □

Theorem 3.6.3. *If P is any point, then there are at least two distinct lines ℓ and m such that P lies on both ℓ and m.*

Proof. Exercise 3.9 □

Theorem 3.6.4. *If ℓ is any line, then there exist lines m and n such that ℓ, m, and n are distinct and both m and n intersect ℓ.*

Proof. Exercise 3.10 □

Theorem 3.6.5. *If P is any point, then there exists at least one line ℓ such that P does not lie on ℓ.*

Proof. Exercise 3.11 □

Theorem 3.6.6. *There exist three distinct lines such that no point lies on all three of the lines.*

Proof. Exercise 3.12 □

Theorem 3.6.7. *If P is any point, then there exist points Q and R such that P, Q, and R are noncollinear.*

Proof. Exercise 3.13 □

Theorem 3.6.8. *If P and Q are two points such that P ≠ Q, then there exists a point R such that P, Q, and R are noncollinear.*

Proof. Exercise 3.14 □

This is a good point at which to remind the reader that every theorem has a proper context. That context is some axiomatic system. Thus every theorem has unstated hypotheses, namely that certain axioms are assumed true. For example, the theorems above are theorems in incidence geometry. This means that every one of them includes the unstated hypothesis that the three incidence axioms are assumed true. In the case of Theorem 3.6.6, the unstated hypotheses are the only hypotheses.

One final remark about writing proofs: Except for the gaps we have discussed, Euclid's proofs serve as excellent models for you to follow. Euclid usually includes just the right amount of detail and clearly states his reasons for each step in exactly the way that is advocated in this chapter. He also includes helpful explanations of where the proof is going, so that the reader has a better chance of understanding the big picture. In learning to write good proofs you can do no better than to study Euclid's proofs, especially those from Book I of the *Elements*. As you come to master those proofs you will begin to appreciate them more and more. You will eventually find yourself reading and enjoying not just the proofs themselves, but also Heath's commentary [30] on the proofs. Heath often explains why Euclid did things as he did and also indicates how other geometers have proved the same theorem. Of course Euclid uses language quite differently from the way we do. His theorem statements themselves do not serve as good models of the kind of precise statements that modern standards of rigor demand.

EXERCISES

3.1. Negate each of he following statements.
   (a) There exists a model for incidence geometry in which the Euclidean Parallel Postulate holds.
   (b) In every model for incidence geometry there are exactly seven points.
   (c) Every triangle has an angle sum of $180°$.
   (d) It is hot and humid outside.
   (e) My favorite color is red or green.
   (f) If the sun shines, then we go hiking.
   (g) All geometry students know how to write proofs.
3.2. Negate each of the parallel postulates stated in Chapter 2.
3.3. Construct truth tables that illustrate De Morgan's laws (page 34).
3.4. Identify the hypothesis and conclusion of each of the following statements.
   (a) If it rains, then I get wet.
   (b) If the sun shines, then we go hiking and biking.
   (c) If $x > 0$, then there exists a y such that $y^2 = 0$.
   (d) If $2x + 1 = 5$, then either $x = 2$ or $x = 3$.
   (e) Every triangle has angle sum less than or equal to $180°$.
3.5. Identify the hypothesis and conclusion of each of the following statements.
   (a) I can take topology if I pass geometry.
   (b) I get wet whenever it rains.
   (c) A number is divisible by 4 only if it is even.
3.6. State the converse and contrapositive of each of the statements in Exercise 3.4.
3.7. State the converse of Theorem 3.6.1.
3.8. Prove Theorem 3.6.2.
3.9. Prove Theorem 3.6.3.
3.10. Prove Theorem 3.6.4.
3.11. Prove Theorem 3.6.5.
3.12. Prove Theorem 3.6.6.
3.13. Prove Theorem 3.6.7.
3.14. Prove Theorem 3.6.8.

CHAPTER 4

Set Notation and the Real Numbers

4.1 SOME ELEMENTARY SET THEORY
4.2 PROPERTIES OF THE REAL NUMBERS
4.3 FUNCTIONS
4.4 THE FOUNDATIONS OF MATHEMATICS

In this chapter we briefly review two topics that will be used extensively in our exposition of geometry. Rather than treating geometry as an isolated subject, we will formulate the axioms of geometry in terms of set-theoretic terminology and base them on properties of the real number system.

4.1 SOME ELEMENTARY SET THEORY

Modern geometry, like most branches of modern mathematics, is formulated using set-theoretic terminology. In order to study the geometry presented in this book, you don't need to know a lot of set theory, but you do need to know at least the basic terminology and notation associated with the subject.

The words *set* and *element* are undefined terms in set theory, so we do not attempt to give precise definitions. The following informal definition will suffice: A *set* is a collection of objects, called *elements* of the set. In order to describe a set, it is necessary to specify the elements.

One way to specify the elements of a set is with a list or roster. Here are some examples.

$$A = \{3, 6, 29, -99\}$$
$$B = \{\text{Ed}, \text{Mary}\}$$
$$C = \{3, 6, 9, 12, \dots\}$$

The set A contains four elements, namely the numbers 3, 6, 29, and -99. The statement "6 is an element of A" is written $6 \in A$. The set B has two elements, Ed and Mary. The set C consists of all positive multiples of 3. It is not necessary (or possible) to list all multiples of 3, but just enough of them are listed so that any reasonable person will know what the dots represent.

Another way to specify the elements of a set is by means of a rule. For example,

$$D = \{x \mid x \text{ is a real number and } x > 0\}.$$

The set D consists of all real numbers greater than 0. So $3 \in D$ and $\pi \in D$, but $0 \notin D$. The symbol "$\mid$" is read "such that."

One set that has a special name is the *empty set*, denoted $\emptyset$, which is the set with no elements in it. The empty set is also called the *null set*. Two other sets that have special importance are the set of *natural numbers* (denoted $\mathbb{N}$) and the set of *integers* (denoted $\mathbb{Z}$).

$$\mathbb{N} = \{1, 2, 3, 4, 5, \dots\}$$
$$\mathbb{Z} = \{0, \pm 1, \pm 2, \pm 3, \dots\}$$

Note that the set C, defined above, can also be defined by a rule.

$$C = \{3n \mid n \in \mathbb{N}\}$$

We say that the a set S *is contained in* set T, written $S \subseteq T$, if every element of S is an element of T (i.e., $x \in S \Rightarrow x \in T$). In the examples above we have $C \subseteq D$, for example.

A set is completely determined by its elements, and the only thing that really matters about a set is what its elements are. Thus the assertion that $S = T$ means that S and T have the same elements. Here is an example of two sets E and F for which $E = F$.

$$E = \{3, 6, 9, 12\}$$
$$F = \{12, 3, 6, 3, 9, 3\}$$

The statement $S = T$ is equivalent to the compound statement $S \subseteq T$ and $T \subseteq S$. Any proof that $S = T$ should therefore have two parts, one proving that $x \in S \Rightarrow x \in T$ and the second proving that $x \in T \Rightarrow x \in S$.

Given two sets S and T, we can form new sets by combining S and T using the operations of union and intersection. The *intersection* of S and T, written $S \cap T$, and the *union* of S and T, written $S \cup T$, are defined as follows.

$$S \cap T = \{x \mid x \in S \text{ and } x \in T\}$$
$$S \cup T = \{x \mid x \in S \text{ or } x \in T\}$$

For example, if A and C are as above, then $A \cap C = \{3, 6\}$. Also, $\{2, 4, 6\} \cup \{2, 3, 4\} = \{2, 3, 4, 6\}$.

Another new set that can be formed out of sets S and T is the *set difference*, $S \setminus T$. The definition is this:

$$S \setminus T = \{x \mid x \in S \text{ but } x \notin T\}.$$

For example, if $E = \{2, 4, 6, 8, 10\}$ and $F = \{4, 8, 12, 16\}$, then $E \setminus F = \{2, 6, 10\}$ and $F \setminus E = \{12, 16\}$.

Let A and B be sets. We say that A *intersects* B (or that A and B have nonempty intersection) if $A \cap B \neq \emptyset$. We say that A and B are *disjoint* if $A \cap B = \emptyset$.

4.2 PROPERTIES OF THE REAL NUMBERS

This course will make extensive use of the real number system. In fact our entire development of geometry will be based on that system. You have surely had lots of experience with real numbers in high school and in calculus and, as a result, it should not be necessary to say a great deal about the real numbers here. In particular, it will be assumed that you know all the usual facts about the algebra of real numbers, such as the associative and distributive laws. The remainder of this chapter contains a few selected facts about real numbers that may not have been emphasized in your previous courses. These are presented as review material and no attempt is made to give complete proofs of all the results. Two sample proofs are included only because they serve as nice illustrations of some of the proof techniques that were discussed in the preceding chapter.

One of the most delicate aspects of working with real numbers is the problem of pinning down just exactly what a real number is. We will simply say that a real number is a number that can be represented by a (possibly infinite) decimal. This neatly solves the problem of defining what a real number is, but the price that must be paid is that it makes it surprisingly difficult to give a precise definition of such things as addition and multiplication. We do not intend to get into that problem here, but Exercise 4.10 does give some hint of the kind of difficulties that arise.

An important property of real numbers is the fact that they are ordered from smaller to larger. Here is a property of the ordering that we often take for granted but that will be important to us in geometry. Because of its importance it has been given a special name.

Axiom 4.2.1 (The Trichotomy Postulate). *If x and y are two real numbers, then exactly one of the following possibilities must hold:*

(1) $x < y$,

(2) $x = y$, or

(3) $x > y$.

Real numbers come in two types: rational and irrational. A *rational number* is a real number that can be written as a quotient of two integers. It is not too difficult to see that a real number is rational if and only if it has a decimal expansion that either terminates or repeats. An *irrational number* is any real number that is not rational. An irrational number has an infinite, nonrepeating decimal representation and cannot be written as a quotient of two integers. We use $\mathbb{R}$ to denote the set of real numbers and $\mathbb{Q}$ to denote the set of rational numbers. We do not have a special symbol for the set of irrational numbers, although $\mathbb{R} \smallsetminus \mathbb{Q}$ works.

It is easy to give examples of irrational numbers. Here is one.

$$x = 0.101001000100001000001\ldots$$

Even though there is a pattern in the digits, the expansion does not end with one block of digits that is repeated infinitely often.

Two irrational numbers that play important roles in geometry are π and $\sqrt{2}$. The number π is defined to be the ratio of the circumference of a circle to its diameter. The first rigorous proof that π is irrational was given by Johann Lambert in 1768. The reason $\sqrt{2}$ appears prominently in geometry is that it is the length of the diagonal of a square whose side has length 1. The Pythagoreans used a nice *reductio ad absurdum* argument to prove that $\sqrt{2}$ must be irrational.[1]

Proposition 4.2.2. *If x is a real number such that $x^2 = 2$, then x is irrational.*

Proof. Let x be a real number such that $x^2 = 2$ (hypothesis). Suppose x is rational (RAA hypothesis). Then x can be written as p/q, where p and q are both integers (definition of rational). Reduce the fraction p/q to lowest terms. Then p and q have no common factor; in particular, p and q are not both even. Since $p^2 = 2q^2$, p^2 must be even and so p is even.[2] Write $p = 2s$. Then $(2s)^2 = 2q^2$, so $q^2 = 2s^2$ and we see that q is even as well. We have now arrived at a contradiction: we started with p and q not both even and have concluded that they are both even. This contradiction forces us to reject the RAA hypothesis and conclude that x is irrational. $\square$

While most numbers we encounter in everyday life are rational, there is a technical sense in which there are far more irrational numbers than rational ones. We do not need to know that technical fact for this course, but we do need to know that both the rationals and the irrationals are *dense* in the set of real numbers. Informally this means that there are lots of numbers of each kind and wherever we look on the real line we will see both kinds. Here is a technical statement of what it means for these two sets of numbers to be dense.

Axiom 4.2.3 (The Density Postulate). *If a and b are real numbers such that $a < b$, then there exists a rational number x such that $a < x < b$ and there exists an irrational number y such that $a < y < b$.*

The axiom may be summarized this way: Between any two rational numbers there is an irrational number and between any two irrational numbers there is a rational number. This fact about rational and irrational numbers will play an important role in geometry. In particular, it will allow us to prove certain theorems about all real numbers by just proving that they hold for all rational numbers. The following theorem makes that principle precise.

---

[1]The argument in the proof is very old and is probably due to the followers of Pythagoras. They would not have stated the theorem as we have, however, since the concept of an irrational number would have been unknown to them. They would more likely have thought of this as a proof that the length of the diagonal of the unit square is not represented by a number—see [30, volume 2, page 112].

[2]We are using a nontrivial property of prime numbers (such as 2). The fact that p^2 is even means that 2 divides p^2. Since 2 is prime, this implies that 2 must divide p. More generally, if r is prime and r divides ab, then either r divides a or r divides b.

Theorem 4.2.4 (The Comparison Theorem). *If x and y are any real numbers such that*

(1) *every rational number that is less than x is also less than y, and*

(2) *every rational number that is less than y is also less than x,*

then $x = y$.

It is quite easy to give a proof of the Comparison Theorem based on trichotomy and the Density Postulate. The structure of the proof is a compound RAA argument.

Proof of Theorem 4.2.4. Let x and y be real numbers such that (1) every rational number that is less than x is also less than y, and (2) every rational number that is less than y is also less than x (hypothesis). We must prove that $x = y$. Either $x < y$, $x = y$, or $x > y$ (trichotomy). Suppose $x < y$ (RAA hypothesis). Then there exists a rational number r such that $x < r < y$ (Theorem 4.2.3). But this contradicts hypothesis (2). As a result, we must reject the RAA hypothesis $x < y$ and conclude that $x \geqslant y$. Suppose, now, that $x > y$ (another RAA hypothesis). Then there exists a rational number s such that $x > s > y$ (Theorem 4.2.3). But this contradicts hypothesis (1). Because of this contradiction we must also reject the RAA hypothesis $x > y$ and can conclude that $x = y$.   □

There are two more axioms regarding real numbers that will be important in our treatment of geometry. Here is the first.

Axiom 4.2.5 (The Archimedean Property of Real Numbers). *If M and ϵ are any two positive real numbers, then there exists a positive integer n such that $n\epsilon > M$.*

In order to appreciate the Archimedean Property, you should think of ϵ as being very small and M as being very large. It is a property of the real numbers that seems intuitively obvious to most people since it simply asserts that there are arbitrarily large natural numbers. (It can be restated to say that for every real number K there is a natural number n such that n is larger than K. This is an equivalent statement because we can take $K = M/\epsilon$.) From another point of view, however, it is deep; it is an assertion about how the inductively defined positive integers fit into the natural ordering of the real numbers, which are defined in an entirely different way.

The Archimedean Property is named after Archimedes of Syracuse, who lived soon after Euclid and who developed many of the infinite limiting constructions we now associate with calculus. The Archimedean Property will be used many times in this book, each time in a situation in which we are making a construction that is repeated over and over again indefinitely. Since this is exactly the kind of construction that Archimedes often used in his work, it is fitting that the property should be named after him.[3]

---

[3] Although Archimedes perfected such repeated constructions, they are already found in Euclid's *Elements*, and the property that is named after Archimedes is implicit in Definition 4 in Book V of the *Elements*.

There is one last postulate regarding the real numbers that is needed in geometry. It captures the deepest properties that distinguish the system of real numbers from other sets of numbers such as the rationals. Before we can state it we need a definition.

Definition 4.2.6. Let A be a set of real numbers. A number b is called an *upper bound* for A if $x \leq b$ for every $x \in A$. The number b_0 is called the *least upper bound* for A if b_0 is an upper bound for A and $b_0 \leq b$ for every b that is an upper bound for A.

Axiom 4.2.7 (The Least Upper Bound Postulate). *If A is any nonempty set of real numbers that has an upper bound, then A has a least upper bound.*

The Least Upper Bound Postulate may be familiar to you from high school algebra or from calculus. If not, you will certainly study it in your first course in real analysis, where it is used to give rigorous proofs of some of the basic theorems regarding continuous functions of real numbers (such as the Intermediate Value Theorem and the Maximum Value Theorem). Exercise 4.4.15 shows that the Least Upper Bound Postulate implies the Archimedean Property of real numbers. The Least Upper Bound Postulate is stronger than the Archimedean property in the sense that the rational numbers satisfy the Archimedean Property but do not satisfy the Least Upper Bound Postulate.

Our study of geometry will often lead us into repeated constructions that involve some sort of limiting process. In most such instances we will be able to use the Archimedean Property to reach the desired conclusion. But there will be times when we will need a stronger property of the real numbers. These situations usually involve some form of continuity. Whenever possible we will frame our proofs of continuity results in terms of the Intermediate Value Theorem since that theorem is familiar from calculus. There is only one point in this book where it is necessary to apply the Least Upper Bound Postulate directly; that point comes in the construction of limiting parallel rays in Chapter 8.

4.3 FUNCTIONS

In this section we briefly review a few facts about functions that will be used in the remainder of the book. The main goal is to define a special type of function, called a one-to-one correspondence, that will be important in our statements of the axioms for geometry.

Definition 4.3.1. Let A and B be sets. A *function from A to B* is a rule f that assigns to each element x of A a unique element $f(x)$ in B. The set A is called the *domain* of the function and the set B is called the *range* of the function.

Note that the domain and range are part of the definition of the function. We use the notation $f : A \rightarrow B$ to indicate that f is a function whose domain is A and whose range is B. The statement "f is a function whose domain is A and whose range is B" is usually shortened to "f is a function from A to B."

Definition 4.3.2. A function $f : A \rightarrow B$ is *one-to-one* (abbreviated 1-1) if $x_1 \neq x_2$ implies $f(x_1) \neq f(x_2)$. The function f is *onto* if for every $y \in B$ there exists an $x \in A$ such that $f(x) = y$.

Functions that are both one-to-one and onto will play an important role in our formulation of the axioms of geometry. They have been given a special name.

Definition 4.3.3. Let A and B be sets. A *one-to-one correspondence* from A to B is a function $f : A \rightarrow B$ that is both one-to-one and onto.

An example of a one-to-one correspondence from calculus is the function $f : \mathbb{R} \rightarrow (-\pi/2, \pi/2)$ defined by $f(x) = \arctan x$. The function $g : (-1, 1) \rightarrow [0, 1)$ defined by $g(x) = x^2$ is onto but not 1-1.[4] The function $h : \mathbb{R} \rightarrow \mathbb{R}$ defined by $h(x) = \sin x$ is neither 1-1 nor onto. On the other hand, the function $k : \mathbb{R} \rightarrow [-1, 1]$ defined by $k(x) = \sin x$ is onto. (It's still not 1-1.) Because of this difference we do not consider h and k to be the same function.

The kinds of distinctions made in the previous paragraph are often not important in calculus and precalculus courses. It is customary in calculus to speak of the function f defined by the rule $f(x) = \sqrt{x}$. The usual convention in calculus is that the domain is understood to be the largest set of real numbers for which the formula makes sense while the range is assumed to be the set of function values. Thus it is understood in calculus that the domain of the function f defined by the rule $f(x) = \sqrt{x}$ is $[0, \infty)$ and that the range consists of the same set.

In this course, by contrast, we will want to distinguish between the function $g : \mathbb{R} \rightarrow \mathbb{R}$ defined by $g(x) = x^2$ and the function $h : \mathbb{R} \rightarrow [0, \infty)$ defined by $h(x) = x^2$. A specific difference that is important here is the fact that h is onto but g is not. Neither g nor h is one-to-one. On the other hand, if we define $k : [0, \infty) \rightarrow [0, \infty)$ by $k(x) = x^2$, we obtain a function that is both one-to-one and onto. The same rule is used to define all three functions, but the functions are different because they have different domains and ranges.

You may find it necessary to adjust your thinking a bit to remember that the specification of the domain and range are part of the definition of the function. These distinctions are important to us because the definitions of one-to-one and onto only make sense if the domain and range are part of the definition of the function.

4.4   THE FOUNDATIONS OF MATHEMATICS

This is a course on the foundations of geometry. We intend to present geometry as an axiomatic system whose foundations are built on the rudiments of set theory and properties of the real numbers. One could, of course, take a broader view and examine the foundations of all of mathematics.

One way to organize the foundations of mathematics is to treat the various branches of mathematics separately. In particular, we could formulate axioms for geometry that do not rely on the real numbers at all. Such an axiomatic system for

---

[4]We are using the standard calculus notation for intervals of real numbers. For example, $[0, 1) = \{x \in \mathbb{R} \mid 0 \leqslant x < 1\}$.

geometry was proposed by David Hilbert and will be discussed in the next chapter. For reasons that will be explained in that chapter, we have chosen a different approach.

The way we are formulating the foundations, all of mathematics is ultimately built on set theory. Set theory itself is treated as an axiomatic system. As mentioned earlier, the words *set* and *element* are taken as undefined terms. Examples of standard axioms for set theory are the assertion that an empty set exists (Null Set Axiom) and the assertion that two sets are equal if and only if they contain the same elements (Axiom of Extensibility). All other branches of mathematics are built on set theory. In particular, the real numbers are constructed from set theory. (The natural numbers and the rational numbers are constructed first.)

Rather than constructing the real numbers, it is equally valid to treat the real numbers axiomatically. Many high school textbooks and real analysis textbooks give an axiomatic description of the real numbers, so you may already be familiar with such a treatment. That is in effect what we have done in this chapter by stating various axioms for the real numbers. In this course we will not concern ourselves with the construction of the real numbers; we will simply accept the real numbers as given and build geometry from there.

SUGGESTED READING

1. Chapter 3 of *Journey Through Genius*, [23].
2. Part II (pages 51–92) of *Euclid's Window*, [50].
3. Chapter 5 of *Geometry: Our Cultural Heritage*, [35].
4. Chapter XII of *Mathematics in Western Culture*, [40].

EXERCISES

4.1. Let $A = \{3, 6, 9, 12\}$ and $B = \{2, 4, 6, 8, 10, 12\}$. Find each of the following sets.
   (a) $A \cup B$.
   (b) $A \cap B$.
   (c) $A \smallsetminus B$.
   (d) $B \smallsetminus A$.

4.2. Write 24/7 as a decimal.

4.3. Convert $2.3\overline{571} = 2.3571571571571\ldots$ to a fraction.

4.4. Find an irrational number between 2.5834556 and 2.5834557.

4.5. Find a rational number between π and $\pi + 0.00001$.

4.6. Find functions $f : \mathbb{N} \to \mathbb{N}$ that have the following properties.
   (a) f is onto, but not one-to-one.
   (b) f is one-to-one, but not onto.
   (c) f is neither one-to-one nor onto.
   (d) f is a one-to-one correspondence.

4.7. Find a function $f : \{1, 3, 5\} \to \{2, 4, 6, 8\}$ that is one-to-one. Find a function $f : \{1, 3, 5, 7\} \to \{2, 4, 6\}$ that is onto.

4.8. Is there a function $f : \{1, 3, 5, 7\} \to \{2, 4, 6, 8\}$ that is onto but not one-to-one? Explain. Is there a function $f : \{1, 3, 5, 7\} \to \{2, 4, 6, 8\}$ that is one-to-one but not onto? Explain.

4.9. Verify that $f : (0, 1) \to (1, \infty)$ defined by $f(x) = 1/x$ is a one-to-one correspondence.

4.10. Let x and y be real numbers. This means that each of them has a decimal expansion. Write a rule for determining the sixth decimal digit of $x + y$ in terms of the digits in the decimal expansions of x and y. Try your rule on the numbers $x = 5.888888888888$ and $y = 1.111111111111$. Now try your rule on the numbers $x = 5.888888888888$ and $y = 1.111111111112$. How many of the digits in the expansions of x and y do you need to know in order to be certain of the sixth digit in $x + y$?

4.11. Use the technique of proof of Proposition 4.2.2 to prove that $\sqrt{3}$ is irrational. Be sure you understand why this works for $\sqrt{3}$ but does not work for $\sqrt{4}$.

4.12. Find the least upper bounds of the following sets of real numbers.

(a) $(0, 3)$.

(b) $[0, 3)$.

(c) $[0, 3]$.

(d) $[0, 3) \cup \{4\}$.

(e) $\mathbb{N}$.

4.13. Does the empty set have an upper bound? Does the empty set have a least upper bound? Explain.

4.14. Show that the rational numbers do not satisfy the Least Upper Bound Postulate.

4.15. The Least Upper Bound Postulate can be used to prove the Archimedean Property of Real Numbers. Here is an outline of the proof; your job is to fill in the details. Let $\epsilon > 0$ be given and define $A = \{n\epsilon \mid n \in \mathbb{N}\}$. It would suffice to prove that A does not have an upper bound (explain). Suppose A has an upper bound (RAA hypothesis). Let b be the least upper bound for A. Use the fact that b is the *least* upper bound for A to prove that there must be $a \in A$ such that $a > b - \epsilon$. Prove that $x \in A$ implies $x + \epsilon \in A$. But $a + \epsilon > b$, which contradicts the fact that b is an upper bound for A.

CHAPTER 5

The Axioms of Plane Geometry

5.1 SYSTEMS OF AXIOMS FOR GEOMETRY
5.2 THE UNDEFINED TERMS
5.3 EXISTENCE AND INCIDENCE
5.4 DISTANCE
5.5 PLANE SEPARATION
5.6 ANGLE MEASURE
5.7 BETWEENNESS AND THE CROSSBAR THEOREM
5.8 SIDE-ANGLE-SIDE
5.9 THE PARALLEL POSTULATES
5.10 MODELS

This is a course on the *foundations* of geometry and it is in the present chapter that we finally begin to lay those foundations. The preceding chapters explained the necessary background so that the reader can now appreciate what it means to organize geometry as an axiomatic system. The next task is to list the undefined terms and to state carefully the axioms that spell out what we assume about those terms.

The axioms are presented in two groups. The first six axioms will be common to all the geometries studied in this book. The geometry that can be done using only those six axioms is called *neutral geometry*, and we will prove our theorems in that setting whenever possible. At the end of this chapter we will state three postulates regarding parallelism. Much of the remainder of the course will be devoted to an exploration of the logical relationships between the neutral axioms and the various parallel postulates.

In addition to statements of the axioms themselves, the chapter includes a number of theorems. The theorems in this chapter are generally foundational in the sense that they spell out the kinds of details that Euclid tacitly assumed but did not prove. This chapter also establishes the basic terminology that will be used throughout the remainder of the course. As a result, the chapter contains a great many definitions. The good news is that most of them will already be familiar to you from your high school geometry course.

It should be noted that the axioms in this chapter are axioms for two-dimensional *plane geometry*. Additional axioms would be needed to describe the geometry of three-dimensional space.

5.1 SYSTEMS OF AXIOMS FOR GEOMETRY

The first question we must confront is the question of which axioms to use as we lay the foundations for the remainder of the course. We know that Euclid's postulates do not contain all the necessary assumptions, so we do not consider using them. But there are many alternative systems of axioms that can be used for geometry courses such as this and we must choose from among them.

One commonly used set of axioms is due to David Hilbert [34], who in the late nineteenth century made a special project of finding a good set of axioms for geometry. Hilbert's axioms are very much in the spirit of Euclid's original work in that they are purely geometric and do not make use of coordinates or the real numbers. There are a total of 20 axioms that carefully spell out exactly what is assumed in ordinary geometry. Their statements may be found in Appendix B. There are many good arguments in favor of using Hilbert's axioms for a college-level course such as this, and in the past it has been fairly standard to do so. The primary reason for this is the fact that Hilbert's axioms most faithfully preserve the spirit of Euclid's original work while building geometry on a completely rigorous foundation. For those who want to study geometry based on Hilbert's axioms, the treatments in the books by Greenberg [27] and Hartshorne [29] are highly recommended.

We will choose to use a different set of axioms, one based on the real number system. One reason we do not use Hilbert's axioms is the very fact that those axioms spell out in detail all of the geometric assumptions without making use of our prior knowledge of other parts of mathematics. The number of axioms is large and, even so, the development of geometry must begin with the proofs of many highly technical results that appear to be intuitively obvious to most people. A second reason we choose not to use Hilbert's axioms is the fact that all contemporary high school textbooks use axiom systems for geometry that are based on measurement and the real numbers. We use a similar system in this course in order to facilitate making direct connections with the high school course. This allows preservice teachers to learn geometry using axioms that are quite similar to those they will encounter when they enter the high school classroom. Since this course is designed to serve the needs of future high school geometry teachers, the latter consideration outweighs the arguments in favor of using Hilbert's axioms.[1]

---

[1] The main argument against the choice made here is the contention that geometry should stand alone as an axiomatic system. It is asserted that the real numbers and set theory contain many deep problems and that it is better to keep those complications out of geometry. This argument is not convincing, however, because a development of geometry based on Hilbert's axioms is forced to include axioms that have implicit in them the deepest properties of the real number system. For example, the development in [27] contains several axioms of continuity, one of which is at least as substantial as the Least Upper Bound Postulate. Even so, it soon becomes necessary to use real numbers to measure distances and angles. The measurement functions could in theory be constructed from Hilbert's axioms, but that construction is too lengthy and complicated for a course at this level; instead their existence is assumed. For example, Theorem 4.3 of [27], which asserts the existence of the measurement functions and lists their properties, is stated and used extensively but never proved. We prefer to include

In the early seventeenth century a great revolution took place in geometry when coordinates were introduced into the plane. The two mathematicians most closely identified with this development are René Descartes and Pierre de Fermat. The use of coordinates made it possible to bring powerful algebraic techniques to bear on geometric problems and led eventually to the development of the calculus. That perspective, in turn, allowed geometric problems to be analyzed using methods other than the deductive, axiomatic method. It is difficult for most of us to truly appreciate how revolutionary this approach was and to understand how foreign it would have seemed to Euclid. We have been taught since childhood to identify points in the plane with pairs of numbers and curves with equations, so we find it hard even to imagine the plane without a pair of coordinate axes imposed on it.

The approach to geometry in which points are identified with pairs of real numbers and geometric properties are studied via the use of algebraic operations and symbolism is called *analytic geometry*. One of the characteristic features of analytic geometry is the identification of a curve with an algebraic equation. The kind of geometry done by Euclid, in which points are viewed as pure geometric objects, coordinates are not used, and geometric truths are developed by deductive reasoning from axioms, is called *synthetic geometry*. Given how pervasive the analytic way of thinking is and how useful it has proved to be, it seems reasonable to incorporate coordinates into the foundations of geometry, and that is what we intend to do. In this approach, geometry is not treated as an isolated subject, but, like many other branches of mathematics, is built on the real number system.

The kind of axiom system we will use, which is based on the real numbers and measurement, was originated by George David Birkhoff in 1932 [7]. Birkhoff's system brings together the two different approaches to geometry: The axiomatic, synthetic approach of Euclid and Hilbert is to some extent merged with the analytic approach of Descartes and Fermat. Real numbers and coordinates had for centuries been freely used in analytic geometry, but it was Birkhoff who took the step of introducing them into the foundations by incorporating the real numbers directly into the axioms of geometry. He recognized that this idea would make geometry more accessible to students and he published a high school textbook [8] that was based on his axioms and is still in print today.

Birkhoff's system has just four simple postulates from which he deduces all of Euclidean geometry. One postulate is the usual assumption that two points determine a unique line. His Postulate of Line Measure is where the real numbers first enter; he assumes that the points on a line can be put into one-to-one correspondence with the real numbers in a way that is compatible with distance measurement. His third postulate asserts that angles can be measured by real numbers. His final postulate makes assumptions regarding similar triangles. Complete statements of Birkhoff's axioms may be found in Appendix B.

The question of which axioms to use in school geometry courses was revisited in the 1960s when the "new math" curricula were being developed. Saunders MacLane [46] proposed a set of axioms that are like Birkhoff's in that they are

---

any unproven assumptions in the axioms.

based on the real numbers. MacLane stated his axioms regarding lines in terms of a function that measures distances (a *metric*). MacLane also introduced an axiom, which he named the Continuity Axiom, that incorporates what we will call the Crossbar Theorem into the foundations.

Soon thereafter the School Mathematics Study Group (SMSG) developed a system of axioms that was specifically designed for use in high school geometry courses. This system based geometry on axioms that are like those of Birkhoff and MacLane in that they emphasize measurement and build on the real numbers but are different in that they greatly expand on the austere, minimalist axioms of Birkhoff. The SMSG system includes a fairly large number of axioms, several of them not strictly necessary in the sense that they are logically implied by other axioms. The idea is that, in the high school setting, it is better to risk introducing some redundancy into the axioms by assuming more than is absolutely necessary than it is to concentrate excessively on proving all the basic theorems from a minimal set of axioms. This approach allows the geometry course to move almost immediately to more interesting and less intuitively obvious results. Statements of the SMSG postulates may be found in Appendix B. The SMSG geometry textbook [64] was only published in mimeographed form but Edwin E. Moise published two books based on modified versions of the SMSG axioms and those books are still in print today. Moise collaborated with Floyd Downs on a high school-level textbook [52] and also wrote a college-level textbook [51] that continues to be quite influential.

The axioms used in current high school textbooks are very close to those of SMSG. The main difference between the currently used axioms and the SMSG axioms is that the transformational approach has been incorporated into the axioms. A good example of a high school textbook that is in wide use at the present time is the geometry textbook [71] in the University of Chicago School Mathematics Project (UCSMP) textbook series. It includes a "Reflection Postulate," which asserts that certain transformations exist and have specified properties.

We will take a middle course by adopting a system of six axioms that combines features of the axiom systems of Birkhoff, MacLane, Moise, SMSG, and UCSMP. A major consideration in the design of the axioms used in this book is the goal of keeping the axioms as close as possible to those used in contemporary high school textbooks since that facilitates making connections between this course and the high school course. Here is an example of one implication of that choice: We will use the Plane Separation Postulate of Hilbert and SMSG as one of our axioms even though some of the proofs that appear later in this chapter could be simplified or eliminated if we adopted MacLane's Continuity Axiom instead. We will see that the Plane Separation Axiom is strong enough to prove what we need, but our treatment will also show that such consequences as the Crossbar Theorem require proof.

The axioms in this chapter will be stated in language that is as close as possible to that used in UCSMP [71]. The language cannot be exactly the same as UCSMP because we want to eliminate some of the redundancy and because we are aiming for a more careful, rigorous, and logical development of the subject than is appropriate in a high school course. We also do not follow UCSMP in taking transformation as a foundational concept. We make this choice because understanding geometric relationships in terms of functions of points and angles involves a somewhat higher level of thinking than does working with points and angles directly. For that reason we postpone a consideration of the transformational approach to the foundations until later in the book.

The axioms that will be used in this book are stated and explained in the remainder of the chapter.

5.2   THE UNDEFINED TERMS

For now there are five undefined terms, namely *point*, *line*, *distance*, *half-plane*, and *angle measure*. In a later chapter we will add *area* to the list, for a total of six undefined terms. In the remainder of this chapter we will explain exactly what is to be assumed about the first five undefined terms. For each of the five terms there is an axiom that spells out what we need to know about that term. There is one additional axiom that specifies how distance and angle measure interact with each other.

5.3   EXISTENCE AND INCIDENCE

The first axiom states our rudimentary assumptions about points. The main thing we need to assume about them is that they exist. We will seldom find it necessary to invoke this postulate explicitly, but it is reassuring to have it around. Without it we run the risk of making complicated statements about nothing.

It is enough to assume the existence of two points because the existence of more points follows from that assumption together with the other postulates stated later in the chapter. Such relationships will be explored in the exercises.

Axiom 5.3.1 (The Existence Postulate). *The collection of all points forms a nonempty set. There is more than one point in that set.*

Definition 5.3.2. The set of all points is called *the plane* and is denoted by $\mathbb{P}$.

The second axiom states our most basic assumptions regarding lines. It is the same as the first axiom of incidence geometry and is also essentially the same as Euclid's first postulate. In his postulate Euclid only asserted that two points determine a line and did not explicitly say that the line is unique, but it is clear from his proofs that he meant it to be unique. As mentioned in an earlier chapter, the word *distinct* simply means "not equal."

Axiom 5.3.3 (The Incidence Postulate). *Every line is a set of points. For every pair of distinct points A and B there is exactly one line ℓ such that $A \in \ell$ and $B \in \ell$.*

FIGURE 5.1: Two points determine a unique line

Notation. We use the notation $\overleftrightarrow{AB}$ to indicate the line determined by A and B.

Since a line is a set of points, we can define what it means for a point to lie on a line and it is not necessary to take "lie on" as one of the undefined terms.

Definition 5.3.4. A point P is said to *lie on* line ℓ if $P \in \ell$. The statements "P is incident with ℓ" and "ℓ is incident with P" are also used to specify the same relationship.

It is convenient to have a name for the opposite of lie on.

Definition 5.3.5. A point Q is called an *external point* for line ℓ if Q does not lie on ℓ.

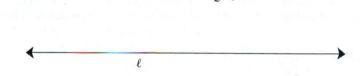

FIGURE 5.2: The point Q is an external point for line ℓ

This is a good time to repeat a definition that was made earlier.

Definition 5.3.6. Two lines ℓ and m are said to be *parallel*, written $\ell \parallel m$, if there is no point P such that P lies on both ℓ and m (i.e., if $\ell \cap m = \emptyset$).

Be sure to notice that, according to this definition, a line is not parallel to itself. This may conflict with the way the word *parallel* was used in your high school geometry course.

We are now ready for our first theorem.

Theorem 5.3.7. *If ℓ and m are two distinct, nonparallel lines, then there exists exactly one point P such that P lies on both ℓ and m.*

Proof. This is a restatement of the first theorem of incidence geometry, Theorem 3.6.1. The proof given earlier is still valid since it was based on just the Incidence Postulate. □

The theorem can be restated as a type of trichotomy for pairs of lines. If ℓ and m are two lines, then exactly one of the following conditions will hold: Either $\ell = m$, $\ell \parallel m$, or $\ell \cap m$ consists of precisely one point.

5.4 DISTANCE

The third axiom spells out what we assume regarding the undefined term *distance*. It is one of the ways in which the algebra of real numbers is brought into the axioms of geometry.

Axiom 5.4.1 (The Ruler Postulate). *For every pair of points P and Q there exists a real number PQ, called the distance from P to Q. For each line ℓ there is a one-to-one correspondence from ℓ to ℝ such that if P and Q are points on the line that correspond to the real numbers x and y, respectively, then PQ = |x − y|.*

There are several additional terms that can be defined using the undefined term *distance*. The next few definitions are fundamental to our development of geometry.

Definition 5.4.2. Let A, B, and C be three distinct points. The point C is *between A and B*, written $A * C * B$, if $C \in \overleftrightarrow{AB}$ and $AC + CB = AB$.

We will see in the next chapter that the condition $AC + CB = AB$ implies that $C \in \overleftrightarrow{AB}$, so we will be able to drop the assumption $C \in \overleftrightarrow{AB}$ from the definition of between. But that proof relies on axioms we have not yet stated. Without additional assumptions both conditions are necessary; see Exercises 5.24 and 5.25.

Later in this section we will prove that $A * C * B$ if and only if $B * C * A$. In Section 5.7 we will study betweenness in more depth.

Definition 5.4.3. Define the *segment* $\overline{AB}$ by

$$\overline{AB} = \{A, B\} \cup \{P \mid A * P * B\}$$

and the *ray* $\overrightarrow{AB}$ by

$$\overrightarrow{AB} = \overline{AB} \cup \{P \mid A * B * P\}.$$

The segment $\overline{AB}$ consists of the two points A and B together with all the points between A and B. The ray $\overrightarrow{AB}$ contains the segment $\overline{AB}$ together with all the points that are "beyond" B in the sense that B is between A and the point. The notation used is consistent with the notation $\overleftrightarrow{AB}$ for the line determined by A and B. In each case the notation suggests the shape of the set. (See Fig. 5.3.)

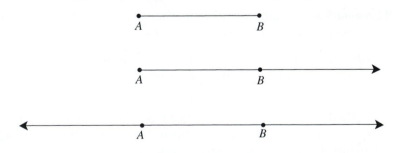

FIGURE 5.3: Segment, ray, line

Definition 5.4.4. The *length* of segment $\overline{AB}$ is AB, the distance from A to B. Two segments $\overline{AB}$ and $\overline{CD}$ are said to be *congruent*, written $\overline{AB} \cong \overline{CD}$, if the segments have the same length.

Definition 5.4.5. The points A and B are the *endpoints* of the segment $\overline{AB}$; all other points of $\overline{AB}$ are *interior points*. The point A is the *endpoint* of the ray $\overrightarrow{AB}$.

Before moving on to statements of the remaining axioms, we pause to examine the Ruler Postulate in more depth. The Ruler Postulate really asserts the existence of two different kinds of functions. There is one function that measures distances between points. In addition, for each line there is a function that assigns coordinates to points on that line in a way that is consistent with the overall distance-measuring function. We will examine each of these aspects of the postulate separately.

The next theorem lists the basic properties of the function that measures distance. The proof illustrates the way in which the postulate allows us to bring facts about the algebra of real numbers to bear on geometry.

Theorem 5.4.6. *If P and Q are any two points, then*

1. *$PQ = QP$,*
2. *$PQ \geqslant 0$, and*
3. *$PQ = 0$ if and only if $P = Q$.*

Proof. Let P and Q be two points (hypothesis). We will first show that there is a line ℓ such that both P and Q lie on ℓ. Either $P = Q$ or $P \neq Q$ and we consider each case separately. Suppose $P \neq Q$. Then there is exactly one line $\ell = \overleftrightarrow{PQ}$ such that P and Q lie on ℓ (Incidence Postulate). If $P = Q$, there must be another point $R \neq P$ (Existence Postulate). In that case we take ℓ to be the unique line such that P and R lie on ℓ (Incidence Postulate). In either case ℓ is a line such that both P and Q lie on ℓ.

There exists a one-to-one correspondence from ℓ to $\mathbb{R}$ having the properties specified in the Ruler Postulate. In particular, P and Q correspond to real numbers x and y, respectively, with $PQ = |x - y|$ and $QP = |y - x|$ (Ruler Postulate). But $|x - y| = |y - x|$ (algebra), so $PQ = QP$. Also $|x - y| \geqslant 0$ (algebra), so $PQ \geqslant 0$. This proves conclusions (1) and (2) of the theorem.

Suppose $PQ = 0$ (hypothesis). Then $|x - y| = 0$, so $x = y$ (algebra). Therefore, $P = Q$ (the correspondence is one-to-one). Finally, if $P = Q$, then $x = y$ (the correspondence is a function), so $PQ = |x - x| = 0$. This completes the proof of both parts of (3). $\qquad\square$

Corollary 5.4.7. *$A * C * B$ if and only if $B * C * A$.*

Proof. Let A, B, and C be three points such that $C \in \overleftrightarrow{AB}$ (hypothesis). If $A * C * B$, then $AC + CB = AB$ (definition). Since $AB = BA$, $AC = CA$, and $CB = BC$, it is also the case that $BC + CA = BA$. Therefore, $B * C * A$. The proof of the converse is similar. $\qquad\square$

Any function that has the properties spelled out in Theorem 5.4.6 can be used to measure distances. Such a function is called a *metric*.

Definition 5.4.8. A *metric* is a function $D : \mathbb{P} \times \mathbb{P} \to \mathbb{R}$ such that

1. $D(P, Q) = D(Q, P)$ for every P and Q,
2. $D(P, Q) \geq 0$ for every P and Q, and
3. $D(P, Q) = 0$ if and only if $P = Q$.

A version of the triangle inequality

$$D(P, Q) \leq D(P, R) + D(R, Q)$$

is usually included as part of the definition of metric, but we do not include it in the definition because we will prove the triangle inequality as a theorem in the next chapter.[2]

The familiar distance formula from high school geometry is an example of a metric.

EXAMPLE 5.4.9   The Euclidean metric

Define the distance between points (x_1, y_1) and (x_2, y_2) in the Cartesian plane by

$$d((x_1, y_1), (x_2, y_2)) = \sqrt{(x_2 - x_1)^2 + (y_2 - y_1)^2}.$$

This metric is called the *Euclidean metric*. The verification that d satisfies the conditions in the definition of metric is left as an exercise (Exercise 5.1). ∎

There are other ways in which to measure distances in the Cartesian plane. Here is an example that will be important later in the chapter.

EXAMPLE 5.4.10   The taxicab metric

Define the distance between points (x_1, y_1) and (x_2, y_2) in the Cartesian plane by

$$\rho((x_1, y_1), (x_2, y_2)) = |x_2 - x_1| + |y_2 - y_1|.$$

This metric is called the *taxicab metric*. The verification that ρ satisfies the conditions in the definition of metric is left as an exercise (Exercise 5.2).

The name *taxicab* is given to this metric because the distance from (x_1, y_1) to (x_2, y_2) is measured by traveling along a line parallel to the x-axis and then along one parallel to the y-axis in much the same way that a taxicab in Manhattan must follow a rectangular grid of streets and cannot travel directly from point A to point B. ∎

---

[2] What we are calling a metric should more properly be called a *semimetric*. However, all the examples of metrics that we will consider satisfy the triangle inequality and are metrics according to the standard definition. We omit the triangle inequality from the definition only because it is an unnecessary assumption in this context.

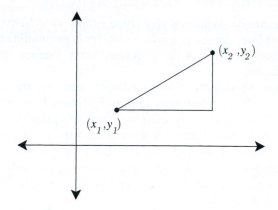

FIGURE 5.4: The Euclidean distance from (x_1, y_1) to (x_2, y_2) is the length of the "straight" line segment while the taxicab distance is the sum of the horizontal and vertical lengths

We now turn our attention to the second part of the Ruler Postulate. The assertion that there is a one-to-one correspondence from ℓ to $\mathbb{R}$ means that there is a function $f : \ell \to \mathbb{R}$ such that f is both one-to-one and onto. The postulate says that for any two points P and Q that lie on ℓ, $PQ = |f(P) - f(Q)|$. Such functions are important because they facilitate the transition from geometry to algebra.

Definition 5.4.11. Let ℓ be a line. A one-to-one correspondence $f : \ell \to \mathbb{R}$ such that $PQ = |f(P) - f(Q)|$ for every P and Q on ℓ is called a *coordinate function* for the line ℓ and the number $f(P)$ is called the *coordinate of the point P*.

It is possible to construct coordinate functions associated with each of the metrics we have defined.

EXAMPLE 5.4.12   Coordinate functions in the Euclidean metric

If ℓ is a nonvertical line in the Cartesian plane, then ℓ has an equation of the form $y = mx + b$. In that case we define $f : \ell \to \mathbb{R}$ by $f(x, y) = x\sqrt{1 + m^2}$. If ℓ is a vertical line, then it has an equation of the form $x = a$. In that case we define f by $f(a, y) = y$. The verification that these functions are coordinate functions in the Euclidean metric is left as an exercise (Exercise 5.3). ∎

EXAMPLE 5.4.13   Coordinate functions in the taxicab metric

If ℓ is a nonvertical line in the Cartesian plane with equation $y = mx + b$, define $f : \ell \to \mathbb{R}$ by $f(x, y) = x(1 + |m|)$. If ℓ is a vertical line with equation $x = a$, define f by $f(a, y) = y$. The verification that these functions are coordinate functions in the taxicab metric is left as an exercise (Exercise 5.4). ∎

It should be emphasized that these coordinate functions on the Cartesian plane are merely examples. We are not restricting ourselves to the Cartesian plane in our study of geometry. Later we will construct coordinate functions on lines in the Klein disk, for example. While we give these examples to help clarify the meaning of the Ruler Postulate, our approach to geometry is axiomatic and we remain open to any possible interpretation of the undefined term *distance* that could serve as a model for the axioms.

Using the definition of coordinate function, the second part of the Ruler Postulate can be restated quite simply: *For every line there is a coordinate function.* The postulate asserts only existence, not uniqueness; each line can have many different coordinate functions associated with it.   Given two points on ℓ, it is often convenient to arrange that the first is assigned the coordinate 0 and and the second has positive coordinate. The following refinement of the Ruler Postulate is an example of one of the redundant postulates in the SMSG system (see SMSG Postulate 4, Appendix B).

Theorem 5.4.14 (The Ruler Placement Postulate). *For every pair of distinct points P and Q, there is a coordinate function* $f : \overleftrightarrow{PQ} \to \mathbb{R}$ *such that* $f(P) = 0$ *and* $f(Q) > 0.$

Proof. Exercise 5.6.   □

The final aspect of the Ruler Postulate that we wish to examine is the fact that it specifies that a coordinate function is a one-to-one correspondence between points on a line and the *real* numbers. The fact that the real numbers are used, rather than some other system of numbers, ensures that there are no "holes" or "gaps" in a line or a circle. There is nothing in the other postulates we have stated that ensures this, but it is a property we need.

EXAMPLE 5.4.15   The rational plane

Interpret *point* to mean an ordered pair of rational numbers. The collection of such ordered pairs is called the *rational plane* and a point in the rational plane is called a *rational point*. Since every rational number is also a real number, the rational plane is a subset of the Cartesian plane. We interpret *line* to mean all the rational points on a Cartesian line that has rational slope and intercept. In other words, a line in the rational plane is a set of the form

$$\ell = \{(r, s) \mid r \text{ and } s \text{ are rational numbers and } ar + bs + c = 0\}$$

for some fixed rational numbers a, b, and c with a and b not both 0. We measure distances in the rational plane using the Euclidean metric.

The Euclidean metric is a metric on the rational plane, so the rational plane satisfies the first part of the Ruler Postulate. It also satisfies the Existence and Incidence Postulates. The rational plane does not satisfy the second part of the Ruler Postulate because there is no one-to-one correspondence between the points on a line in the rational plane and the real numbers. It is easy to see that there is a one-to-one correspondence between points on a rational line and the rational numbers. But a theorem from set theory asserts that there is no one-to-one correspondence from $\mathbb{Q}$ to $\mathbb{R}$, so there can be no one-to-one correspondence between the points on a rational line and the real numbers.   ■

The rational plane satisfies all five of Euclid's postulates. However, the proof of Euclid's very first proposition, which we examined in Chapter 1, breaks down in the rational plane.

EXAMPLE 5.4.16   Circles in the rational plane

Consider the rational points $A = (0,0)$ and $B = (2,0)$. The circles of radius 2 centered at A and B do not intersect in the rational plane. There are two points in the Cartesian plane at which the corresponding circles intersect, namely $(1, \pm\sqrt{3})$, but those points are not points in the rational plane. ∎

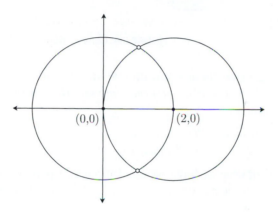

FIGURE 5.5: Two circles in the rational plane

The rational plane shows that Euclid was indeed using unstated hypotheses in his proofs. Since the rational plane satisfies all of Euclid's postulates, Euclid's implicit assertion in the proof of Proposition 1 that the two circles he constructs must intersect is based on assumptions that are not made explicit in the postulates. According to modern standards of rigor, this means that there is a gap in Euclid's proof. Our assumption that coordinate functions use real numbers will allow us to fill that gap. Specifically, we will prove that if one circle contains a point that is inside a second circle and another point that is outside the second circle, then the two circles must intersect. Even using the powerful axioms in this chapter the proof is not easy, however, and we will not attempt that particular proof until Chapter 10 (see Theorem 10.5.4).

5.5 PLANE SEPARATION

The fourth axiom tells us what we need to know about half-planes. It guarantees that the plane is two-dimensional in the sense that a line separates it into two disjoint half-planes. This axiom also allows us to define angle, the interior of an angle, and triangle.

Definition 5.5.1. A set of points S is said to be a *convex set* if for every pair of points A and B in S, the entire segment $\overline{AB}$ is contained in S.

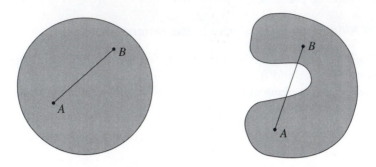

FIGURE 5.6: Convex, not convex

Axiom 5.5.2 (The Plane Separation Postulate). *For every line ℓ, the points that do not lie on ℓ form two disjoint, nonempty sets H_1 and H_2, called half-planes bounded by ℓ, such that the following conditions are satisfied.*

1. *Each of H_1 and H_2 is convex.*
2. *If $P \in H_1$ and $Q \in H_2$, then $\overline{PQ}$ intersects ℓ.*

Some of what we have assumed about half-planes can be more succinctly stated using set theoretic notation. Here, in symbols, is what the postulate says about H_1 and H_2.

- $H_1 \cup H_2 = \mathbb{P} \smallsetminus \ell$.
- $H_1 \cap H_2 = \emptyset$.
- $H_1 \neq \emptyset$ and $H_2 \neq \emptyset$.
- If $A \in H_1$ and $B \in H_1$, then $\overline{AB} \subseteq H_1$ and $\overline{AB} \cap \ell = \emptyset$.
- If $A \in H_2$ and $B \in H_2$, then $\overline{AB} \subseteq H_2$ and $\overline{AB} \cap \ell = \emptyset$.
- If $A \in H_1$ and $B \in H_2$, then $\overline{AB} \cap \ell \neq \emptyset$.

Notation. Let ℓ be a line and let A be an external point. We use the notation H_A to denote the half-plane bounded by ℓ that contains A.

Definition 5.5.3. Let ℓ be a line and let A and B be two external points. We say that A and B are *on the same side of ℓ* if they are both in H_1 or both in H_2. The points A and B are *on opposite sides of ℓ* if one is in H_1 and the other is in H_2.

Equivalently, A and B are on opposite sides of ℓ if and only if $B \notin H_A$ and A and B are on the same side of ℓ if and only if $B \in H_A$. Using the definitions above, Axiom 5.5.2 can be restated as follows.

Proposition 5.5.4. *Let ℓ be a line and let A and B be points that do not lie on ℓ. The points A and B are on the same side of ℓ if and only if $\overline{AB} \cap \ell = \emptyset$. The points A and B are on opposite sides of ℓ if and only if $\overline{AB} \cap \ell \neq \emptyset$.*

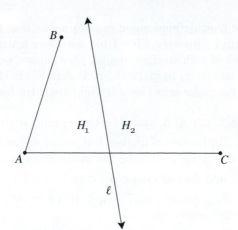

FIGURE 5.7: *A* and *B* are on the same side of ℓ; *A* and *C* are on opposite sides of ℓ

Definition 5.5.5. Two rays $\overrightarrow{AB}$ and $\overrightarrow{AC}$ having the same endpoint are *opposite* rays if the two rays are unequal but $\overleftrightarrow{AB} = \overleftrightarrow{AC}$. Otherwise they are *nonopposite*.

Another way to state the definition is this: $\overrightarrow{AB}$ and $\overrightarrow{AC}$ are opposite rays if $B * A * C$.

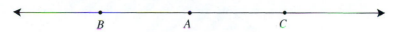

FIGURE 5.8: Opposite rays

Definition 5.5.6. An *angle* is the union of two nonopposite rays $\overrightarrow{AB}$ and $\overrightarrow{AC}$ sharing the same endpoint. The angle is denoted by either $\angle BAC$ or $\angle CAB$. The point *A* is called the *vertex* of the angle and the rays $\overrightarrow{AB}$ and $\overrightarrow{AC}$ are called the *sides* of the angle.

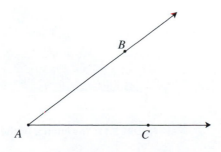

FIGURE 5.9: An angle

Notice that the angle $\angle BAC$ is the same as the angle $\angle CAB$.

Remark. We have defined angle in such a way that there is no angle determined by opposite rays. This may differ from what you learned in high school, where you probably worked with straight angles. One reason we define angle the way we do is so that we can go on to make the next definition. The definition of interior of an angle would not make sense for a straight angle (or for larger angles).

Definition 5.5.7. Let A, B, and C be three points such that the rays $\overrightarrow{AB}$ and $\overrightarrow{AC}$ are nonopposite. The *interior* of angle $\angle BAC$ is defined as follows. If $\overrightarrow{AB} \neq \overrightarrow{AC}$, then the interior of $\angle BAC$ is defined to be the intersection of the half-plane H_B determined by B and $\overleftrightarrow{AC}$ and the half-plane H_C determined by C and $\overleftrightarrow{AB}$ (i.e., the interior of $\angle BAC$ is the set of points $H_B \cap H_C$). If $\overrightarrow{AB} = \overrightarrow{AC}$, then the interior of $\angle BAC$ is defined to be the empty set.

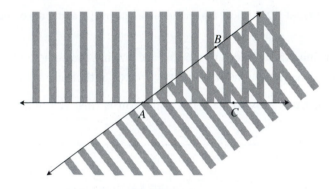

FIGURE 5.10: The two half-planes determined by an angle

The interior of any angle is a convex set; see Exercise 5.9. Another way to state the definition is this: A point P is in the interior of $\angle BAC$ if P is on the same side of $\overleftrightarrow{AB}$ as C and on the same side of $\overleftrightarrow{AC}$ as B.

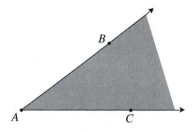

FIGURE 5.11: The interior of the angle

Definition 5.5.8. Three points A, B, and C are *collinear* if there exists one line ℓ such that A, B, and C all lie on ℓ. The points are *noncollinear* otherwise.

Observation. If A, B, and C are noncollinear, then rays $\overrightarrow{AB}$ and $\overrightarrow{AC}$ are neither opposite nor equal.

Definition 5.5.9. Let A, B, and C be three noncollinear points. The *triangle* $\triangle ABC$ consists of the union of the three segments $\overline{AB}$, $\overline{BC}$, and $\overline{AC}$; that is,

$$\triangle ABC = \overline{AB} \cup \overline{BC} \cup \overline{AC}.$$

The points A, B, and C are called the *vertices* of the triangle and the segments $\overline{AB}$, $\overline{BC}$, and $\overline{AC}$ are called the *sides* of the triangle.

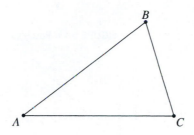

FIGURE 5.12: A triangle

Notice that we have only defined $\triangle ABC$ in case A, B, and C are noncollinear points. Thus the statement "$\triangle ABC$ is a triangle" is to be interpreted to mean, among other things, that the points A, B, and C are noncollinear.

The following theorem is sometimes taken as an axiom. It is a consequence of our Plane Separation Postulate. In fact it is really just a restatement of the postulate in different terminology.

Theorem 5.5.10 (Pasch's Axiom). *Let $\triangle ABC$ be a triangle and let ℓ be a line such that none of A, B, and C lies on ℓ. If ℓ intersects $\overline{AB}$, then ℓ also intersects either $\overline{AC}$ or $\overline{BC}$.*

Proof. Let $\triangle ABC$ be a triangle and ℓ be a line such that ℓ intersects $\overline{AB}$ and none of the points A, B, and C lies on ℓ (hypothesis). Let H_1 and H_2 be the two half-planes determined by ℓ (Axiom 5.5.2). The points A and B are in opposite half-planes (hypothesis and Proposition 5.5.4). Let us say that $A \in H_1$ and $B \in H_2$ (notation). It must be the case that either C is in H_1 or C is in H_2 (Axiom 5.5.2). If $C \in H_2$, then $\overline{AC} \cap \ell \neq \varnothing$ (Axiom 5.5.2, Part 2). If $C \in H_1$, then $\overline{BC} \cap \ell \neq \varnothing$ (Axiom 5.5.2, Part 2). $\qquad \square$

5.6 ANGLE MEASURE

The fifth axiom spells out the properties of angle measure (the last undefined term). First we need a definition.

Definition 5.6.1. Ray $\overrightarrow{AD}$ is *between* rays $\overrightarrow{AB}$ and $\overrightarrow{AC}$ if D is in the interior of $\angle BAC$.

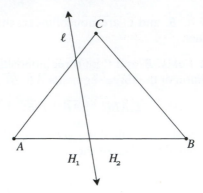

FIGURE 5.13: Pasch's Axiom

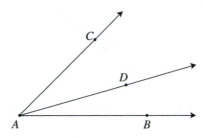

FIGURE 5.14: Betweenness for rays

Notice that the definition of betweenness for rays appears to depend on which point D is used to define the ray. However, it follows from a theorem to be proved later (Theorem 5.7.7) that if $\overrightarrow{AD} = \overrightarrow{AD'}$, then D is in the interior of $\angle BAC$ if and only if D' is in the interior of $\angle BAC$. Therefore, it does not matter which point D is used to describe the ray.

Axiom 5.6.2 (The Protractor Postulate). *For every angle $\angle BAC$ there is a real number $\mu(\angle BAC)$, called the measure of $\angle BAC$, such that the following conditions are satisfied.*

1. $0° \leqslant \mu(\angle BAC) < 180°$ *for every angle $\angle BAC$.*
2. $\mu(\angle BAC) = 0°$ *if and only if $\overrightarrow{AB} = \overrightarrow{AC}$.*
3. (Angle Construction Postulate) *For each real number r, $0 < r < 180$, and for each half-plane H bounded by $\overleftrightarrow{AB}$ there exists a unique ray $\overrightarrow{AE}$ such that E is in H and $\mu(\angle BAE) = r°$.*
4. (Angle Addition Postulate) *If the ray $\overrightarrow{AD}$ is between rays $\overrightarrow{AB}$ and $\overrightarrow{AC}$, then*

$$\mu(\angle BAD) + \mu(\angle DAC) = \mu(\angle BAC).$$

In Part 3 of the postulate, it is the *ray* that is unique, not the point E. Many different points determine the same ray.

We define congruence of angles using angle measure in much the same way as we defined segment congruence using distance.

Definition 5.6.3. Two angles $\angle BAC$ and $\angle EDF$ are said to be *congruent*, written $\angle BAC \cong \angle EDF$, if $\mu(\angle BAC) = \mu(\angle EDF)$.

The Protractor Postulate tells us that every angle has a measure that is a real number less than $180°$. We will divide angles into three types, depending on how the measure compares with $90°$.

Definition 5.6.4. Angle $\angle BAC$ is a *right angle* if $\mu(\angle BAC) = 90°$, $\angle BAC$ is an *acute angle* if $\mu(\angle BAC) < 90°$, and $\angle BAC$ is an *obtuse angle* if $\mu(\angle BAC) > 90°$.

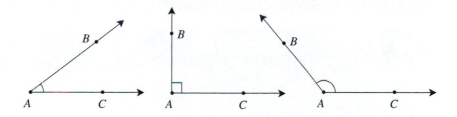

FIGURE 5.15: Types of angles: acute, right, and obtuse

We are familiar with the measurement of angles from previous mathematics courses and it is intuitively clear that a measurement function like that described in the Protractor Postulate exists in the Cartesian plane. It is not easy to write down a precise definition, however, and we make no attempt to do so at this point. Most of this book is devoted to an axiomatic investigation of geometry; so it is not necessary to have a definition of angle measure, we simply assume that angle measure exists and work from there. In Chapter 13 we will come back to the question of the existence of models and at that time we will define angle measure in the Cartesian plane.

Even though we do not give a definition of angle measure, we can draw diagrams that illustrate our intuitive understanding. For example, in Fig. 5.16 we have $\mu(\angle BAC) = 20°$, $\mu(\angle BAD) = 45°$, $\mu(\angle BAE) = 140°$, and (by the Angle Addition Postulate) $\mu(\angle CAD) = 25°$.

The units of angle measure that we use are called *degrees* and are denoted by the symbol $°$. Other units can be used; for example, radian measure is used in many mathematics courses and you are familiar with it from your study of calculus. It is traditional to use degree measure in a geometry course such as this. Degrees are dimensionless and the measure is really just a real number. We will sometimes omit the symbol $°$ and just use the number.

The Protractor Postulate can be used to establish a one-to-one correspondence between angles based on one side of the ray $\overrightarrow{AB}$ and the real numbers in the interval

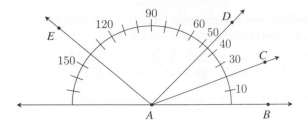

FIGURE 5.16: Measuring angles

[0, 180). In fact, Birkhoff first stated the Protractor Postulate in terms of just such a one-to-one correspondence (see Appendix B). We choose to state the postulate the way we do because we want to stay as close as possible to the statements of the postulates in the high school texts [64] and [71].

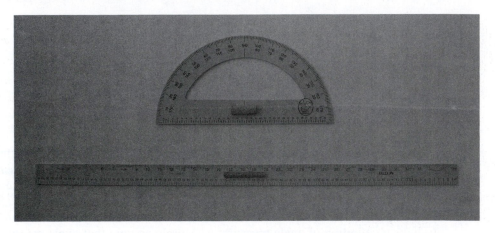

FIGURE 5.17: A protractor and a ruler

It is worth noting that there are many parallels between the Ruler Postulate and the Protractor Postulate. Both postulates assert that certain geometric properties can be quantified and measured. In fact the postulates are named after the everyday tools we use to measure those quantities: the ruler in case of distance and the protractor in case of angles. Euclid's tools, the compass and the straightedge, have been replaced by new ones. The new tools are like the old ones in that one of them is used on lines and the other on angles (or circles), but they are very unlike the old ones in that both include numerical scales.

5.7 BETWEENNESS AND THE CROSSBAR THEOREM

The postulates used in this book are meant to be powerful enough to allow us to move quickly past some of the early technical results in the foundations of geometry to the more interesting, nonintuitive results that will be discussed in the next chapter.

It is not possible, however, to avoid foundational technicalities altogether. Such technical results are collected in this section in order to make it clear just what is implicit in the axioms we have listed so far but is not stated directly in them. The results in this section will be used repeatedly in the remainder of the course. Their usual function is to allow us to fill in the gaps in Euclid's proofs.

We begin with a discussion of betweenness. We study betweenness separately for points and for rays. We also prove a theorem which shows that the use of the term *between* for points is consistent and compatible with the use of the same term for rays. The main theorem in the section is the Crossbar Theorem. We will prove it and then apply it to prove several other useful relationships.

In Chapter 12 we will revisit the foundations, and it will be important at that time to know that we were able to prove the theorems in this section using only the five axioms that have been stated so far and that the proofs do not require the final neutral axiom that is to be stated in the next section.

The first theorem clarifies what it means for one point to be between two others.

Theorem 5.7.1 (Betweenness Theorem for Points). *Let ℓ be a line; let A, B, and C be three distinct points that all lie on ℓ; and let $f : \ell \to \mathbb{R}$ be a coordinate function for ℓ. The point C is between A and B if and only if either $f(A) < f(C) < f(B)$ or $f(A) > f(C) > f(B)$.*

Proof. Let ℓ be a line, let A, B, and C be three distinct points that all lie on ℓ, and let $f : \ell \to \mathbb{R}$ be a coordinate function for ℓ (hypothesis). If $f(A) < f(C) < f(B)$, then

$$|f(C) - f(A)| + |f(B) - f(C)| = |f(B) - f(A)|$$

(algebra), so C is between A and B (definition of between). In a similar way it follows that C is between A and B if $f(A) > f(C) > f(B)$. This completes the proof of one half of the theorem.

Now suppose C is between A and B (hypothesis). Then

$$|f(C) - f(A)| + |f(B) - f(C)| = |f(B) - f(A)|$$

(definition of between). It is also the case that $(f(C) - f(A)) + (f(B) - f(C)) = (f(B) - f(A))$ (no absolute values), so $f(C) - f(A)$ and $f(B) - f(C)$ have the same sign; that is, either both are positive or both are negative (algebra).[3] In case both are positive, $f(A) < f(C) < f(B)$ (algebra). In case both are negative, $f(A) > f(C) > f(B)$ (algebra). This completes the proof. □

Corollary 5.7.2. *Let A, B, and C be three points such that B lies on $\overrightarrow{AC}$. Then $A * B * C$ if and only if $AB < AC$.*

---

[3]The fact from high school algebra being used is this: If x and y are two nonzero real numbers such that $|x| + |y| = |x + y|$, then either both x and y are positive or both x and y are negative (Exercise 5.28).

Corollary 5.7.3. *If A, B, and C are three distinct collinear points, then exactly one of them lies between the other two.*

Proof. Under any coordinate function, the three distinct points correspond to three distinct real numbers x, y, and z (the coordinate function is one-to-one). The numbers x, y, and z can be ordered from smallest to largest and, by the theorem, this uniquely determines the order of the points on the line.   □

This is a good time to reflect on the definition of betweenness for points. It would be natural to use the ordering of the real numbers to define betweenness. In other words, we could define B to be between A and C if the coordinate of B is between the coordinates of A and C on the real line. This is, in fact, the way betweenness is defined in most high school geometry books. (See [71], page 46, for example.) The problem with this definition is that it then becomes necessary to prove that betweenness is well defined, that it does not depend on which coordinate function is used for the line containing the three points. In order to avoid that minor technicality, we have followed the majority of college-level geometry books in defining betweenness using the equation $AB + BC = AC$. Either way it is necessary to prove a betweenness theorem that relates the two possible definitions. A complete proof of such a theorem is included in the margin of the teachers' edition of [71] (see page 50).

Definition 5.7.4. Let A and B be two distinct points. The point M is a *midpoint* of $\overline{AB}$ if M is between A and B and $AM = MB$.

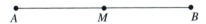

$$A \qquad\qquad M \qquad\qquad B$$

FIGURE 5.18: M is the midpoint of $\overline{AB}$

The midpoint of $\overline{AB}$ is between A and B, so $AM + MB = AB$. Simple algebra shows that $AM = (1/2)AB = MB$.

The next theorem asserts that midpoints exist and are unique. For us the proof is an exercise in the use of coordinate functions. Euclid also proved that midpoints exist (Proposition I.10), but for Euclid this meant proving that the midpoint can be constructed using compass and straightedge.

Theorem 5.7.5 (Existence and Uniqueness of Midpoints). *If A and B are distinct points, then there exists a unique point M such that M is the midpoint of $\overline{AB}$.*

Proof. Exercise 5.29.   □

You might wonder why we prove that a midpoint exists when we did not stop to prove that other objects we defined earlier (such as segments, angles, triangles, etc.) exist. It is easy to define something like a midpoint, but the act of making the definition does not by itself guarantee that there is anything that satisfies the definition. For example, we could define "the first point in the interior of $\overline{AB}$" to be the point F in the interior of $\overline{AB}$ such that $F * C * B$ for every point $C \neq F$ in the interior of $\overline{AB}$. Even though this reads like a reasonable definition, every calculus student knows that no such point exists. For this reason it is a good idea to prove that something like a midpoint exists. At the same time this should not be taken to an unreasonable extreme; it is not necessary, for example, to prove that the segment determined by two points A and B exists. The segment was defined to be the set of points that satisfy a certain condition. This set exists whether or not there are any points that satisfy the condition. (Of course it would be easy to use the Ruler Postulate to prove that there are lots of points in $\overline{AB}$.) Similar comments apply to objects like angles, triangles, and so on.

We now prove a simple theorem that relates betweenness of points to plane separation.

Theorem 5.7.6. *Let ℓ be a line, let A be a point on ℓ, and let B be an external point for ℓ. If C is a point between A and B, then B and C are on the same side of ℓ.*

Proof. Let ℓ be a line, let A be a point on ℓ, let B be an external point for ℓ, and let C be a point between A and B (hypothesis). We must prove that $\overline{BC} \cap \ell = \emptyset$ (definition of same side).

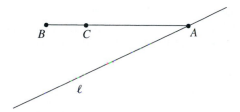

FIGURE 5.19: B and C are on the same side of ℓ

Note, first, that A is not between B and C (Corollary 5.7.3) so A is not on the segment $\overline{BC}$. The lines ℓ and $\overleftrightarrow{AB}$ have only one point in common (Theorem 5.3.7) and that point must be A, which is not on $\overline{BC}$ (previous statement), so $\overline{BC} \cap \ell = \emptyset$. This completes the proof. ☐

The theorem has several corollaries that will be important later in the section. The third one is known as the "Z-Theorem" because of the shape of the diagram that accompanies it. The Z-Theorem is the main ingredient in the proof of the Crossbar Theorem, below.

Corollary 5.7.7. *Let ℓ be a line, let A be a point on ℓ, and let B be an external point for ℓ. If C is a point on $\overrightarrow{AB}$ and $C \neq A$, then B and C are on the same side of ℓ.*

Proof. Let ℓ be a line, let A be a point on ℓ, let B be an external point for ℓ, and let C be a point on $\overrightarrow{AB}$ such that $C \neq A$ (hypothesis). If $C = B$, then the conclusion is obvious. Otherwise, C is on $\overrightarrow{AB}$ and is not equal to A or B, so either $A * C * B$ or

$A * B * C$ (definition of ray). In either case, B and C are on the same side of ℓ by Theorem 5.7.6. (In the second application of the theorem, the roles of B and C are reversed.)   □

Corollary 5.7.8. *Betweenness for rays is well defined (i.e., if D is in interior of $\angle BAC$, then every point on $\overrightarrow{AD}$ (except for A) is in the interior of $\angle BAC$).*

Proof. Let A, B, and C be three noncollinear points and let D be a point in the interior of $\angle BAC$ (hypothesis). Then D is on the same side of $\overleftrightarrow{AB}$ as C (definition of interior of angle), so every point on $\overrightarrow{AD}$ (except for A) is on the same side of $\overleftrightarrow{AB}$ as C (Corollary 5.7.7). In the same way, D is on the same side of $\overleftrightarrow{AC}$ as B (definition of interior of angle) and so every point on $\overrightarrow{AD}$ (except for A) is on the same side of $\overleftrightarrow{AC}$ as B. Therefore, every point on $\overrightarrow{AD}$ is in the interior of $\angle BAC$ (definition of interior of angle).   □

Corollary 5.7.9 (The Z-Theorem). *Let ℓ be a line and let A and D be distinct points on ℓ. If B and E are points on opposite sides of ℓ, then $\overrightarrow{AB} \cap \overrightarrow{DE} = \emptyset$.*

Proof. Except for endpoints, all the points of $\overrightarrow{AB}$ lie in one half-plane and all the points of $\overrightarrow{DE}$ lie in the other half-plane (Theorem 5.7.7). The half-planes are disjoint by the Plane Separation Postulate. Thus the only place the rays could intersect is in their endpoints. But the endpoints are distinct by hypothesis.   □

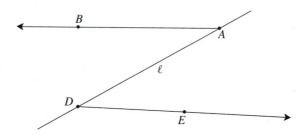

FIGURE 5.20: The rays are disjoint

We have defined *between* for both points and rays. The next theorem asserts that those definitions are consistent with each other.

Theorem 5.7.10. *Let A, B, and C be three noncollinear points and let D be a point on the line $\overleftrightarrow{BC}$. The point D is between the points B and C if and only if the ray $\overrightarrow{AD}$ is between the rays $\overrightarrow{AB}$ and $\overrightarrow{AC}$.*

Proof. Let A, B, and C be three noncollinear points and let D be a point on $\overleftrightarrow{BC}$ (hypothesis). Suppose, first, that D is between B and C (hypothesis). Then C and D are on the same side of $\overleftrightarrow{AB}$ (Theorem 5.7.6) and B and D are on the same side of $\overleftrightarrow{AC}$ (Theorem 5.7.6 again). Therefore, D is in the interior of $\angle BAC$ (definition of interior of angle) and $\overrightarrow{AD}$ is between $\overrightarrow{AB}$ and $\overrightarrow{AC}$ (definition of betweenness for rays). This completes the proof of the first half of the theorem.

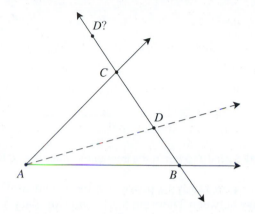

FIGURE 5.21: The point D is between B and C if and only if D is in the interior of $\angle BAC$

Now suppose that the ray $\overrightarrow{AD}$ is between the rays $\overrightarrow{AB}$ and $\overrightarrow{AC}$ (hypothesis). Then D is in the interior of $\angle BAC$ (Corollary 5.7.8). Therefore, B and D are on the same side of $\overleftrightarrow{AC}$ (definition of interior of angle), so C is not on segment $\overline{BD}$ (Plane Separation Postulate). In a similar way we can prove that B is not on segment $\overline{CD}$. Thus B, C, and D are three collinear points such that B is not between C and D and C is not between B and D. It follows that D is between B and C (Corollary 5.7.3). This completes the proof of the second half of the theorem. □

The next lemma is a technical fact that we need in the proof of the Betweenness Theorem for Rays.

Lemma 5.7.11. *If A, B, C, and D are four distinct points such that C and D are on the same side of $\overleftrightarrow{AB}$ and D is not on $\overrightarrow{AC}$, then either C is in the interior of $\angle BAD$ or D is in the interior of $\angle BAC$.*

Proof. Let A, B, C, and D be four distinct points such that C and D are on the same side of $\overleftrightarrow{AB}$ and D is not on $\overrightarrow{AC}$ (hypothesis). We will assume that D is not in the interior of $\angle BAC$ (hypothesis) and use that assumption to prove that C is in the interior of $\angle BAD$.

We know that C and D are on the same side of $\overleftrightarrow{AB}$ (hypothesis), so B and D must lie on opposite sides of $\overleftrightarrow{AC}$ (negation of definition of angle interior). Thus $\overline{BD} \cap \overleftrightarrow{AC} \neq \emptyset$ (Plane Separation Postulate). Let C' be the unique point at which $\overline{BD}$ intersects $\overleftrightarrow{AC}$. (See Fig. 5.22.) By Theorem 5.7.10, C' is in the interior of $\angle BAD$. In particular, D and C' lie on the same side of $\overleftrightarrow{AB}$. Since C and C' both lie on the same side of $\overleftrightarrow{AB}$ as D, A cannot be between C and C', so $\overrightarrow{AC}$ and $\overrightarrow{AC'}$ cannot be opposite rays. Therefore, $\overrightarrow{AC} = \overrightarrow{AC'}$. It follows that C is in the interior of $\angle BAD$ (Corollary 5.7.8). □

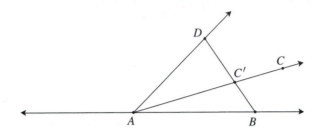

FIGURE 5.22: If D is not in the interior of $\angle BAC$, then C is in the interior of $\angle BAD$

We are now ready to prove our Betweenness Theorem for Rays. Note the close analogy between Theorem 5.7.12 and Corollary 5.7.2.

Theorem 5.7.12 (Betweenness Theorem for Rays). *Let A, B, C, and D be four distinct points such that C and D lie on the same side of $\overleftrightarrow{AB}$. Then $\mu(\angle BAD) < \mu(\angle BAC)$ if and only if $\overrightarrow{AD}$ is between rays $\overrightarrow{AB}$ and $\overrightarrow{AC}$.*

Proof. Let A, B, C, and D be four distinct points such that C and D lie on the same side of $\overleftrightarrow{AB}$ (hypothesis). Assume, first, that $\overrightarrow{AD}$ is between rays $\overrightarrow{AB}$ and $\overrightarrow{AC}$ (hypothesis). Then D is in the interior of $\angle BAC$ (Corollary 5.7.8). Thus $\mu(\angle BAD) + \mu(\angle DAC) = \mu(\angle BAC)$ (Protractor Postulate, Part 4) and $\mu(\angle DAC) > 0$ (Protractor Postulate, Parts 1 and 2), so $\mu(\angle BAD) < \mu(\angle BAC)$ (algebra). This completes the proof of the first half of the theorem.

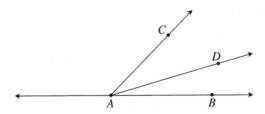

FIGURE 5.23: $\mu(\angle BAD) < \mu(\angle BAC)$ if and only if D is in the interior of $\angle BAC$

We will prove the contrapositive of the second half of the theorem. Suppose $\overrightarrow{AD}$ is not between rays $\overrightarrow{AB}$ and $\overrightarrow{AC}$ (hypothesis). We must prove that $\mu(\angle BAD) \geqslant \mu(\angle BAC)$. If D lies on $\overrightarrow{AC}$, then $\mu(\angle BAD) = \mu(\angle BAC)$. Otherwise, C is in the interior of $\angle BAD$ (Lemma 5.7.11). Therefore, $\mu(\angle BAD) > \mu(\angle BAC)$ (by the first half of the theorem, which was proved in the previous paragraph). This completes the proof of the theorem. □

We can use the Betweenness Theorem for Rays to prove the existence of angle bisectors.

Definition 5.7.13. Let A, B, and C be three noncollinear points. A ray $\overrightarrow{AD}$ is an *angle bisector* of $\angle BAC$ if D is in the interior of $\angle BAC$ and $\mu(\angle BAD) = \mu(\angle DAC)$.

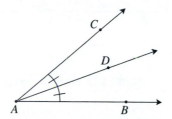

FIGURE 5.24: The angle bisector

Theorem 5.7.14 (Existence and Uniqueness of Angle Bisectors). *If A, B, and C are three noncollinear points, then there exists a unique angle bisector for $\angle BAC$.*

Proof. Exercise 5.30. ☐

We now have all the machinery in place to prove the Crossbar Theorem. The Crossbar Theorem asserts that if a ray starts out in the interior of one of the angles in a triangle, then it must intersect the opposite side of the triangle, which is called a "crossbar" for the angle.

Theorem 5.7.15 (The Crossbar Theorem). *If $\triangle ABC$ is a triangle and D is a point in the interior of $\angle BAC$, then there is a point G such that G lies on both $\overrightarrow{AD}$ and $\overline{BC}$.*

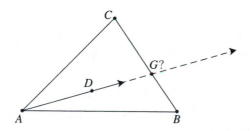

FIGURE 5.25: Statement of Crossbar Theorem

Proof. Let $\triangle ABC$ be a triangle and let D be a point in the interior of $\angle BAC$ (hypothesis). Choose points E and F such that $E * A * B$ and $F * A * D$ (Ruler Postulate) and let $\ell = \overleftrightarrow{AD}$.

Since D is in the interior of $\angle BAC$, neither B nor C lies on ℓ. Thus we can apply Pasch's Axiom (Theorem 5.5.10) to the triangle $\triangle EBC$ to conclude that ℓ must

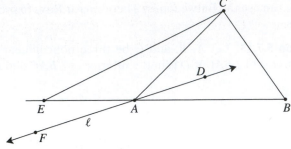

FIGURE 5.26: Proof of the Crossbar Theorem

intersect either $\overline{EC}$ or $\overline{BC}$. In order to complete the proof, we must show that it is the ray $\overrightarrow{AD}$ that intersects either $\overline{EC}$ or $\overline{BC}$ (and not the opposite ray $\overrightarrow{AF}$) and that $\overrightarrow{AD}$ does not intersect $\overline{EC}$. In symbols, we must show $\overrightarrow{AF} \cap \overline{EC} = \emptyset$, $\overrightarrow{AF} \cap \overline{BC} = \emptyset$, and $\overrightarrow{AD} \cap \overline{EC} = \emptyset$. This will be accomplished by three applications of the Z-Theorem (Corollary 5.7.9).

Since A is between F and D, F and D lie on opposite sides of $\overleftrightarrow{AB}$ (Plane Separation Postulate). On the other hand, C and D are on the same side of $\overleftrightarrow{AB}$ (D is in the interior of $\angle BAC$), so C and F are on opposite sides of $\overleftrightarrow{AB}$ (Plane Separation Postulate). Thus $\overrightarrow{EC} \cap \overrightarrow{AF} = \emptyset$ (Z-Theorem). Since $\overline{EC}$ is a subset of $\overrightarrow{EC}$, we have $\overrightarrow{AF} \cap \overline{EC} = \emptyset$.

By the previous paragraph, C and F are on opposite sides of $\overleftrightarrow{AB}$. Thus $\overrightarrow{BC} \cap \overrightarrow{AF} = \emptyset$ (the Z-Theorem again). Since $\overline{BC}$ is a subset of $\overrightarrow{BC}$, we have $\overrightarrow{AF} \cap \overline{BC} = \emptyset$.

Since A is between E and B, E and B lie on opposite sides of $\overleftrightarrow{AC}$ (Plane Separation Postulate). On the other hand, B and D are on the same side of $\overleftrightarrow{AC}$ (D is in the interior of $\angle BAC$), so E and D are on opposite sides of $\overleftrightarrow{AC}$ (Plane Separation Postulate). Thus $\overrightarrow{CE} \cap \overrightarrow{AD} = \emptyset$ (the Z-Theorem once again). Since $\overline{EC}$ is a subset of $\overrightarrow{CE}$, we have $\overrightarrow{AD} \cap \overline{EC} = \emptyset$. This completes the proof of the Crossbar Theorem. ☐

Theorem 5.7.10 asserts that if the ray $\overrightarrow{AD}$ intersects the interior of the crossbar $\overline{BC}$, then D is in the interior of $\angle BAC$. Thus there is a sense in which Theorem 5.7.10 is a converse to the Crossbar Theorem. The next theorem combines them into one theorem.

Theorem 5.7.16. *A point D is in the interior of the angle $\angle BAC$ if and only if the ray $\overrightarrow{AD}$ intersects the interior of the segment $\overline{BC}$.*

Proof. Suppose, first, that D is in the interior of $\angle BAC$. Then $\overrightarrow{AD}$ intersects the interior of $\overline{BC}$ by the Crossbar Theorem. Next suppose that $\overrightarrow{AD}$ intersects the interior of $\overline{BC}$. Let E be a point of intersection. Then E is in the interior of $\angle BAC$ by Theorem 5.7.10. By Corollary 5.7.8, D is also in the interior of $\angle BAC$.   □

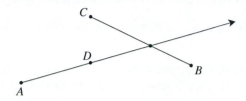

FIGURE 5.27: D is in the interior of $\angle BAC$

Remark. Theorem 5.7.16 is taken as an axiom by MacLane [46]. He calls it the "Continuity Axiom" because it implies the Continuity Axiom of Birkhoff—see Theorem 5.7.27. In MacLane's formulation of the axioms, Theorem 5.7.16 is assumed in place of the Plane Separation Postulate.

We now wish to use the results of this section to prove a sort of generalization of the Angle Addition Postulate (Part 4 of the Protractor Postulate). First we need a definition.

Definition 5.7.17. Two angles $\angle BAD$ and $\angle DAC$ form a *linear pair* if $\overrightarrow{AB}$ and $\overrightarrow{AC}$ are opposite rays.

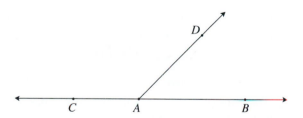

FIGURE 5.28: Angles $\angle BAD$ and $\angle DAC$ form a linear pair

The following theorem is often taken as an axiom (see [51], [64], [38], and [72], for example). A common name for it is the *Supplement Postulate*. It is another example of a redundant axiom in the SMSG system.

Theorem 5.7.18 (Linear Pair Theorem). *If angles $\angle BAD$ and $\angle DAC$ form a linear pair, then $\mu(\angle BAD) + \mu(\angle DAC) = 180°$.*

Definition 5.7.19. Two angles $\angle BAC$ and $\angle EDF$ are *supplementary* (or *supplements*) if $\mu(\angle BAC) + \mu(\angle EDF) = 180°$.

The Linear Pair Theorem can be restated as follows: *If two angles form a linear pair, then they are supplements.* In order to be a linear pair, two angles must share a side, so supplementary angles do not usually form a linear pair.

Before tackling the full proof of the Linear Pair Theorem, we prove a lemma. The lemma really constitutes part of the proof of the Linear Pair Theorem, but the proof is easier to understand if this technical fact is separated out.

Lemma 5.7.20. *If $C * A * B$ and D is in the interior of $\angle BAE$, then E is in the interior of $\angle DAC$.*

Proof. Let A, B, C, D, and E be five points such that $C * A * B$ and D is in the interior of $\angle BAE$ (hypothesis). In order to show that E is in the interior of $\angle DAC$, we must show that E and D are on the same side of $\overleftrightarrow{AC}$ and that E and C are on the same side of $\overrightarrow{AD}$ (definition of angle interior).

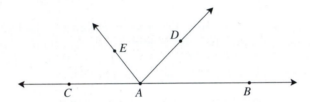

FIGURE 5.29: If D is in the interior of $\angle BAE$, then E is in the interior of $\angle DAC$.

Since D is in the interior of $\angle BAE$, E and D are on the same side of $\overleftrightarrow{AB}$ (definition of angle interior). But $\overleftrightarrow{AB} = \overleftrightarrow{AC}$, so E and D are on the same side of $\overleftrightarrow{AC}$. In addition, $\overrightarrow{AD}$ must intersect $\overline{BE}$ (Crossbar Theorem). Therefore, E and B lie on opposite sides of $\overleftrightarrow{AD}$ (Plane Separation Postulate). The fact that A is between C and B means that C and B are on opposite sides of $\overleftrightarrow{AD}$ and so we can conclude that C and E are on the same side of $\overleftrightarrow{AD}$ (Plane Separation Postulate). □

Proof of the Linear Pair Theorem. Let $\angle BAD$ and $\angle DAC$ be two angles that form a linear pair (hypothesis). Then $\overrightarrow{AB}$ and $\overrightarrow{AC}$ are opposite rays. In order to simplify notation in the proof, we will drop the degree notation and use α to denote $\mu(\angle BAD)$ and β to denote $\mu(\angle DAC)$. We must prove that $\alpha + \beta = 180$. By trichotomy, either $\alpha + \beta < 180$, $\alpha + \beta = 180$, or $\alpha + \beta > 180$. We will show that the first and last possibilities lead to contradictions, and this will allow us to conclude that $\alpha + \beta = 180$.

Suppose, first, that $\alpha + \beta < 180$. By the Angle Construction Postulate (Protractor Postulate, Part 3), there is a point E, on the same side of $\overleftrightarrow{AB}$ as D, such that $\mu(\angle BAE) = \alpha + \beta$. By the Betweenness Theorem for Rays (Theorem 5.7.12), D is in the interior of $\angle BAE$. Therefore, $\mu(\angle BAD) + \mu(\angle DAE) = \mu(\angle BAE)$ (Angle Addition Postulate) and so $\mu(\angle DAE) = \beta$ (algebra). Now E is in the interior of $\angle DAC$ (Lemma 5.7.20), so $\mu(\angle DAE) + \mu(\angle EAC) = \mu(\angle DAC)$ (Angle

Addition Postulate). It follows that $\mu(\angle EAC) = 0$ (algebra). This contradicts Parts 1 and 2 of the Protractor Postulate. Thus we conclude that $\alpha + \beta < 180$ is impossible.

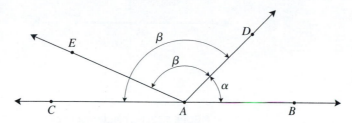

FIGURE 5.30: Proof of the Linear Pair Theorem in case $\alpha + \beta < 180$

Suppose, next, that $\alpha + \beta > 180$. Choose a point F, on the same side of $\overleftrightarrow{AB}$ as D, such that $\mu(\angle BAF) = (\alpha + \beta) - 180$ (Protractor Postulate, Part 3). Now $\beta < 180$ (Protractor Postulate, Part 1), so $\alpha + \beta - 180 < \alpha$ (algebra). By the Betweenness Theorem for Rays (Theorem 5.7.12), F is in the interior of $\angle BAD$. Therefore $\mu(\angle BAF) + \mu(\angle FAD) = \mu(\angle BAD)$ (Angle Addition Postulate) and so $\mu(\angle FAD) = 180 - \beta$ (algebra). By Lemma 5.7.20, D is in the interior of $\angle FAC$. Thus $\mu(\angle FAD) + \mu(\angle DAC) = \mu(\angle FAC)$ (Angle Addition Postulate). It follows that $\mu(\angle FAC) = 180°$ (algebra). This contradicts Part 1 of the Protractor Postulate and so we conclude that $\alpha + \beta > 180$ is impossible. This completes the proof of the Linear Pair Theorem. $\square$

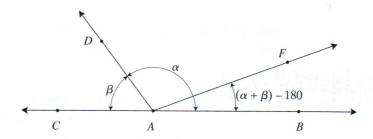

FIGURE 5.31: Proof of the Linear Pair Theorem in case $\alpha + \beta > 180$

Definition 5.7.21. Two lines ℓ and m are *perpendicular* if there exists a point A that lies on both ℓ and m and there exist points $B \in \ell$ and $C \in m$ such that $\angle BAC$ is a right angle. Notation: $\ell \perp m$.

It is easy to see, using the Linear Pair Theorem, that if the lines ℓ and m are perpendicular, then they contain rays that make four different right angles (Exercise 5.32).

Definition 5.7.22. Let A and B be two distinct points. A *perpendicular bisector* of $\overline{AB}$ is a line ℓ such that the midpoint of $\overline{AB}$ lies on ℓ and $\ell \perp \overleftrightarrow{AB}$.

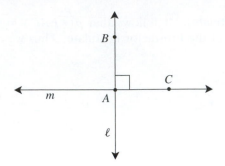

FIGURE 5.32: Perpendicular lines

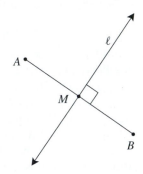

FIGURE 5.33: ℓ is the perpendicular bisector of $\overline{AB}$

Theorem 5.7.23 (Existence and Uniqueness of Perpendicular Bisectors). *If A and B are two distinct points, then there exists a unique perpendicular bisector for $\overline{AB}$.*

Proof. Exercise 5.33. □

Definition 5.7.24. Angles $\angle BAC$ and $\angle DAE$ form a *vertical pair* (or are *vertical angles*) if rays $\overrightarrow{AB}$ and $\overrightarrow{AE}$ are opposite and rays $\overrightarrow{AC}$ and $\overrightarrow{AD}$ are opposite or if rays $\overrightarrow{AB}$ and $\overrightarrow{AD}$ are opposite and rays $\overrightarrow{AC}$ and $\overrightarrow{AE}$ are opposite.

Theorem 5.7.25 (Vertical Angles Theorem). *Vertical angles are congruent.*

Proof. Exercise 5.35. □

We conclude this section with an application of the Crossbar Theorem. The final theorem in this section will not be used again until we prove circular continuity in Section 10.5, so the remainder of the section can be omitted for now without serious consequence.

The theorem we will prove is called the *Continuity Axiom*. It asserts that the relationship between angle measure and distance is a continuous one. It is called an axiom because Birkhoff stated this fact as part of his version of the Protractor

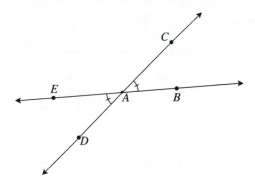

FIGURE 5.34: $\angle BAC$ and $\angle DAE$ are vertical angles

Postulate. It is included here because it adds interest to the statement of the Crossbar Theorem and because it relates the Crossbar Theorem to the more familiar concept of continuity.

We first need a lemma about functions defined on intervals of real numbers. The lemma should seem intuitively plausible: If a function from the real numbers to the real numbers is strictly increasing, then a jump in the graph would result in a gap in the range of the function.

Lemma 5.7.26. *Let $[a, b]$ and $[c, d]$ be closed intervals of real numbers and let $f : [a, b] \to [c, d]$ be a function. If f is strictly increasing and onto, then f is continuous.*

Proof. Let $f : [a, b] \to [c, d]$ be a function that is both increasing and onto (hypothesis) and let $x \in [a, b]$. We must show that f is continuous at x. Let us assume that x is in the interior of $[a, b]$. The proof of the case in which x is an endpoint is similar.

Let $\epsilon > 0$ be given. We must show that there exists a positive number δ such that if $|x - y| < \delta$, then $|f(x) - f(y)| < \epsilon$ (definition of continuous at x).

Since f is strictly increasing, $c < f(x) < d$. We may assume that $c < f(x) - \epsilon$ and $f(x) + \epsilon < d$. (If this is not the case, simply replace ϵ with a smaller number.) Since f is onto, there exist numbers x_1 and x_2 in $[a, b]$ such that $f(x_1) = f(x) - \epsilon$ and $f(x_2) = f(x) + \epsilon$. Choose $\delta > 0$ to be small enough so that $x_1 < x - \delta$ and $x_2 > x + \delta$.

Suppose $|x - y| < \delta$. Then $x - \delta < y < x + \delta$. Since f is increasing, $f(x_1) < f(x - \delta) < f(y) < f(x + \delta) < f(x_2)$. Therefore, $f(x) - \epsilon < f(y) < f(x) + \epsilon$ and so $|f(x) - f(y)| < \epsilon$ (algebra). $\qquad\square$

Setting for the Continuity Axiom. Let A, B, and C be three noncollinear points. For each point D on $\overline{BC}$ there is an angle $\angle CAD$ and there is a distance CD. We will define a function that relates the distance and the angle measure. Let $d = BC$. The Ruler Placement Postulate (Theorem 5.4.14) gives a one-to-one

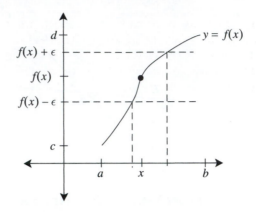

FIGURE 5.35: Proof of Lemma 5.7.26

correspondence from the interval $[0, d]$ to points on $\overline{BC}$ such that C corresponds to 0 and B corresponds to d. Let D_x be the point that corresponds to the number x; i.e., D_x is the point on $\overline{BC}$ such that $CD_x = x$. Define a function $f : [0, d] \to \mathbb{R}$ by $f(x) = \mu(\angle D_x AB)$.

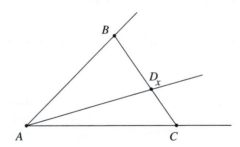

FIGURE 5.36: Setting for the Continuity Axiom

Theorem 5.7.27 (The Continuity Axiom). *The function f described in the previous paragraph is a continuous function, as is the inverse of f.*

Proof. Let f be the function described above. By Theorems 5.7.10 and 5.7.12, f is a strictly increasing function. By the Crossbar Theorem and Theorem 5.7.12, f is onto. Therefore, f is continuous (Lemma 5.7.26). It is obvious that the inverse of f is increasing and onto, so the inverse is also continuous. ☐

5.8 THE SIDE-ANGLE-SIDE POSTULATE

So far we have formulated one axiom for each of the undefined terms. It would be reasonable to expect this to be enough axioms since we now know the basic properties of each of the undefined terms. But there is still something missing: The postulates stated so far do not tell us quite enough about how distance (or length of segments) and angle measure interact with each other. In this section we will give an example that illustrates the need for additional information and then state one final axiom to complete the picture.

The simplest objects that combine both segments and angles are triangles. We have defined what it means for two segments to be congruent and what it means for two angles to be congruent. We now extend that definition to triangles, where the two types of congruence are combined.

Definition 5.8.1. Two triangles are *congruent* if there is a correspondence between the vertices of the first triangle and the vertices of the second triangle such that corresponding angles are congruent and corresponding sides are congruent. Thus the assertion that two triangles are congruent is really the assertion that there are six congruences, three angle congruences and three segment congruences.

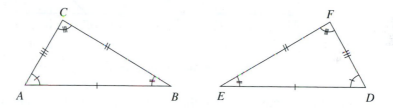

FIGURE 5.37: Congruent triangles

Notation. We use the symbol $\cong$ to indicate congruence of triangles. It is understood that the notation $\triangle ABC \cong \triangle DEF$ means that the two triangles are congruent under the correspondence $A \leftrightarrow D, B \leftrightarrow E$, and $C \leftrightarrow F$. Specifically, $\triangle ABC \cong \triangle DEF$ means $\overline{AB} \cong \overline{DE}, \overline{BC} \cong \overline{EF}, \overline{AC} \cong \overline{DF}, \angle ABC \cong \angle DEF, \angle BCA \cong \angle EFD$, and $\angle CAB \cong \angle FDE$.

One of the most basic theorems in geometry is the Side-Angle-Side Triangle Congruence Condition (abbreviated SAS). It states that if two sides and the included angle of one triangle are congruent to two sides and included angle of a second triangle, then the triangles are congruent. This theorem is Euclid's Proposition 4 and Euclid's proof of the proposition was discussed in Chapter 1. SAS is one of the essential ingredients in the proofs of more complicated theorems of geometry, so it is a key theorem we need in our development of the subject. But the axioms we have stated so far are not strong enough to allow us to prove this theorem.

We have seen that the Cartesian plane with the taxicab metric satisfies the Ruler Postulate. While distances are measured in an unusual way in taxicab geometry, the lines are the standard Euclidean lines and so we can use the ordinary Euclidean angle measure to measure angles in taxicab geometry. The Cartesian plane equipped with the taxicab metric and the Euclidean angle measure is a model for geometry that satisfies all five of the axioms stated so far in this chapter. The next example shows that SAS fails in that model.

EXAMPLE 5.8.2 SAS fails in taxicab geometry

Consider the triangles $\triangle ABC$ and $\triangle DEF$ in the Cartesian plane that have vertices at $A = (0,0)$, $B = (0,2)$, $C = (2,0)$, $D = (0,1)$, $E = (-1,0)$, and $F = (1,0)$. In the taxicab metric, $AB = AC = DE = DF = 2$. Furthermore, both the angles $\angle BAC$ and $\angle EDF$ are right angles. Nonetheless, $\triangle ABC \not\cong \triangle DEF$ because $BC = 4$ while $EF = 2$. ∎

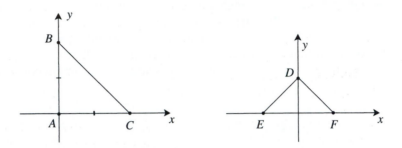

FIGURE 5.38: Two triangles in taxicab geometry

From this example we see that the postulates we have stated thus far are not powerful enough to prove the SAS theorem. We will fix this by simply assuming Side-Angle-Side as an axiom. In doing so we are following both Hilbert and SMSG. Since SAS is true in the Cartesian plane with the Euclidean metric but fails in the Cartesian plane with the taxicab metric, the SAS Postulate is independent of the other postulates of neutral geometry.

Axiom 5.8.3 (The Side-Angle-Side Postulate or SAS). *If $\triangle ABC$ and $\triangle DEF$ are two triangles such that $\overline{AB} \cong \overline{DE}$, $\angle ABC \cong \angle DEF$, and $\overline{BC} \cong \overline{EF}$, then $\triangle ABC \cong \triangle DEF$.*

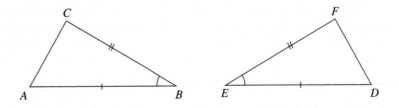

FIGURE 5.39: Side-Angle-Side

Euclid proved SAS using his method of superposition, so Euclid's "proof" depends on the ability to move triangles around in the plane without distorting them. Examples such as the one above show that we cannot take this for granted. Even though taxicab geometry satisfies all the axioms stated earlier in this chapter, it allows only a limited number of rigid motions (motions that preserve both distances and angle measures).

Some high school textbooks (such as [71]) replace the SAS Postulate with an axiom which states that certain kinds of transformations are allowed and then prove SAS as a theorem. We will discuss that transformational approach in Chapter 12. It is interesting to note that the latter approach is in some ways closer to Euclid's original treatment than is the Hilbert/SMSG approach of taking SAS as an axiom. Euclid gives Common Notion 4 ("Things which coincide with one another are equal to one another") as the justification for the method of superposition that he uses to prove SAS. It is not unreasonable to interpret Common Notion 4 to be an axiom that permits rigid motions in the plane.

The transformational approach has implications for the definition of congruence as well. This will also be discussed in more detail in Chapter 12. For now we will simply point out that we have defined congruence of triangles to mean that corresponding parts are congruent. (A triangle has six "parts," three sides and three angles.) In high school textbooks congruence of triangles is defined differently and so the statement "corresponding parts of congruent triangles are congruent" becomes a theorem. That theorem is abbreviated CPCT, or more generally, CPCF, which stands for "corresponding parts of congruent figures." For us, CPCT is a definition not a theorem.

In order to illustrate the use of SAS, we now give two proofs of the familiar Isosceles Triangle Theorem. The example proofs also serve to illustrate the use of the theorems from the previous section.

Definition 5.8.4. A triangle is called *isosceles* if it has a pair of congruent sides.

Theorem 5.8.5 (Isosceles Triangle Theorem). *If $\triangle ABC$ is a triangle and $\overline{AB} \cong \overline{AC}$, then $\angle ABC \cong \angle ACB$.*

First proof of Theorem 5.8.5. Let $\triangle ABC$ be a triangle such that $\overline{AB} \cong \overline{AC}$ (hypothesis). We must prove that $\angle ABC \cong \angle ACB$. Let D be a point in the interior of $\angle BAC$ such that $\overrightarrow{AD}$ is the bisector of $\angle BAC$ (Theorem 5.7.14). There is a point E at which the ray $\overrightarrow{AD}$ intersects the segments $\overline{BC}$ (Crossbar Theorem 5.7.15). Then $\triangle BAE \cong \triangle CAE$ (SAS) and so $\angle ABE \cong \angle ACE$ (definition of congruent triangles). $\square$

The proof above, which is essentially the same as the proof given in a typical high school geometry textbook, has two drawbacks from our point of view. One is the fact that it requires the use of the Crossbar Theorem. Many lower-level textbooks give this proof and simply ignore the fact that the Crossbar Theorem is needed. (The use of the Crossbar Theorem can be suppressed by saying something like this: "Let E be the point at which the angle bisector intersects the side $\overline{BC}$." It is manifestly obvious from the diagram that such a point exists, so most readers do not ask for a justification.) A second drawback to the proof is the fact that it is unnecessarily complicated. There is a much more elegant way to prove the theorem that does not make use of the Crossbar Theorem at all.

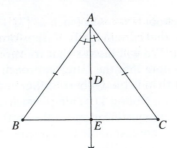

FIGURE 5.40: First proof of the Isosceles Triangle Theorem

Second proof of Theorem 5.8.5. Let $\triangle ABC$ be a triangle such that $\overline{AB} \cong \overline{AC}$ (hypothesis). Then $\triangle BAC \cong \triangle CAB$ (SAS), so $\angle ABC \cong \angle ACB$ (definition of congruent triangles). □

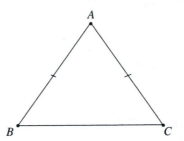

FIGURE 5.41: Second proof of the Isosceles Triangle Theorem

We can understand the second proof visually by imagining that the triangle is picked up, turned over, and placed back on top of itself, bringing A back to where it was but interchanging B and C. This beautifully simple proof is attributed to Pappus of Alexandria who lived about 600 years after Euclid and was the last great ancient Greek geometer. So why do high school textbooks give the longer, more complicated, proof? Mostly because of the transformational approach that is emphasized in those texts. The concept of a reflection across a line is taken to be a more basic concept than the abstract congruence we use in the second proof. The second proof also requires more sophisticated thinking in that we must think of one triangle as two and imagine a correspondence of a triangle to itself that is not the identity. We somehow feel more comfortable thinking of the two triangles in the SAS Postulate as two completely separate triangles, even though there is nothing in the statement that actually requires this.

The Isosceles Triangle Theorem is Euclid's Proposition 5, so we are following in the tradition of Euclid when we give this theorem as the very first application of SAS. Euclid himself gives a proof that is different from either of the two proofs discussed here. Euclid extends the sides $\overline{AB}$ and $\overline{AC}$ first and then works with two triangles that lie below the original triangle. It seems likely that he made this (unnecessary) construction in order to have two distinct triangles to which he could apply SAS.

Euclid's Proposition 5 was known in medieval universities as the *pons asinorum* (the bridge of asses), probably because the diagram that accompanies Euclid's proof looks like a bridge. Another possible explanation for the name is the fact that this proof is a narrow bridge that separates those who can understand and appreciate Euclid's work from those who cannot.

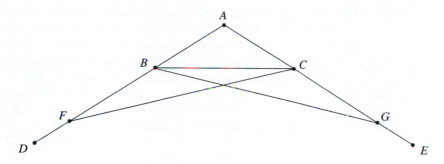

FIGURE 5.42: The *pons asinorum*

5.9 THE PARALLEL POSTULATES

The geometry that can be done using only the six postulates stated earlier in this chapter is called *neutral geometry*. As we will see in the next chapter, a great many of the standard theorems of geometry can be proved in neutral geometry. But we will also find that the neutral postulates are not strong enough to allow us to prove much about parallel lines. In Chapter 2 we stated three postulates regarding parallelism. We repeat them here for completeness.

Euclidean Parallel Postulate. *For every line ℓ and for every point P that does not lie on ℓ, there is exactly one line m such that P lies on m and m ∥ ℓ.*

Elliptic Parallel Postulate. *For every line ℓ and for every point P that does not lie on ℓ, there is no line m such that P lies on m and m ∥ ℓ.*

Hyperbolic Parallel Postulate. *For every line ℓ and for every point P that does not lie on ℓ, there are at least two lines m and n such that P lies on both m and n and both m and n are parallel to ℓ.*

We will see in the next chapter that it is possible to prove the existence of parallel lines in neutral geometry. It follows that the Elliptic Parallel Postulate is not consistent with the axioms of neutral geometry. Later in the course we will prove that each of the other two parallel postulates is consistent with the axioms of neutral geometry. We will use the term *Euclidean geometry* to refer to the geometry we can do if we add the Euclidean Parallel Postulate as a seventh postulate and *hyperbolic geometry* to refer to the geometry we can do if we instead add the Hyperbolic Parallel Postulate as a seventh postulate.

5.10 MODELS

The method we will use to prove that the Euclidean and hyperbolic parallel postulates are both consistent with the axioms of neutral geometry is to construct models. We need to develop many more tools before we can construct those models, but we should at least briefly consider the question of the existence of models here.

It is clear from the Ruler Postulate that any model for neutral geometry must contain an infinite number of points, so we can rule out all of the finite geometries of Chapter 2 as possible models for neutral geometry. We can also rule out the sphere $\mathbb{S}^2$ since it does not satisfy either the Incidence Postulate or the Ruler Postulate. The rational plane fails as a model for neutral geometry because it does not satisfy the Ruler Postulate and taxicab geometry fails to satisfy SAS.

This leaves only the Cartesian plane and the Klein disk as possible candidates from among the previously discussed examples. The Cartesian plane with the Euclidean metric certainly is a model for neutral geometry and it is the model you studied in high school. Although it is far from obvious, the Klein disk can also be made into a model for neutral geometry. The undefined terms *point* and *line* have already been interpreted in the Klein disk; in order to see that it is a model for neutral geometry it is necessary to interpret the terms *distance* and *angle measure* and then to verify that all the axioms of neutral geometry are valid in that interpretation. The natural way to measure distances in the Klein disk will certainly not work since there is a finite upper bound on distances and that contradicts the Ruler Postulate. One of the major tasks of the second half of this course is to define a metric in the Klein disk in such a way that the result is a model for hyperbolic geometry.

The question of the existence of models will be set aside for now. First we will give axiomatic treatments of neutral, Euclidean, and hyperbolic geometries.

Even though we will treat these three geometries axiomatically, we will still want to draw diagrams that illustrate the relationships being discussed. A diagram should be viewed as one possible interpretation of a theorem. Both the theorem statements and the proofs should stand alone and should not depend on the diagrams. The purpose of a diagram is to help us to visually organize the information being discussed. Diagrams add tremendously to our intuitive understanding, but they are only meant to illustrate, not to narrow the range of possible interpretations.

EXERCISES

5.1. Show that the Euclidean metric defined in Example 5.4.9 is a metric (i.e., verify that the function d satisfies the three conditions in the definition of metric on page 60).

5.2. Show that the taxicab metric defined in Example 5.4.10 is a metric (i.e., verify that the function ρ satisfies the three conditions in the definition of metric on page 60).

5.3. Show that the functions defined in Example 5.4.12 are coordinate functions.

5.4. Show that the functions defined in Example 5.4.13 are coordinate functions.

5.5. Assume that $f : \ell \to \mathbb{R}$ is a coordinate function for ℓ
   (a) Prove that $-f$ is also a coordinate function for ℓ.
   (b) Prove that $g : \ell \to \mathbb{R}$ defined by $g(P) = f(P) + c$ for some constant c is also a coordinate function for ℓ.
   (c) Prove that if $h : \ell \to \mathbb{R}$ is any coordinate function for ℓ then there must exist a constant c such that either $h(P) = f(P) + c$ or $h(P) = -f(P) + c$.

5.6. Prove the Ruler Placement Postulate (Theorem 5.4.14).

5.7. Suppose f is a coordinate function for $\ell = \overleftrightarrow{AB}$ such that $f(A) = 0$ and $f(B) > 0$. Prove that $\overrightarrow{AB} = \{P \in \ell \mid f(P) \geqslant 0\}$.

5.8. Let ℓ be a line and let $f : \ell \to \mathbb{R}$ be a function such that $PQ = |f(P) - f(Q)|$ for every $P, Q \in \ell$. Prove that f is a coordinate function for ℓ.

5.9. Prove that the intersection of two convex sets is convex. Show by example that the union of two convex sets need not be convex. Is the empty set convex?

5.10. Check that the trivial geometry containing just one point and no lines satisfies all the postulates of neutral geometry except the Existence Postulate. Which parallel postulate is satisfied by this geometry? (This exercise is supposed to convince you of the need for the Existence Postulate.)

5.11. Relationship between neutral geometry and incidence geometry.
   (a) Explain how the axioms stated in this chapter imply that there exists at least one line.
   (b) Explain how the existence of three noncollinear points (Incidence Axiom I-2 or SMSG Postulate 5, Part 2) can be deduced from the axioms stated in this chapter.
   (c) Explain why every model for neutral geometry must have infinitely many points.
   (d) Explain why every model for neutral geometry must have infinitely many lines.
   (e) Explain why every model for neutral geometry must also be a model for incidence geometry.
   (f) Explain why all the theorems of incidence geometry must also be theorems in neutral geometry.

5.12. Prove: If ℓ and m are two lines, the number of points in $\ell \cap m$ is either 0, 1, or ∞.

5.13. Let A and B be two distinct points. Prove that $\overline{AB} = \overline{BA}$.

5.14. Prove that the endpoints of a segment are well defined (i.e., if $\overline{AB} = \overline{CD}$, then either $A = C$ and $B = D$ or $A = D$ and $B = C$).

5.15. Prove that the vertices and edges of a triangle are well defined (i.e., if $\triangle ABC = \triangle DEF$, then $\{A, B, C\} = \{D, E, F\}$).

5.16. Let A and B be two distinct points. Prove each of the following equalities.
   (a) $\overrightarrow{AB} \cup \overrightarrow{BA} = \overleftrightarrow{AB}$.
   (b) $\overrightarrow{AB} \cap \overrightarrow{BA} = \overline{AB}$.

5.17. Let A and B be two distinct points. Prove that each of the sets $\{A\}$, $\overline{AB}$, $\overrightarrow{AB}$, and $\overleftrightarrow{AB}$ is a convex set.

5.18. Let ℓ be a line and let H be one of the half-planes bounded by ℓ. Prove that $H \cup \ell$ is a convex set.

5.19. Let A, B, and C be three collinear points such that $A * B * C$. Prove each of the following set equalities.
   (a) $\overrightarrow{BA} \cup \overrightarrow{BC} = \overleftrightarrow{AB}$,
   (b) $\overrightarrow{BA} \cap \overrightarrow{BC} = \{B\}$,
   (c) $\overline{AB} \cup \overline{BC} = \overline{AC}$,
   (d) $\overline{AB} \cap \overline{BC} = \{B\}$, and
   (e) $\overrightarrow{AB} = \overrightarrow{AC}$.

5.20. (Segment Construction Theorem) Prove the following theorem. If $\overline{AB}$ is a segment and $\overrightarrow{CD}$ is a ray, then there is a unique point E on $\overrightarrow{CD}$ such that $\overline{AB} \cong \overline{CE}$.

5.21. (Segment Addition Theorem) Prove the following theorem. If $A * B * C$, $D * E * F$, $\overline{AB} \cong \overline{DE}$, and $\overline{BC} \cong \overline{EF}$, then $\overline{AC} \cong \overline{DF}$.

5.22. (Segment Subtraction Theorem) Prove the following theorem. If $A * B * C$, $D * E * F$, $\overline{AB} \cong \overline{DE}$, and $\overline{AC} \cong \overline{DF}$, then $\overline{BC} \cong \overline{EF}$.

5.23. Suppose $\triangle ABC$ is a triangle and ℓ is a line such that none of the vertices A, B, or C lies on ℓ. Prove that ℓ cannot intersect all three sides of $\triangle ABC$. Is it possible for a line to intersect all three sides of a triangle?

5.24. Betweenness in taxicab geometry. Let $A = (0,0)$, $B = (1,0)$ and $C = (1,1)$ and let ρ denote the taxicab metric (Example 5.4.10).
   (a) Find all points P such that $\rho(A, P) + \rho(P, B) = \rho(A, B)$. Draw a sketch in the Cartesian plane.
   (b) Find all points P such that $\rho(A, P) + \rho(P, C) = \rho(A, C)$. Draw a sketch in the Cartesian plane.[4]

5.25. Betweenness on the sphere. Let A and B be two points on the sphere $\mathbb{S}^2$ (see Example 2.5.8). Define $D(A, B)$, the *distance from A to B*, to be the length of the shortest arc of a great circle containing A and B. More specifically, if A and B are antipodal points, define $D(A, B) = \pi$. If A and B are not antipodal points, then they lie on a unique great circle and they are the endpoints of two subarcs of that great circle. The distance from A to B is the length of the shorter arc. As usual, define C to be between A and B if A, B, and C are collinear and $D(A, C) + D(C, B) = D(A, B)$. Also define *segment* and *ray* in the usual way. A great circle divides the sphere into two hemispheres, which we could think of as half-planes determined by the great circle.
   (a) Find all points that are between A and C in case A and C are nonantipodal points. Sketch the segment from A to C.
   (b) Find all points that are between A and C in case A and C are antipodal points. Sketch the segment from A to C.
   (c) Does the Plane Separation Postulate hold in this setting? Explain.
   (d) Let A and B be distinct, nonantipodal points on $\mathbb{S}^2$. Find all points C such that $A * B * C$. Sketch the ray $\overrightarrow{AB}$.
   (e) Does Theorem 5.7.6 hold in this setting? Explain.
   (f) Does Corollary 5.7.7 hold in this setting? Explain.

---

[4]In the next chapter we will prove the triangle inequality for neutral geometry. One consequence is the fact that if $AB + BC = AC$, then A, B, and C are collinear. The triangle inequality does not hold in taxicab geometry because its proof relies on SAS.

5.26. A circle in taxicab geometry. Sketch all the points (x, y) in $\mathbb{R}^2$ such that $\rho((0, 0), (x, y)) = 1$, where ρ is the taxicab metric.

5.27. The square metric. Define the distance between two points (x_1, y_1) and (x_2, y_2) in $\mathbb{R}^2$ by

$$D((x_1, y_1), (x_2, y_2)) = \max\{|x_2 - x_1|, |y_2 - y_1|\}.^5$$

Sketch all the points (x, y) in $\mathbb{R}^2$ such that $D((0, 0), (x, y)) = 1$. (This should explain the name *square metric*.) Measure angles in the usual way. Show by example that $\mathbb{R}^2$ with the square metric does not satisfy the SAS Postulate.

5.28. Prove the following fact from high school algebra that was needed in the proof of Theorem 5.7.1: If x and y are two nonzero real numbers such that $|x| + |y| = |x + y|$, then either both x and y are positive or both x and y are negative.

5.29. Prove existence and uniqueness of midpoints (Theorem 5.7.5).

5.30. Prove existence and uniqueness of angle bisectors (Theorem 5.7.14) using the Betweenness Theorem for Rays. Do not use SAS.

5.31. Prove existence and uniqueness of angle bisectors (Theorem 5.7.14) using SAS and the Isosceles Triangle Theorem but not using the Betweenness Theorem for Rays.

5.32. If $\ell \perp m$, then ℓ and m contain rays that make four different right angles.

5.33. Prove existence and uniqueness of perpendicular bisectors (Theorem 5.7.23).

5.34. Prove that supplements of congruent angles are congruent.

5.35. Prove the Vertical Angles Theorem (Theorem 5.7.25).

5.36. Prove the following converse to the Vertical Angles Theorem: *If A, B, C, D, and E are points such that $A * B * C$, D and E are on opposite sides of $\overleftrightarrow{AB}$, and $\angle DBC \cong \angle ABE$, then D, B, and E are collinear.*

5.37. Use the Continuity Axiom and the Intermediate Value Theorem to prove the Crossbar Theorem.

5.38. Show that Hilbert's Axioms of Order (Axioms II-1, II-2, and II-3 in Appendix B) are theorems in neutral geometry.

5.39. Show that Hilbert's Axioms of Congruence (Axioms III-1, III-2, and III-3 in Appendix B) are theorems in neutral geometry.

5.40. Identify Hilbert's Axioms II-4 and III-4 (Appendix B) with axioms or theorems stated in this chapter.

5.41. Explain the relationship between Hilbert's Axiom III-5 (Appendix B) and SAS.

5.42. Explain how Birkhoff's Postulate of Similarity (Appendix B) implies SAS.

---

[5] $\max\{a, b\}$ denotes the larger of the two real numbers a and b.

CHAPTER 6

Neutral Geometry

6.1   GEOMETRY WITHOUT THE PARALLEL POSTULATE

In this chapter we study neutral geometry from an axiomatic point of view (no model is assumed). As explained in the previous chapter, *neutral geometry* is the geometry that is based on the five undefined terms *point*, *line*, *distance*, *half-plane*, and *angle measure* together with the following axioms:

1. The Existence Postulate
2. The Incidence Postulate
3. The Ruler Postulate
4. The Plane Separation Postulate
5. The Protractor Postulate
6. The Side-Angle-Side Postulate

Neutral geometry is "neutral" in the sense that it does not take a stand on the parallel postulate. It is the geometry we can do without any postulate regarding parallelism.

In the preceding chapter we spelled out all the assumptions we will use to replace Euclid's first four postulates together with all the unstated assumptions that Euclid took for granted in his proofs. Now we move on to the propositions that Euclid stated and proved in Book I of the *Elements*. We intend to study those propositions while maintaining a "neutral" stance regarding parallelism, so our treatment will necessarily differ quite substantially from that of Euclid.

By studying neutral geometry as a separate subject, we are able to clarify the role of the parallel postulate in geometry. The first part of our study of neutral geometry consists of proving as many of the theorems of plane geometry as we can without assuming any axiom regarding parallelism. We will see that a surprisingly large number of the standard theorems from high school geometry can be proved without any parallel postulate at all.

There will be other theorems that cannot be proved without some additional axiom regarding parallelism. In the second part of the chapter we will see that many familiar theorems not only require a parallel postulate for their proofs, but are, in fact, logically equivalent to the Euclidean Parallel Postulate (at least in the context of the axioms of neutral geometry). The chapter ends with a theorem, known as the Universal Hyperbolic Theorem, which shows that the negation of the Euclidean Parallel Postulate is equivalent in neutral geometry to the Hyperbolic Parallel Postulate. Thus there are only two possibilities: In any model of neutral geometry, either the Euclidean Parallel Postulate holds or the Hyperbolic Parallel Postulate holds.

There is a sense in which Euclid himself did neutral geometry first. He did not use his Fifth Postulate until it was absolutely necessary to do so. It is clear from his arrangement of the material in Book I of the *Elements* that he recognized that the Fifth Postulate plays a special role in geometry and that it would be good to prove as many theorems as possible without it. In fact, all of the major theorems in the next four sections of this chapter are found in Book I of the *Elements*, and Euclid proves all of them without any mention of his Fifth Postulate.

It is not entirely standard to classify all the material in this chapter as neutral geometry. Many authors include the proofs that various statements are equivalent to the Euclidean Parallel Postulate in their treatment of Euclidean geometry and include the Universal Hyperbolic Theorem in their treatment of hyperbolic geometry. The general organizational principle used in this book is the following: Each theorem has a natural context and that context is an axiomatic system; the axioms of the relevant system are the unstated hypotheses in any theorem. From that perspective it is obvious that all the theorems in this chapter must be classified as theorems in neutral geometry. What they have in common is the fact that the six neutral axioms are implicitly assumed as unstated hypotheses. By way of contrast, when we prove that the Euclidean Parallel Postulate implies some other statement, we must explicitly state, as an additional hypothesis, the fact that the Euclidean Parallel Postulate is being assumed. Similarly, the only unstated hypotheses in the Universal Hyperbolic Theorem are the axioms of neutral geometry.

6.2 ANGLE-SIDE-ANGLE AND ITS CONSEQUENCES

In this section we build on the Side-Angle-Side Postulate, which we have assumed as an axiom, to prove another of the familiar triangle congruence conditions. We also investigate some of the basic constructions of plane geometry, in particular those constructions that allow us to "drop a perpendicular" or to construct a copy of a triangle on a given base.

We begin with the Angle-Side-Angle triangle congruence condition. It is one half of Euclid's Proposition 26; the other half (Angle-Angle-Side) will have to wait until after we have proved the Exterior Angle Theorem in the next section.

Theorem 6.2.1 (ASA). *If $\triangle ABC$ and $\triangle DEF$ are two triangles such that $\angle CAB \cong \angle FDE$, $\overline{AB} \cong \overline{DE}$, and $\angle ABC \cong \angle DEF$, then $\triangle ABC \cong \triangle DEF$.*

Proof. Let $\triangle ABC$ and $\triangle DEF$ be two triangles such that $\angle CAB \cong \angle FDE$, $\overline{AB} \cong \overline{DE}$, and $\angle ABC \cong \angle DEF$ (hypothesis). We must show that $\triangle ABC \cong \triangle DEF$. There exists a point C' on $\overrightarrow{AC}$ such that $\overline{AC'} \cong \overline{DF}$ (Ruler Postulate).

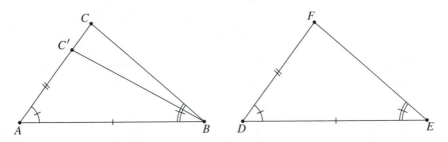

FIGURE 6.1: One possible location for C' in proof of ASA

Now $\triangle ABC' \cong \triangle DEF$ (SAS) and so $\angle ABC' \cong \angle DEF$ (definition of congruent triangles). Since $\angle ABC \cong \angle DEF$ (hypothesis), we can conclude that $\angle ABC \cong \angle ABC'$. Hence $\overrightarrow{BC} = \overrightarrow{BC'}$ (Protractor Postulate, Part 3). But $\overrightarrow{BC}$ can only intersect $\overleftrightarrow{AC}$ in at most one point (Theorem 5.3.7), so $C = C'$ and the proof is complete.   □

The next theorem is the converse to the Isosceles Triangle Theorem. It is Euclid's Proposition 6.

Theorem 6.2.2 (Converse to the Isosceles Triangle Theorem). *If $\triangle ABC$ is a triangle such that $\angle ABC \cong \angle ACB$, then $\overline{AB} \cong \overline{AC}$.*

Proof. Exercise 6.2.   □

The following theorem will be an important tool in our development of geometry. It is closely related to Euclid's Proposition 12, although Euclid meant something quite different by his proposition. Euclid was asserting that the perpendicular could be constructed using straightedge and compass; we are simply saying that the perpendicular line exists.

Theorem 6.2.3 (Existence of Perpendiculars). *For every line ℓ and for every external point P, there exists a line m such that P lies on m and $m \perp \ell$.*

Terminology. When we wish to apply Theorem 6.2.3 in our proofs we will say "drop a perpendicular from P to ℓ." That statement is to be interpreted as an invocation of Theorem 6.2.3. By definition of perpendicular, there is a point F that lies on both ℓ and m; that point is called the *foot* of the perpendicular from P to ℓ.

Proof of theorem 6.2.3. Let ℓ be a line and let P be an external point (hypothesis). We must show that there exists a line m such that $P \in m$ and $m \perp \ell$. There exist distinct points Q and Q' on ℓ (Ruler Postulate). There exists a point R, on the opposite side of ℓ from P, such that $\angle Q'QP \cong \angle Q'QR$ (Protractor Postulate, Part 3). Choose a point P' on $\overrightarrow{QR}$ such that $\overline{QP} \cong \overline{QP'}$ (Ruler Postulate).

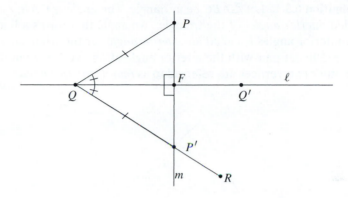

FIGURE 6.2: Proof of existence of perpendiculars

Let $m = \overleftrightarrow{PP'}$; the proof will be complete if we show that $m \perp \ell$. There exists a point $F \in \overline{PP'} \cap \ell$ (Plane Separation Postulate). If $Q = F$, then $\angle Q'FP$ and $\angle Q'FP'$ are supplements. Since they are also congruent (earlier statement), they must both be right angles and so $m \perp \ell$. On the other hand, if $Q \neq F$, then $\triangle FQP \cong \triangle FQP'$ (SAS) and so $\angle QFP \cong \angle QFP'$ (definition of congruent triangles). Since $\angle QFP$ and $\angle QFP'$ are supplements, they must both be right angles. So we conclude that $m \perp \ell$ in this case as well. $\qquad\square$

The perpendicular constructed in the preceding theorem is unique, but we do not yet have the tools to prove that. We will prove uniqueness of perpendiculars in the next section as a corollary to the Exterior Angle Theorem.

The last theorem in the section asserts that it is possible to construct a congruent copy of a triangle on a given base. It is roughly equivalent to Euclid's Proposition 7.

Theorem 6.2.4. *If $\triangle ABC$ is a triangle, $\overline{DE}$ is a segment such that $\overline{DE} \cong \overline{AB}$, and H is a half-plane bounded by $\overleftrightarrow{DE}$, then there is a unique point $F \in H$ such that $\triangle DEF \cong \triangle ABC$.*

Proof. Exercise 6.4. $\qquad\square$

6.3   THE EXTERIOR ANGLE THEOREM

The main result in this section, the Exterior Angle Theorem, is one of the fundamental results of neutral geometry. It is Euclid's Proposition 16. Our "modern" proof is almost exactly the same as Euclid's, right down to the diagram we use to understand the relationships in the proof. There is one important difference however: we use the axioms and theorems of Chapter 5 to fill the gap in Euclid's proof. When you read the proof below, be sure to notice how the theorems of Chapter 5 provide exactly the information we need in order to give a complete proof.

Definition 6.3.1. Let $\triangle ABC$ be a triangle. The angles $\angle CAB$, $\angle ABC$, and $\angle BCA$ are called *interior angles* of the triangle. An angle that forms a linear pair with one of the interior angles is called an *exterior angle* for the triangle. If the exterior angle forms a linear pair with the interior angle at one vertex, then the interior angles at the other two vertices are referred to as *remote interior angles*.

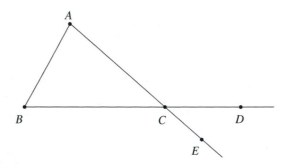

FIGURE 6.3: $\angle ACD$ and $\angle BCE$ are exterior angles for $\triangle ABC$

At each vertex of the triangle there are two exterior angles. Those two exterior angles form a vertical pair and are therefore congruent (Theorem 5.7.25).

Theorem 6.3.2 (Exterior Angle Theorem). *The measure of an exterior angle for a triangle is strictly greater than the measure of either remote interior angle.*

Proof. Let $\triangle ABC$ be a triangle and let D be a point such that $\overrightarrow{CD}$ is opposite to $\overrightarrow{CB}$ (hypothesis). We must prove that $\mu(\angle DCA) > \mu(\angle BAC)$ and that $\mu(\angle DCA) > \mu(\angle ABC)$.

Let E be the midpoint of $\overline{AC}$ (Theorem 5.7.5) and choose F to be the point on $\overrightarrow{BE}$ such that $\overline{BE} \cong \overline{EF}$ (Ruler Postulate). Notice that $\angle BEA \cong \angle FEC$ (Vertical Angle Theorem). Hence $\triangle BEA \cong \triangle FEC$ (SAS) and so $\angle FCA \cong \angle BAC$ (definition of congruent triangles).

Now F and B are on opposite sides of $\overleftrightarrow{AC}$ and B and D are on opposite sides of $\overleftrightarrow{AC}$, so F and D are on the same side of $\overleftrightarrow{AC}$ (Plane Separation Postulate). Also A and E are on the same side of $\overleftrightarrow{CD}$ (Theorem 5.7.6) and E and F are on the same side of $\overleftrightarrow{CD}$ (Theorem 5.7.6 again), so A and F are on the same side of $\overleftrightarrow{CD}$ (Plane Separation Postulate). The last two sentences show that F is in the interior of $\angle ACD$ (definition of angle interior). Hence $\mu(\angle DCA) > \mu(\angle FCA)$

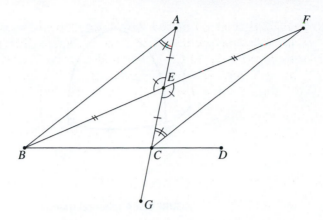

FIGURE 6.4: Proof of the Exterior Angle Theorem

(Betweenness Theorem for Rays, Theorem 5.7.12). Combining this last statement with the previous paragraph gives $\mu(\angle DCA) > \mu(\angle BAC)$ as required.

To prove $\mu(\angle DCA) > \mu(\angle ABC)$ we proceed in exactly the same way as for the other inequality. The only difference is that we replace the exterior angle $\angle DCA$ with the vertical exterior angle $\angle GCB$, where $\overrightarrow{CG}$ is opposite to $\overrightarrow{CA}$.   □

Remark on the form of theorems and proofs. Notice that the theorem was not stated in the standard *if ... then* form. The theorem is easier to understand and remember when it is stated as above and that is why we do so. Such theorem statements make it even more important that the proof begin with a clear identification of the precise hypotheses and conclusions. Perhaps this example will help the reader understand why we have insisted that proofs begin with such apparently unnecessary elements.

The Exterior Angle Theorem is our first major theorem in neutral geometry. Like all the theorems in this chapter, it relies on the postulates of neutral geometry. The theorem does not hold on the sphere. Consider, for example, the triangle $\triangle ABC$ shown in Fig. 6.5. In that triangle, the measure of the angle at vertex A can be much larger than $90°$ while the angles at the other two vertices measure exactly $90°$. As a result, the measure of the exterior angle at B or C can be smaller than the measure of the remote interior angle at A.

It is interesting to follow the steps of Euclid's proof on the sphere. All the steps work there except for the one that Euclid did not justify. If the construction in the proof of Theorem 6.3.2 is carried out on the sphere, starting with the triangle pictured, the point F constructed is in the interior of $\angle ACD$ if and only if $\mu(\angle BAC) < 90°$ (Exercise 6.5).

The Exterior Angle Theorem shows that there can be no triangle in neutral geometry that contains two right angles. We use that result to prove the important fact that perpendiculars are unique.

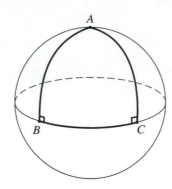

FIGURE 6.5: A spherical triangle

Corollary 6.3.3 (Uniqueness of Perpendiculars). *For every line ℓ and for every external point P, there exists exactly one line m such that P lies on m and m ⊥ ℓ.*

Proof. Let ℓ be a line and let P be an external point (hypothesis). By Theorem 6.2.3 there exists a line m such that P lies on m and $m \perp \ell$. Let Q be the point at which ℓ and m intersect.

   Suppose there exists a line m', different from m, such that P lies on m' and $m' \perp \ell$ (RAA hypothesis). Let Q' be the point at which ℓ and m' intersect. Then $Q' \neq Q$ (Theorem 5.3.7). Now $\triangle PQQ'$ has an exterior angle at Q' that measures $90°$ and an interior angle at Q of the same measure. This contradicts the Exterior Angle Theorem, which says that the measure of an exterior angle must be strictly greater than the measure of a remote interior angle. Thus we must reject the RAA hypothesis and conclude that no such line m' exists. □

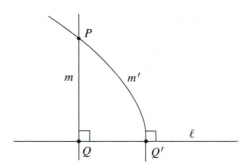

FIGURE 6.6: The existence of two perpendiculars would violate the Exterior Angle Theorem

   Another important corollary of the Exterior Angle Theorem is the Angle-Angle-Side triangle congruence condition. It is the second half of Euclid's Proposition 26.

Theorem 6.3.4 (AAS). *If △ABC and △DEF are two triangles such that ∠ABC ≅ ∠DEF, ∠BCA ≅ ∠EFD, and $\overline{AC}$ ≅ $\overline{DF}$, then △ABC ≅ △DEF.*

Proof. Exercise 6.6.   □

We have now proved three triangle congruence conditions: SAS, ASA, and AAS. The pattern is that three of the six parts of a triangle determine the other three. One might ask whether that is always true. The two conditions ASA and AAS show that any two angles plus a side determine the entire triangle. Two sides plus one angle may or may not determine the triangle. If the angle is included between the two sides, then SAS shows that the triangle is determined; if the angle is not included between the sides, then the triangle is not determined and therefore ASS is not a valid congruence condition. (See Fig. 6.7.)

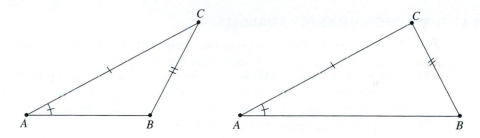

FIGURE 6.7: There is no ASS Theorem

There is one significant special case in which ASS does hold; that is the case in which the given angle is a right angle. The theorem is known as the Hypotenuse-Leg Theorem.

Definition 6.3.5. A triangle is a *right triangle* if one of the interior angles is a right angle. The side opposite the right angle is called the *hypotenuse* and the two sides adjacent to the right angle are called *legs*.

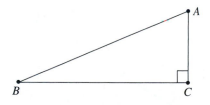

FIGURE 6.8: △ABC is a right triangle; $\overline{AB}$ is the hypotenuse; $\overline{AC}$ and $\overline{BC}$ are legs

Theorem 6.3.6 (Hypotenuse-Leg Theorem). *If △ABC and △DEF are two right triangles with right angles at the vertices C and F, respectively, $\overline{AB}$ ≅ $\overline{DE}$, and $\overline{BC}$ ≅ $\overline{EF}$, then △ABC ≅ △DEF.*

Proof. Exercise 6.7.   □

Three sides always determine the triangle. This is the Side-Side-Side triangle congruence condition, which is Euclid's Proposition 8.

Theorem 6.3.7 (SSS). *If $\triangle ABC$ and $\triangle DEF$ are two triangles such that $\overline{AB} \cong \overline{DE}$, $\overline{BC} \cong \overline{EF}$, and $\overline{CA} \cong \overline{FD}$, then $\triangle ABC \cong \triangle DEF$.*

Proof. Exercise 6.8. □

Before leaving the subject of triangle congruence conditions we should mention that in neutral geometry the three angles of a triangle do not determine the triangle. This may seem obvious to you since you remember from high school that similar triangles are usually not congruent. However, that is a special situation in Euclidean geometry. One of the surprising results we will prove in hyperbolic geometry is that AAA is a valid triangle congruence condition!

6.4   THREE INEQUALITIES FOR TRIANGLES

The Exterior Angle Theorem gives one inequality that is always satisfied by the measures of the angles of a triangle. In this section we use the Exterior Angle Theorem to prove three additional inequalities that will be useful in our study of triangles. The first of these theorems, the Scalene Inequality, extends the Isosceles Triangle Theorem and its converse. It combines Euclid's Propositions 18 and 19. The result can be paraphrased this way: *In any triangle, the greater side lies opposite the greater angle and the greater angle lies opposite the greater side.* The word *scalene* means "unequal" or "uneven." A *scalene triangle* is a triangle that has sides of three different lengths.

Theorem 6.4.1 (Scalene Inequality). *Let A, B, and C be three noncollinear points. Then $AB > BC$ if and only if $\mu(\angle ACB) > \mu(\angle BAC)$.*

Proof. Let A, B, and C be three noncollinear points (hypothesis). We will first assume the hypothesis $AB > BC$ and prove that $\mu(\angle ACB) > \mu(\angle BAC)$. Since $AB > BC$, there exists a point D between A and B such that $\overline{BD} \cong \overline{BC}$ (Ruler Postulate).

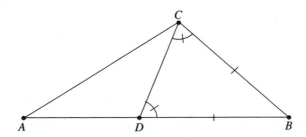

FIGURE 6.9: The greater side lies opposite greater angle

Now $\mu(\angle ACB) > \mu(\angle DCB)$ (Protractor Postulate, Part 4, and Theorem 5.7.10) and $\angle DCB \cong \angle CDB$ (Isosceles Triangle Theorem). But $\angle CDB$ is an exterior angle for $\triangle ADC$ (see Figure 6.9), so $\mu(\angle CDB) > \mu(\angle CAB)$ (Exterior Angle Theorem). The conclusion follows from those inequalities.

The proof of the converse is left as an exercise (Exercise 6.10). $\qquad\square$

The second inequality is the familiar Triangle Inequality. It is Euclid's Proposition 20.

Theorem 6.4.2 (Triangle Inequality). *If A, B, and C are three noncollinear points, then $AC < AB + BC$.*

Proof. Exercise 6.11. $\qquad\square$

The third inequality is Euclid's Proposition 24. It is known as the Hinge Theorem. The reason for the name is the fact that the two sides of the triangle are fixed but the angle is allowed to vary. This means that the triangle can open and close much like a door hinge. The theorem is a generalization of SAS and is sometimes called the SAS Inequality.

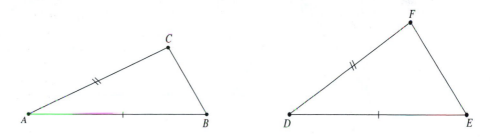

FIGURE 6.10: If $\mu(\angle BAC) < \mu(\angle EDF)$, then $BC < EF$

Theorem 6.4.3 (Hinge Theorem). *If $\triangle ABC$ and $\triangle DEF$ are two triangles such that $AB = DE$, $AC = DF$, and $\mu(\angle BAC) < \mu(\angle EDF)$, then $BC < EF$.*

Proof. Let $\triangle ABC$ and $\triangle DEF$ be two triangles such that $AB = DE$, $AC = DF$, and $\mu(\angle BAC) < \mu(\angle EDF)$. We must show that $BC < EF$. Find a point G, on the same side of $\overleftrightarrow{AB}$ as C, such that $\triangle ABG \cong \triangle DEF$ (Theorem 6.2.4). The proof will be complete if we show that $BG > BC$.

Since C is in the interior of $\angle BAG$ (Theorem 5.7.12), $\overrightarrow{AC}$ must intersect $\overline{BG}$ in a point J (Crossbar Theorem). Let $\overrightarrow{AH}$ be the bisector of $\angle CAG$ (Theorem 5.7.23). Then H is in the interior of $\angle CAG = \angle JAG$ (definition of angle bisector), so $\overrightarrow{AH}$ must intersect $\overline{JG}$ (Crossbar Theorem). In order to simplify the notation, we will assume that H lies on $\overline{JG}$.

Note that $\triangle AHG \cong \triangle AHC$ (SAS). Hence $HG = HC$. Now $BG = BH + HG$ (H is between B and G by Theorem 5.7.10), so $BG = BH + HC$. If H lies on $\overleftrightarrow{BC}$, then $BC = BH - HC$ and $BG = BH + HC$, so $BC < BG$. If H does not lie on

$\overleftrightarrow{BC}$, then $BC < BH + HC$ by the Triangle Inequality. In either case the proof is complete.   □

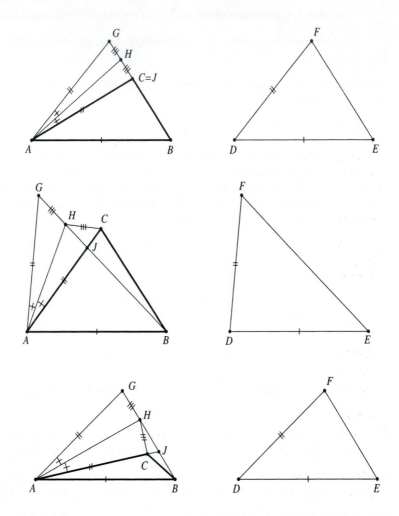

FIGURE 6.11: Three possible diagrams of the proof of the Hinge Theorem

Remark. Euclid's proof of the Hinge Theorem is quite complicated and so we have given a different proof.[1] One reason Euclid's proof is so complicated is the fact that it has three cases that must be considered. In proofs that involved multiple cases, Euclid followed the practice of including the details of only one case, usually the most difficult one. As a result, his written proof cannot really be considered complete by modern standards. It would be easy to fill in the omitted details, but

----

[1]The proof included here is the one Heath calls a "modern" proof of the theorem [30, page 298].

the result is a rather intricate and complicated proof. The nice thing about the proof we gave above is the fact that it is only necessary to consider two different cases. It should be noted, however, that three different diagrams are possible. (See Fig. 6.11.)

The Scalene Inequality allows us to prove that the perpendicular is the shortest line segment joining an external point to a line.

Theorem 6.4.4. *Let ℓ be a line, let P be an external point, and let F be the foot of the perpendicular from P to ℓ. If R is any point on ℓ that is different from F, then PR > PF.*

Proof. Exercise 6.14 (see Fig. 6.12.) □

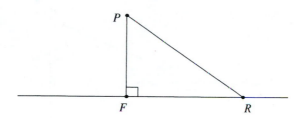

FIGURE 6.12: $PR > PF$

The theorem motivates the following definition.

Definition 6.4.5. If ℓ is a line and P is a point, the *distance from P to ℓ*, denoted $d(P, ℓ)$, is defined to be the distance from P to the foot of the perpendicular from P to ℓ.

We can use that definition to characterize the points that lie on an angle bisector.

Theorem 6.4.6 (Pointwise Characterization of Angle Bisector). *Let A, B, and C be three noncollinear points and let P be a point in the interior of ∠BAC. Then P lies on the angle bisector of ∠BAC if and only if $d(P, \overleftrightarrow{AB}) = d(P, \overleftrightarrow{AC})$.*

Proof. Exercise 6.15 (see Fig. 6.13). □

There is a similar characterization of points that lie on a perpendicular bisector.

Theorem 6.4.7 (Pointwise Characterization of Perpendicular Bisector). *Let A and B be distinct points. A point P lies on the perpendicular bisector of $\overline{AB}$ if and only if PA = PB.*

Proof. Exercise 6.16 (see Fig. 6.14). □

We end this section with an application of the Triangle Inequality that is very different from anything in Euclid. We will prove that the function that measures distances is continuous in a certain technical sense. This theorem will allow us

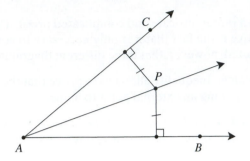

FIGURE 6.13: Points on the angle bisector

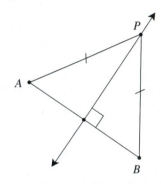

FIGURE 6.14: Points on the perpendicular bisector

to bring results from calculus and real analysis (such as the Intermediate Value Theorem) to bear on geometry. The theorem is not used until Chapters 9 and 10, so most readers will probably want to omit it for now and refer back to it only when it is needed.

Setting for the Continuity Theorem. Let A, B, and C be three noncollinear points. Let $d = AB$. For each $x \in [0, d]$ there exists a unique point $D_x \in \overline{AB}$ such that $AD_x = x$ (Ruler Postulate). Define a function $f : [0, d] \to [0, \infty)$ by $f(x) = CD_x$.

Theorem 6.4.8 (Continuity of Distance). *The function f defined in the previous paragraph is continuous.*

Proof. Let A, B, C, and f be as in the setting for the theorem. We will show that f is continuous at x for $0 < x < d$. The proof of continuity at an endpoint of $[0, d]$ is similar. Let $\epsilon > 0$ be given. We must show that there exists $\delta > 0$ such that if $|x - y| < \delta$ then $|f(x) - f(y)| < \epsilon$ (definition of continuous at x). We claim that $\delta = \epsilon$ works. Suppose y is a number in $[0, d]$ such that $|x - y| < \epsilon$. By the Triangle Inequality, $CD_x < CD_y + D_xD_y < CD_y + \epsilon$. Applying the Triangle Inequality again gives $CD_y < CD_x + D_xD_y < CD_x + \epsilon$. Combining the two inequalities yields $CD_x - \epsilon < CD_y < CD_x + \epsilon$. Therefore $|f(x) - f(y)| = |CD_x - CD_y| < \epsilon$ as required. $\square$

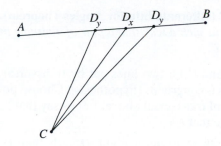

FIGURE 6.15: Two possible locations for D_y in the proof of continuity

6.5 THE ALTERNATE INTERIOR ANGLES THEOREM

We come now to our first theorem about parallel lines. The Alternate Interior Angles Theorem is one of the major theorems of neutral geometry. It is also Euclid's Proposition 27. Before we can state the theorem we need some definitions and notation.

Definition 6.5.1. Let ℓ and ℓ' be two distinct lines. A third line t is called a *transversal* for ℓ and ℓ' if t intersects ℓ in one point B and t intersects ℓ' in one point B' with $B' \neq B$. Notice that there are four angles with vertex B that are formed by rays on t and ℓ and there are four angles with vertex B' that are formed by rays on t and ℓ'. We will give names to some of these eight angles. In order to do so, choose points A and C on ℓ and A' and C' on ℓ' so that $A * B * C$, $A' * B' * C'$, and A and A' are on the same side of t. (Such points exist by the Ruler and Plane Separation Postulates.) The four angles $\angle ABB'$, $\angle A'B'B$, $\angle CBB'$, and $\angle C'B'B$ are called *interior angles* for ℓ and ℓ' with transversal t. The two pairs $\{\angle ABB', \angle BB'C'\}$ and $\{\angle A'B'B, \angle B'BC\}$ are called *alternate interior angles*. See Fig. 6.16.

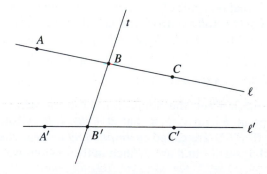

FIGURE 6.16: Definition of transversal

When t is a transversal for lines ℓ and ℓ', we say that "ℓ and ℓ' are *cut* by transversal t." This means that t intersects both lines and that it does so at distinct points.

Theorem 6.5.2 (Alternate Interior Angles Theorem). *If ℓ and ℓ′ are two lines cut by a transversal t in such a way that a pair of alternate interior angles is congruent, then ℓ is parallel to ℓ′.*

Proof. Let ℓ and ℓ′ be two lines cut by transversal t such that a pair of alternate interior angles is congruent (hypothesis). Choose points A, B, C and $A′$, $B′$, $C′$ as in the definition of transversal above. Let us say that $\angle A′B′B \cong \angle B′BC$ (hypothesis). We must prove that $ℓ \parallel ℓ′$.

Suppose there exists a point D such that D lies on both ℓ and ℓ′ (RAA hypothesis). If D lies on the same side of t as C, then $\angle A′B′B$ is an exterior angle for $\triangle BB′D$ while $\angle B′BC$ is a remote interior angle for that same triangle (definitions). This contradicts the Exterior Angles Theorem (Fig. 6.17).

In case D lies on the same side of t as A, then $\angle B′BC$ is an exterior angle and $\angle A′B′B$ is a remote interior angle for $\triangle BB′D$ and we again reach a contradiction. Since D must lie on one of the two sides of t (Plane Separation Postulate), we are forced to reject the RAA hypothesis and the theorem is proved. □

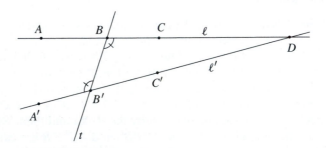

FIGURE 6.17: Impossible alternate interior angles

Notice that one pair of alternate interior angles is congruent if and only if the other one is. This follows from the fact that supplements of congruent angles are congruent, which was an exercise in the previous chapter.

It is also common to state the theorem in terms of corresponding angles. This version is Euclid's Proposition 28.

Definition 6.5.3. Keep the notation used in the definition of interior angles and choose a point $B″$ on t such that $B * B′ * B″$ (Ruler Postulate). The angles $\{\angle B′BC, \angle B″B′C′\}$ are called *corresponding angles*. There are three other pairs of corresponding angles that are defined in the obvious way.

Corollary 6.5.4 (Corresponding Angles Theorem). *If ℓ and ℓ′ are lines cut by a transversal t in such a way that two corresponding angles are congruent, then ℓ is parallel to ℓ′.*

Proof. Exercise 6.17. □

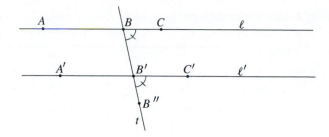

FIGURE 6.18: Congruent corresponding angles

The Alternate Interior Angles Theorem has several other interesting and important corollaries. The first is a partial converse to Euclid's Fifth Postulate.

Corollary 6.5.5. *If ℓ and ℓ' are lines cut by a transversal t in such a way that two nonalternating interior angles on the same side of t are supplements, then ℓ is parallel to ℓ'.*

Proof. Exercise 6.18. □

The next corollary, which asserts that parallel lines exist, is one of the most important consequences of the Alternate Interior Angles Theorem. The construction in the proof is simple, but it is one that we will use over and over in the remainder of the course. In fact, most applications of the corollary will use the construction in the proof rather than the statement of the corollary itself. Be sure to notice that the there is no claim that the parallel constructed in the proof is unique. As we will see later, the uniqueness of parallels cannot be proved in neutral geometry. The existence of parallels is Euclid's Proposition 31; as usual Euclid was really asserting that the parallel could be constructed with straightedge and compass.

Corollary 6.5.6 (Existence of Parallels). *If ℓ is a line and P is an external point, then there is a line m such that P lies on m and m is parallel to ℓ.*

Proof. Let ℓ be a line and let P be an external point (hypothesis). Drop a perpendicular from P to ℓ (Theorem 6.2.3). Call the foot of that perpendicular Q and let $t = \overleftrightarrow{PQ}$. Next construct a line m through P that is perpendicular to t (Protractor Postulate, Part 3, or Exercise 6.6.3). Then $\ell \parallel m$ by the Alternate Interior Angles Theorem (all the interior angles are right angles). □

For future reference we add some of the additional properties of the parallel to the statement.

Addendum. *If ℓ is a line and P is an external point, then there are lines t and m such that P lies on both t and m, m is parallel to ℓ, and t is a transversal for ℓ and m such that t is perpendicular to both ℓ and m.*

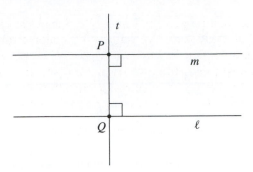

FIGURE 6.19: Proof of the existence of parallels

The existence of parallels implies, in particular, that the Elliptic Parallel Postulate is not consistent with the axioms of neutral geometry. Thus we can state the following corollary.

Corollary 6.5.7. *The Elliptic Parallel Postulate is false in any model for neutral geometry.*

The following special case of the Alternate Interior Angles Theorem is often useful.

Corollary 6.5.8. *If ℓ, m, and n are three lines such that $m \perp \ell$ and $n \perp \ell$, then either $m = n$ or $m \parallel n$.*

Proof. Exercise 6.19. □

6.6 THE SACCHERI-LEGENDRE THEOREM

The next theorem tells us about angle sums for triangles. The theorem is named after two mathematicians who contributed to our understanding of the theorem's place in neutral geometry. It is our first major departure from Euclid's propositions and also the first time we use the Archimedean Property of Real Numbers.

Definition 6.6.1. Let A, B, and C be three noncollinear points. The *angle sum* for $\triangle ABC$ is the sum of the measures of the three interior angles of $\triangle ABC$. More specifically, the angle sum is defined by the equation

$$\sigma(\triangle ABC) = \mu(\angle CAB) + \mu(\angle ABC) + \mu(\angle BCA).$$

It is clear from the definition that congruent triangles have equal angle sums.

We will prove that the angle sum of any triangle in neutral geometry is less than or equal to $180°$. The theorem is not one of the familiar theorems from high school geometry nor is it one of Euclid's propositions. Euclid proves the well-known "fact" that the angle sum of any triangle is exactly equal to $180°$. His proof uses his fifth postulate, so it has no place in neutral geometry. We will see later that there are models for neutral geometry in which triangles have angle sums strictly less than 180° and that angle sums are one of the major distinguishing characteristics for Euclidean and non-Euclidean geometries. What the Saccheri-Legendre Theorem rules out is angle sums that are strictly greater than 180°. Be sure to notice that triangles on the sphere $\mathbb{S}^2$ have angle sums that are strictly greater than 180° (see Figure 6.5) and so the theorem does have substantial content.

Theorem 6.6.2 (Saccheri-Legendre Theorem). *If $\triangle ABC$ is any triangle, then $\sigma(\triangle ABC) \leqslant 180°$.*

The proof of the theorem relies on several lemmas. The first tells us that the sum of the measures of any two angles in a triangle is always strictly less than 180° while the second tells us how angle sums add when a triangle is subdivided. The third allows us to replace one triangle with another triangle that has the same angle sum but at least one angle that is smaller by a factor of 2.

Lemma 6.6.3. *If $\triangle ABC$ is any triangle, then*

$$\mu(\angle CAB) + \mu(\angle ABC) < 180°.$$

Proof. Exercise 6.20.   □

Lemma 6.6.4. *If $\triangle ABC$ is a triangle and E is a point in the interior of $\overline{BC}$, then*

$$\sigma(\triangle ABE) + \sigma(\triangle ECA) = \sigma(\triangle ABC) + 180°.$$

Proof. Exercise 6.21 (Fig. 6.20).   □

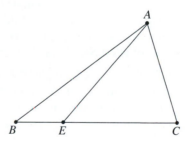

FIGURE 6.20: The triangle $\triangle ABC$ is subdivided into two smaller triangles

Lemma 6.6.5. *If A, B, and C are three noncollinear points, then there exists a point D that does not lie on $\overleftrightarrow{AB}$ such that $\sigma(\triangle ABD) = \sigma(\triangle ABC)$ and the angle measure of one of the interior angles in $\triangle ABD$ is less than or equal to $\frac{1}{2}\mu(\angle CAB)$.*

Proof. Let A, B, and C be three noncollinear points (hypothesis). Let E be the midpoint of $\overline{BC}$ (Theorem 5.7.5). Let D be a point on $\overrightarrow{AE}$ such that $A * E * D$ and

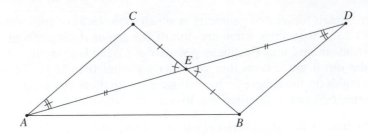

FIGURE 6.21: Proof of Lemma 6.6.5

$AE = ED$ (Ruler Postulate). We will show that $\sigma(\triangle ABD) = \sigma(\triangle ABC)$ and either $\mu(\angle BAD) \leq \frac{1}{2}\mu(\angle BAC)$ or $\mu(\angle ADB) \leq \frac{1}{2}\mu(\angle BAC)$.

Since $\angle AEC \cong \angle DEB$ (Vertical Angles Theorem), we see that $\triangle AEC \cong \triangle DEB$ (SAS) and so $\sigma(\triangle AEC) = \sigma(\triangle DEB)$. Applying Lemma 6.6.4 twice gives

$$\sigma(\triangle ABC) = \sigma(\triangle ABE) + \sigma(\triangle AEC) - 180°$$

and

$$\sigma(\triangle ABD) = \sigma(\triangle ABE) + \sigma(\triangle DEB) - 180°.$$

It follows that $\sigma(\triangle ABD) = \sigma(\triangle ABC)$.

Now $\mu(\angle BAE) + \mu(\angle EAC) = \mu(\angle BAC)$ (Protractor Postulate). Hence either $\mu(\angle BAE) \leq \frac{1}{2}\mu(\angle BAC)$ or $\mu(\angle EAC) \leq \frac{1}{2}\mu(\angle BAC)$ (algebra). Since $\mu(\angle EAC) = \mu(\angle ADB)$ (previous paragraph), the proof is complete.   □

Proof of Theorem 6.6.2. Let $\triangle ABC$ be a triangle (hypothesis). Suppose $\sigma(\triangle ABC) > 180°$ (RAA hypothesis). Let us say that $\sigma(\triangle ABC) = 180° + \epsilon°$, where ϵ is a positive real number. Choose a positive integer n large enough so that $2^n\epsilon° > \mu(\angle CAB)$ (Archimedean Property of Real Numbers, Axiom 4.2.5).

By Lemma 6.6.5, there is a triangle $\triangle A_1 B_1 C_1$ such that $\sigma(\triangle A_1 B_1 C_1) = \sigma(\triangle ABC)$ and one of the angles in $\triangle A_1 B_1 C_1$ has angle measure $\leq \frac{1}{2}\mu(\angle CAB)$. Applying the lemma a second time gives a triangle $\triangle A_2 B_2 C_2$ such that $\sigma(\triangle A_2 B_2 C_2) = \sigma(\triangle ABC)$ and one of the angles in $\triangle A_2 B_2 C_2$ has angle measure less than or equal to $\frac{1}{4}\mu(\angle CAB)$.

Applying Lemma 6.6.5 a total of n times yields a triangle $\triangle A_n B_n C_n$ such that $\sigma(\triangle A_n B_n C_n) = \sigma(\triangle ABC) = 180° + \epsilon°$ and one of the angles in $\triangle A_n B_n C_n$ has measure less than or equal to $\frac{1}{2^n}\mu(\angle CAB) < \epsilon°$. Thus the sum of the measures of the other two angles is greater than $180°$ (algebra). But this contradicts Lemma 6.6.3. Hence we must reject the RAA hypothesis and the proof is complete.   □

Here are two interesting corollaries of the Saccheri-Legendre Theorem. The first improves on the Exterior Angle Theorem. The second is a full converse to Euclid's Fifth Postulate.

Corollary 6.6.6. *The sum of the measures of two interior angles of a triangle is less than or equal to the measure of their remote exterior angle.*

Proof. Exercise 6.22. □

Corollary 6.6.7 (Converse to Euclid's Fifth Postulate). *Let ℓ and ℓ' be two lines cut by a transversal t. If ℓ and ℓ' meet on one side of t, then the sum of the measures of the two interior angles on that side of t is strictly less than $180°$.*

Proof. Exercise 6.23. □

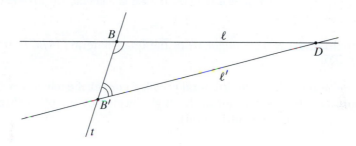

FIGURE 6.22: Converse to Euclid V: $\mu(\angle DBB') + \mu(\angle DB'B) < 180°$

6.7 QUADRILATERALS

In our study of geometry we will want to work with four-sided figures (quadrilaterals) as well as three-sided ones (triangles). In particular, it will be important to understand angle sums for quadrilaterals. This section contains a brief introduction to quadrilaterals and their properties. We begin with some definitions.

Definition 6.7.1. Let A, B, C, and D be four points, no three of which are collinear. Suppose further that any two of the segments $\overline{AB}$, $\overline{BC}$, $\overline{CD}$, and $\overline{DA}$ either have no point in common or have only an endpoint in common. If those conditions are satisfied, then the points A, B, C, and D determine a *quadrilateral*, which we will denote by $\square ABCD$. The quadrilateral is the union of the four segments $\overline{AB}$, $\overline{BC}$, $\overline{CD}$, and $\overline{DA}$. The four segments are called the *sides* of the quadrilateral and the points A, B, C, and D are called the *vertices* of the quadrilateral. The sides $\overline{AB}$ and $\overline{CD}$ are called *opposite sides* of the quadrilateral as are the sides $\overline{BC}$ and $\overline{AD}$. Two quadrilaterals are *congruent* if there is a correspondence between their vertices so that all four corresponding sides are congruent and all four corresponding angles are congruent.

Notice that the order in which the vertices of a quadrilateral are listed is important. In general, $\square ABCD$ is a different quadrilateral from $\square ACBD$. In fact, there may not even be a quadrilateral $\square ACBD$ since the segments $\overline{AC}$ and $\overline{BD}$ may intersect at an interior point.

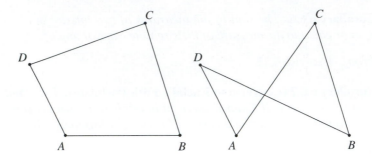

FIGURE 6.23: □$ABCD$ is a quadrilateral, □$ACBD$ is not

Definition 6.7.2. The *diagonals* of the quadrilateral □$ABCD$ are the segments $\overline{AC}$ and $\overline{BD}$.

Definition 6.7.3. A quadrilateral is called *convex* if each vertex of the quadrilateral is contained in the interior of the angle formed by the other three vertices (in their cyclic order on the quadrilateral).

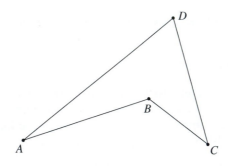

FIGURE 6.24: A nonconvex quadrilateral

As Fig. 6.24 suggests, it is possible to identify an interior for a quadrilateral and the interior is convex according to the earlier definition of convex set if and only if the quadrilateral is convex according to our new definition. Making a precise definition of interior of a quadrilateral involves complications that we prefer to avoid, so we will not make a formal definition.

Definition 6.7.4. If □$ABCD$ is a convex quadrilateral, then the *angle sum* is defined by

$$\sigma(\square ABCD) = \mu(\angle ABC) + \mu(\angle BCD) + \mu(\angle CDA) + \mu(\angle DAB).$$

We will not define angle sums for nonconvex quadrilaterals. The reason should be evident from Figure 6.24. What we intuitively want to call interior angles for the quadrilateral may not be angles at all, at least not according to the way we have defined angles. Each vertex does determine an angle, but one of those angles may be on the "exterior" of the quadrilateral. If we were to ignore that fact and simply use the formula to define an angle sum, we would have trouble proving theorems like the next proposition.

Theorem 6.7.5. *If □ABCD is a convex quadrilateral, then* $\sigma(\Box ABCD) \leq 360°$.

Proof. Exercise 6.25. □

We end this section with several technical facts about convex quadrilaterals that will be needed in later chapters.

Definition 6.7.6. The quadrilateral □*ABCD* is a *parallelogram* if $\overleftrightarrow{AB} \parallel \overleftrightarrow{CD}$ and $\overleftrightarrow{AD} \parallel \overleftrightarrow{BC}$.

Theorem 6.7.7. *Every parallelogram is a convex quadrilateral.*

Proof. Exercise 6.26. □

Theorem 6.7.8. *If* △*ABC is a triangle, D is between A and B, and E is between A and C, then* □*BCED is a convex quadrilateral.*

Proof. Exercise 6.29. (See Figure 6.25.) □

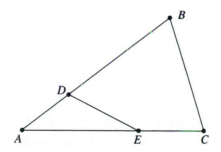

FIGURE 6.25: □*DECB* is a convex quadrilateral

Theorem 6.7.9. *The quadrilateral* □*ABCD is convex if and only if the diagonals* $\overline{AC}$ *and* $\overline{BD}$ *have an interior point in common.*

Proof. Assume, first, that □*ABCD* is a convex quadrilateral (hypothesis). We must prove that the diagonals $\overline{AC}$ and $\overline{BD}$ have a point in common. Then *C* is in the interior of ∠*DAB* (definition of convex quadrilateral). Hence $\overrightarrow{AC}$ intersects $\overline{BD}$ in a point we will call *E* (Crossbar Theorem). It follows in a similar way that $\overrightarrow{BD}$ intersects $\overline{AC}$ in a point *E'*. But $\overleftrightarrow{AC}$ and $\overleftrightarrow{BD}$ meet in at most one point (Incidence Postulate) and so *E* = *E'* and *E* lies on both $\overline{AC}$ and $\overline{BD}$.

The proof of the converse is left as an exercise (Exercise 6.30). □

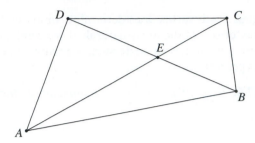

FIGURE 6.26: The diagonals $\overline{AC}$ and $\overline{BD}$ intersect at the point E

Corollary 6.7.10. *If* $\square ABCD$ *and* $\square ACBD$ *are both quadrilaterals, then* $\square ABCD$ *is not convex. If* $\square ABCD$ *is a nonconvex quadrilateral, then* $\square ACBD$ *is a quadrilateral.*

Proof. Exercise 6.31. □

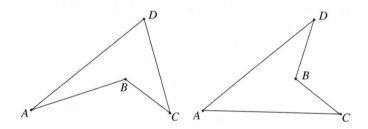

FIGURE 6.27: $\square ABCD$ and $\square ACBD$ are both quadrilaterals

6.8 STATEMENTS EQUIVALENT TO THE EUCLIDEAN PARALLEL POSTULATE

As was pointed out earlier, Euclid delayed using his fifth postulate until he absolutely needed it. Our exploration of neutral geometry paralleled that of Euclid through the Alternate Interior Angles Theorem and its corollaries (Euclid's Propositions 27 and 28). After those theorems our treatment diverged from Euclid and we proved the Saccheri-Legendre Theorem regarding angle sums. In that theorem we avoided the use of any parallel postulate by proving a weaker theorem than Euclid did. We now take a different tack. We return to Euclid and consider his Proposition 29, which is the converse to the Alternate Interior Angles Theorem. We will see that Euclid was exactly right to invoke his Fifth Postulate at that point. We will prove, in fact, that the Fifth Postulate is logically equivalent to Proposition 29.

It is important to understand clearly what it means to say that the two statements are logically equivalent in this context. It means that either one of the statements could be assumed as an axiom and then the other could be proved as a theorem. More specifically, it means that if we start with all the axioms of neutral geometry and add one of the statements as an additional axiom, then the other statement can be proved as a theorem. This is often misunderstood. We are not claiming that these statements are logically equivalent in isolation. For example, Euclid's Fifth Postulate is a correct statement on the sphere, but the Euclidean Parallel Postulate is not. Hence the two statements cannot be logically equivalent by themselves. What we will show in this section is that they are logically equivalent in the presence of the axioms of neutral geometry. Thus adding one of the statements to our existing system of axioms is equivalent to adding the other.

Another way to understand this is in terms of models. When we assert that two statements are equivalent in neutral geometry we are saying that if one of the statements is true in a given model for neutral geometry, then the other must be true in that model as well.

We will prove that a surprisingly large number of the familiar theorems of Euclidean geometry are actually equivalent to the Parallel Postulate in the sense just explained. The approach taken in this section is consistent with what was done historically. The hope of many geometers was to prove Euclid's Fifth Postulate using only Euclid's other postulates. Failing that, the next best thing was to assume a simpler, more intuitively obvious postulate instead and then to demonstrate that Euclid's first four postulates together with the new postulate could be used to prove Euclid's Fifth Postulate. All such attempts to improve on Euclid have come to nothing in the sense that the new postulates invariably turn out to be equivalent to Euclid's postulate.

Here is a statement of Euclid's Proposition 29.

Converse to the Alternate Interior Angles Theorem. *If two parallel lines are cut by a transversal, then both pairs of alternate interior angles are congruent.*

The Converse to the Alternate Interior Angles Theorem is *not* a theorem in neutral geometry. But we do have the following theorem.

Theorem 6.8.1. *The Converse to the Alternate Interior Angles Theorem is equivalent to the Euclidean Parallel Postulate.*

Pay particular attention to the logical structure of the proof of this theorem. The proof itself is part of neutral geometry even though neither of the statements being proved is a theorem in neutral geometry. In the first half of the proof, the entire Converse to the Alternate Interior Angles Theorem is assumed as part of the hypothesis. The "givens" of the Euclidean Parallel Postulate (the line ℓ and the external point P) are also included as hypotheses. In the second half of the proof we prove the Converse to the Alternate Interior Angles Theorem. For that part of the proof, the hypotheses consist of the entire Euclidean Parallel Postulate along with the hypotheses of the Converse to the Alternate Interior Angles Theorem. In both parts of the proof an unstated hypothesis is the assumption that all the axioms of neutral geometry hold.

Proof of Theorem 6.8.1. First assume the Converse to the Alternate Interior Angles Theorem (hypothesis). Let ℓ be a line and let P be an external point (hypothesis). We must prove that there is exactly one line m such that P lies on m and $m \parallel \ell$. First construct a line m using the construction we have used before: Drop a perpendicular t from P to ℓ (Theorem 6.2.3), and then construct a line m such that P lies on m and $m \perp t$. Note that P lies on m (by construction) and $m \parallel \ell$ (Alternate Interior Angles Theorem). We must prove that m is the only such line. Suppose m' is any line such that P lies on m' and $m' \parallel \ell$. Then t is a transversal for ℓ and m' so the angle made by t and m' must be equal to the angle made by t and ℓ (Converse to the Alternate Interior Angles Theorem). Hence $m' \perp t$ and so $m' = m$ (the uniqueness part of Protractor Postulate, Part 3).

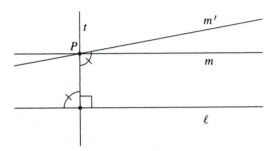

FIGURE 6.28: Converse to the Alternate Interior Angles Theorem implies the Euclidean Parallel Postulate

Now assume the Euclidean Parallel Postulate (hypothesis). Suppose ℓ and ℓ' are parallel lines with transversal t (hypothesis). We must prove that both pairs of alternate interior angles are congruent. Let B' denote the point at which t intersects ℓ'. Let ℓ'' be the line through B' for which the alternate interior angles formed by ℓ and ℓ'' with transversal t are congruent. (This line exists by the Protractor Postulate, Part 3.) By the Alternate Interior Angles Theorem, $\ell'' \parallel \ell$. The uniqueness part of the Euclidean Parallel Postulate implies that $\ell'' = \ell'$. Hence the alternate interior angles formed by ℓ and ℓ' with transversal t are congruent. $\square$

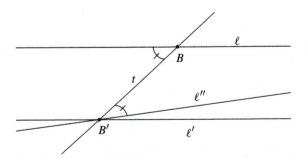

FIGURE 6.29: The Euclidean Parallel Postulate implies the converse to the Alternate Interior Angles Theorem

The axiom we have been calling the Euclidean Parallel Postulate is not exactly the same as Euclid's Fifth Postulate. The axiom we refer to as the Euclidean

Parallel Postulate is often called *Playfair's Postulate* after the eighteenth-century Scottish mathematician John Playfair, even though it had been formulated much earlier by Proclus. Playfair stated the postulate in one of many attempts by various mathematicians to improve on Euclid by replacing Euclid's Fifth Postulate with something simpler. Of course Playfair's Postulate turned out to be logically equivalent to Euclid's. Let us state Euclid's Fifth Postulate in the language of this course.

Euclid's Postulate V. *If ℓ and ℓ' are two lines cut by a transversal t in such a way that the sum of the measures of the two interior angles on one side of t is less than $180°$, then ℓ and ℓ' intersect on that side of t.*

Theorem 6.8.2. *Euclid's Postulate V is equivalent to the Euclidean Parallel Postulate.*

Proof. First assume the Euclidean Parallel Postulate (hypothesis). Let ℓ and ℓ' be two lines crossed by a transversal t such that the sum of the measures of the two interior angles on one side of t is less than $180°$ (hypothesis). Let B and B' be the points where t crosses ℓ and ℓ', respectively. There is a line ℓ'' such that B' lies on ℓ'' and both pairs of nonalternating interior angles formed by ℓ and ℓ'' with transversal t have measures whose sum is $180°$ (by the existence part of Protractor Postulate, Part 3). Note that $\ell'' \neq \ell'$ (by the uniqueness part of Protractor Postulate, Part 3) and $\ell'' \parallel \ell$ (Alternate Interior Angles Theorem). Hence ℓ' is not parallel to ℓ (by the uniqueness part of the Euclidean Parallel Postulate). Thus there exists a point C that lies on both ℓ and ℓ' (negation of the definition of parallel). Now $\triangle BB'C$ is a triangle and so the sum of the measures of any two interior angles must be less than $180°$ (Lemma 6.6.3). It follows that C must be on the side of t where the interior angles formed by ℓ and ℓ' with transversal t have measures whose sum is less than $180°$.

The other half of the proof is left as an exercise (Exercise 6.36).   □

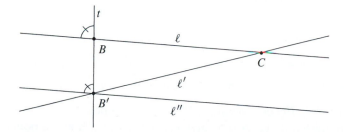

FIGURE 6.30: The Euclidean Parallel Postulate implies Euclid's Postulate V

Hilbert also has a parallel postulate. It is the same as the Euclidean Parallel Postulate except that he only asserts that there is at most one parallel line, rather than exactly one. Since we have already proved that there is at least one parallel line, it is obvious that Hilbert's Parallel Postulate is equivalent to the Euclidean Parallel

Postulate. Hilbert was interested in finding a set of postulates that assumed no more than was absolutely necessary. We prefer to use Playfair's Postulate because it makes a clearer contrast with the Hyperbolic and Elliptic Parallel Postulates.

Hilbert's Parallel Postulate. *For every line ℓ and for every external point P there exists at most one line m such that P lies on m and m $\parallel$ ℓ.*

The next theorem gives several other statements that are equivalent to the Euclidean Parallel Postulate. Be sure to notice the fourth part of the theorem. It is tempting to assume that parallelism is transitive just on the basis of general principles, but that is not the case. Transitivity of Parallelism is Euclid's Proposition 30.

Theorem 6.8.3. *Each of the following statements is equivalent to the Euclidean Parallel Postulate.*

1. (Proclus's Axiom) *If ℓ and ℓ' are parallel lines and $t \neq \ell$ is a line such that t intersects ℓ, then t also intersects ℓ'.*
2. *If ℓ and ℓ' are parallel lines and t is a transversal such that $t \perp \ell$, then $t \perp \ell'$.*
3. *If ℓ, m, n and k are lines such that $k \parallel \ell$, $m \perp k$, and $n \perp \ell$, then either $m = n$ or $m \parallel n$.*
4. (Transitivity of Parallelism) *If ℓ is parallel to m and m is parallel to n, then either $\ell = n$ or $\ell \parallel n$.*

Proof. Exercises 6.37, 6.38, 6.39, and 6.40.   □

There is a close relationship between the parallel postulate and angle sums. We will explore that relationship further in the next section, but for now we prove the basic equivalence. First let us state the usual assumption about angle sums as a postulate.

Angle Sum Postulate. *If $\triangle ABC$ is a triangle, then $\sigma(\triangle ABC) = 180°$.*

Theorem 6.8.4. *The Euclidean Parallel Postulate is equivalent to the Angle Sum Postulate.*

The proof relies on the following lemma.

Lemma 6.8.5. *Suppose $\overline{PQ}$ is a segment and Q' is a point such that $\angle PQQ'$ is a right angle. For every $\epsilon > 0$ there exists a point T on $\overrightarrow{QQ'}$ such that $\mu(\angle PTQ) < \epsilon°$.*

Proof. Let $\overline{PQ}$ be a segment and let Q' be a point such that $\angle PQQ'$ is a right angle (hypothesis). Let $\epsilon > 0$ be given (hypothesis). Choose a point P', on the same side of $\overleftrightarrow{PQ}$ as Q', such that $\overleftrightarrow{PP'} \perp \overleftrightarrow{PQ}$. Observe that every point T on $\overrightarrow{QQ'}$ is in the interior of $\angle QPP'$ (definition of interior). Choose a sequence of points $T_1, T_2, \dots$ on $\overrightarrow{QQ'}$ as follows. First choose T_1 such that $\overline{PQ} \cong \overline{QT_1}$. Then choose T_2 such that $\overline{PT_1} \cong \overline{T_1T_2}$. Inductively choose T_n such that T_{n-1} is between Q and T_n and such that $\overline{PT_{n-1}} \cong \overline{T_{n-1}T_n}$. Points with those properties can be chosen using the Ruler Postulate.

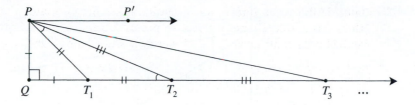

FIGURE 6.31: The construction of $T_1, T_2, \ldots$

Now $\angle QT_nP \cong \angle T_{n-1}PT_n$ for every n (Isosceles Triangle Theorem). For each n, Part 4 of the Protractor Postulate gives

$$\mu(\angle QPT_1) + \mu(\angle T_1PT_2) + \cdots + \mu(\angle T_{n-1}PT_n)$$
$$= \mu(\angle QPT_n) < \mu(\angle QPP') = 90°.$$

Suppose $\mu(\angle T_{i-1}PT_i) \geq \epsilon$ for every i (RAA hypothesis). By the Archimedean Property of Real Numbers (Axiom 4.2.5) there is an n for which $n\epsilon > 90$. It follows that

$$\mu(\angle QPT_1) + \mu(\angle T_1PT_2) + \cdots + \mu(\angle T_{n-1}PT_n) \geq n\epsilon° > 90°.$$

This contradicts the earlier computation. Hence we must reject the RAA hypothesis and can conclude that there exists an i such that $\mu(\angle T_{i-1}PT_i) < \epsilon°$. Therefore, $\mu(\angle QT_iP) < \epsilon°$ and so $T = T_i$ satisfies the conclusion of the lemma.   □

Proof of Theorem 6.8.4 . The proof that the Euclidean Parallel Postulate implies $\sigma(\triangle ABC) = 180°$ for every triangle $\triangle ABC$ is left as an exercise (Exercise 6.41).

For the converse, we must prove that if the angle sum for every triangle is $180°$, then the Euclidean Parallel Postulate holds. We will actually prove the contrapositive; we will prove that if there exists a line ℓ and an external point P through which there are multiple parallel lines, then there is a triangle whose angle sum is different from $180°$.

Let ℓ be a line and let P be an external point such that there is more than one line through P that is parallel to ℓ (hypothesis). Drop a perpendicular from P to ℓ and call the foot of that perpendicular Q. As usual, let m be the line through P that is perpendicular to $\overleftrightarrow{PQ}$. Then $m \parallel \ell$. By our hypothesis on ℓ and P, there exists another line m', different from m, such that P lies on m' and $m' \parallel \ell$. Choose a point S on m' such that S is on the same side of m as Q. Choose a point R on m such that R is on the same side of $\overleftrightarrow{PQ}$ as S. Finally, choose a point T on ℓ such that T lies on the same side of $\overleftrightarrow{PQ}$ as S and $\mu(\angle QTP) < \mu(\angle SPR)$ (Lemma 6.8.5). We will prove that $\sigma(\triangle PQT) < 180°$.

Now T is in the interior of $\angle QPS$ (otherwise $\overrightarrow{PS}$ would have to meet $\overline{QT}$ by the Crossbar Theorem). Therefore, $\mu(\angle QPT) < \mu(\angle QPS)$ (Protractor Postulate).

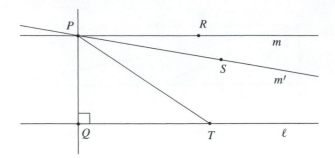

FIGURE 6.32: $\triangle PQT$ has angle sum less than $180°$

The way in which R and S were chosen guarantees that S is in the interior of $\angle QPR$, and so $\mu(\angle SPR) + \mu(\angle SPQ) = \mu(\angle RPQ)$ (Protractor Postulate). Hence

$$\begin{aligned}
\sigma(\triangle QTP) &= \mu(\angle PQT) + \mu(\angle QTP) + \mu(\angle TPQ) \\
&< \mu(\angle PQT) + \mu(\angle SPR) + \mu(\angle QPS) \\
&= \mu(\angle PQT) + \mu(\angle RPQ) \\
&= 180°
\end{aligned}$$

because angles $\angle PQT$ and $\angle RPQ$ are right angles. $\square$

We end this section with an equivalence between the existence of (noncongruent) similar triangles and the Euclidean Parallel Postulate. A mathematician named John Wallis (1616–1703) took the existence of similar triangles as his postulate and used it to prove Euclid's Fifth Postulate. Wallis's Postulate also proved to be equivalent to Euclid's Fifth Postulate.

Definition 6.8.6. Triangles $\triangle ABC$ and $\triangle DEF$ are *similar* if $\angle ABC \cong \angle DEF$, $\angle BCA \cong \angle EFD$, and $\angle CAB \cong \angle FDE$.

Notation. Write $\triangle ABC \sim \triangle DEF$ if $\triangle ABC$ is similar to $\triangle DEF$.

Wallis's Postulate. *If $\triangle ABC$ is a triangle and $\overline{DE}$ is a segment, then there exists a point F such that $\triangle ABC \sim \triangle DEF$.*

Theorem 6.8.7. *Wallis's Postulate is equivalent to the Euclidean Parallel Postulate.*

Proof. First assume the Euclidean Parallel Postulate (hypothesis). Let $\triangle ABC$ be a triangle and let $\overline{DE}$ be a segment (hypothesis). There exists a ray $\overrightarrow{DG}$ such that $\angle EDG \cong \angle BAC$ and there exists a ray $\overrightarrow{EH}$, with H on the same side of $\overleftrightarrow{DE}$ as G, such that $\angle DEH \cong \angle ABC$ (Protractor Postulate, Part 3). There exists a point F where $\overrightarrow{DG}$ and $\overrightarrow{EH}$ meet (Euclid's Postulate V). It follows from the Euclidean Parallel Postulate and Theorem 6.8.4 that $\sigma(\triangle ABC) = \sigma(\triangle DEF) = 180°$. Hence subtraction gives $\mu(\angle EFD) = \mu(\angle BCA)$. Therefore, $\triangle ABC \sim \triangle DEF$ (definition of similar).

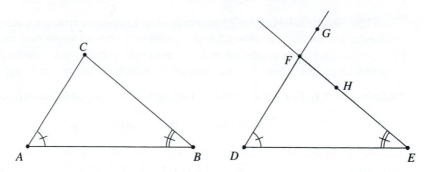

FIGURE 6.33: The Euclidean Parallel Postulate implies Wallis's Postulate

Now assume Wallis's Postulate (hypothesis). Let ℓ be a line and let P be an external point (hypothesis). We must show that there exists exactly one line m such that P lies on m and $m \parallel \ell$. Drop a perpendicular from P to ℓ and call the foot of that perpendicular Q. Let m be the line through P that is perpendicular to $\overleftrightarrow{PQ}$. Let m' be a line such that P lies on m' and $m' \parallel \ell$. Suppose $m' \neq m$ (RAA hypothesis). Choose a point S on m' such that S and Q are on the same side of m. Drop a perpendicular from S to $\overleftrightarrow{PQ}$ and call the foot of that perpendicular R. By Wallis's Postulate there exists a point T such that $\triangle PRS \sim \triangle PQT$. Since $\angle PQT$ is a right angle, T must lie on ℓ (Protractor Postulate, Part 3). Since $\angle QPT \cong \angle RPS$, T must lie on m' (Protractor Postulate, Part 3). Hence T is a point of $m' \cap \ell$ and so $m' \nparallel \ell$ (definition of parallel). This last statement contradicts the choice of m' so we must reject the RAA hypothesis and conclude that the parallel m is unique. $\square$

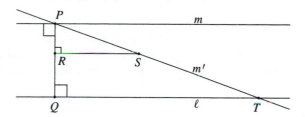

FIGURE 6.34: Wallis's Postulate implies the Euclidean Parallel Postulate

6.9   RECTANGLES AND DEFECT

In this section we explore angle sums for triangles in more depth. We have already seen that the Euclidean Parallel Postulate is equivalent to the assertion that every triangle has angle sum 180°. The main result of this section says that it is all or nothing: If just one triangle has angle sum equal to 180°, then every triangle has angle sum equal to 180°. Along the way toward the proof of that assertion we will be led to consider the question of whether or not a rectangle exists.

The Saccheri-Legendre Theorem asserts that every triangle has angle sum less than or equal to 180°. It allows for the possibility that an angle sum might fall short of the expected amount and we now want to seriously consider that possibility. It is convenient to give a name to the amount by which the angle sum might fall short.

Definition 6.9.1. For any triangle $\triangle ABC$, the *defect* of $\triangle ABC$ is defined by

$$\delta(\triangle ABC) = 180° - \sigma(\triangle ABC).$$

By Saccheri-Legendre Theorem, the defect of every triangle is nonnegative. We will call a triangle *defective* if its defect is positive. Similarly, if $\square ABCD$ is a convex quadrilateral, define its defect by

$$\delta(\square ABCD) = 360° - \sigma(\square ABCD).$$

The next theorem follows easily from Lemma 6.6.4. The simple additive relationship between the defect of a triangle and the defects of the triangles in a subdivision is the main reason that it is easier to work with defects than with angle sums.

Theorem 6.9.2 (Additivity of Defect).

 1. *If $\triangle ABC$ is a triangle and E is a point in the interior of $\overline{BC}$, then*

$$\delta(\triangle ABC) = \delta(\triangle ABE) + \delta(\triangle ECA).$$

 2. *If $\square ABCD$ is a convex quadrilateral, then*

$$\delta(\square ABCD) = \delta(\triangle ABC) + \delta(\triangle ACD).$$

Proof. Exercises 6.42 and 6.43.   □

The following theorem is the main result of the section. The corollary that is stated immediately after the theorem is the result in which we are really interested, but it is convenient to state the theorem this way because it breaks the proof of the corollary down into manageable steps. The theorem as stated also clarifies the relationship between defect and the existence of rectangles.

Definition 6.9.3. A *rectangle* is a quadrilateral each of whose angles is a right angle.

Notice that the angle sum of a rectangle is 360° and so its defect is 0°. It follows from the Alternate Interior Angles Theorem and Theorem 6.7.7 that every rectangle is a convex quadrilateral. The following theorem shows that we cannot take the existence of rectangles for granted.

Theorem 6.9.4. *The following statements are equivalent.*

 1. *There exists a triangle whose defect is $0°$.*
 2. *There exists a right triangle whose defect is $0°$.*
 3. *There exists a rectangle.*

4. *There exist arbitrarily large rectangles.*
5. *The defect of every right triangle is* $0°$.
6. *The defect of every triangle is* $0°$.

Corollary 6.9.5. *In any model for neutral geometry, there exists one triangle whose defect is* $0°$ *if and only if every triangle in that model has defect* $0°$.

The statement "there exist arbitrarily large rectangles" means that for any given positive number M there exists a rectangle each of whose sides has length greater than M.

Before we tackle the proof of Theorem 6.9.4, we need a preliminary lemma.

Lemma 6.9.6. *If* $\triangle ABC$ *is any triangle, then at least two of the interior angles in* $\triangle ABC$ *are acute. If the interior angles at vertices A and B are acute, then the foot of the perpendicular from C to* $\overleftrightarrow{AB}$ *is between A and B.*

Proof. It follows from the Saccheri-Legendre Theorem that at least two of the interior angles must be acute. Let us say that the interior angles at vertices A and B are acute. Drop a perpendicular from C to $\overleftrightarrow{AB}$ and call the foot D. We must prove that D is between A and B.

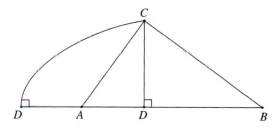

FIGURE 6.35: If D is not between A and B, the Exterior Angle Theorem is violated

First note that D cannot be equal to either A or B because if it were then the interior angle at that vertex would be a right angle. Suppose D is not between A and B (RAA hypothesis). Then either $D * A * B$ or $A * B * D$. If $D * A * B$, then the acute angle $\angle CAB$ is an exterior angle for $\triangle ACD$ and the right angle $\angle CDA$ is a remote interior angle for the same triangle. This contradicts the Exterior Angle Theorem. A similar proof shows that $A * B * D$ is impossible. Thus we can conclude that D is between A and B. □

Proof of Theorem 6.9.4. We will prove that $1 \Rightarrow 2 \Rightarrow 3 \Rightarrow 4 \Rightarrow 5 \Rightarrow 6 \Rightarrow 1$. This cycle of implications shows that each of the statements implies all the others.

$(1 \Rightarrow 2)$ Let $\triangle ABC$ be a triangle such that $\delta(\triangle ABC) = 0°$ (hypothesis). Relabel the vertices, if necessary, so that the interior angles at A and B are acute. By Lemma 6.9.6, there is a point D between A and B such that $\angle CDA$ is a right angle. Then triangles $\triangle ADC$ and $\triangle DBC$ are both right triangles. By Lemma 6.9.2,

$\delta(\triangle ABC) = \delta(\triangle ADC) + \delta(\triangle DBC)$. Since all defects are nonnegative and $\delta(\triangle ABC) = 0°$, it follows that $\delta(\triangle ADC) = \delta(\triangle DBC) = 0°$.

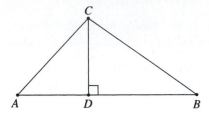

FIGURE 6.36: $(1 \Rightarrow 2)$ Split the given triangle into two right triangles

$(2 \Rightarrow 3)$ Suppose $\triangle ABC$ is a triangle such that $\angle ABC$ is a right angle and $\delta(\triangle ABC) = 0°$ (hypothesis). It follows that $\mu(\angle BAC) + \mu(\angle BCA) = 90°$. There exists a point D, on the opposite side of $\overleftrightarrow{AC}$ from B, such that $\triangle CDA \cong \triangle ABC$ (Theorem 6.2.4). It is easy to check that $\square ABCD$ is a rectangle—see Fig. 6.37.

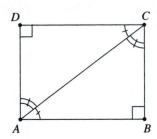

FIGURE 6.37: $(2 \Rightarrow 3)$ Two copies of $\triangle ABC$ make a rectangle

$(3 \Rightarrow 4)$ The idea is to place two copies of the given rectangle next to each other to form a larger rectangle. This process is repeated to produce a rectangle whose dimensions are larger than any specified number—see Fig. 6.38.

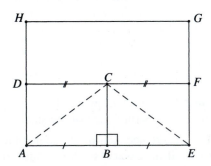

FIGURE 6.38: $(3 \Rightarrow 4)$ Place copies of the rectangle next to each other to form a larger rectangle

Let $\square ABCD$ be a rectangle. Choose a point E on $\overrightarrow{AB}$ such that $A * B * E$ and $AB = BE$; then choose a point F on $\overrightarrow{DC}$ such that $D * C * F$ and $DC = CF$. By SAS, $\triangle ABC \cong \triangle EBC$. In particular, $\mu(\angle BCE) = \mu(\angle BCA) < 90°$, so E is in the interior of $\angle BCF$ (Theorem 5.7.12). The Angle Addition Postulate and subtraction give $\angle ACD \cong \angle ECF$, so a second application of SAS shows that $\triangle ADC \cong \triangle EFC$.

Now $\delta(\triangle ABC) + \delta(\triangle ACD) = \delta(\square ABCD)$ (Theorem 6.9.2), so $\delta(\triangle ABC) = \delta(\triangle ACD) = 0°$. (The defect of any rectangle is zero.) The fact that $\delta(\triangle ABC) = 0°$ means that $\mu(\angle BAC) + \mu(\angle BCA) = 90°$. Since $\square ABCD$ is a parallelogram, it is convex (Theorem 6.7.7). Hence C is in the interior of $\angle BAD$ and therefore $\mu(\angle BAC) + \mu(\angle CAD) = 90°$. Combining the last two statements gives $\angle DAC \cong \angle ACB$. A similar argument shows $\angle ACD \cong \angle CAB$. Therefore $\triangle ABC \cong \triangle CDA$ (ASA).

We now have $\triangle ABC \cong \triangle CDA \cong \triangle CFE$, so $\angle CFE$ is a right angle. A similar argument shows that $\angle BEF$ is a right angle. Therefore, $\square AEFD$ is a rectangle. One pair of sides of $\square AEFD$ is twice as long as the corresponding side of $\square ABCD$; specifically, $AE = 2AB = DF$. As suggested by Figure 6.38, we can apply the construction again to produce a rectangle $\square AEGF$ such that $AH = 2AD = EG$. By the Archimedean Property of Real Numbers, iteration of this process will produce arbitrarily large rectangles.

$(4 \Rightarrow 5)$ Let $\triangle ABC$ be a right triangle with right angle at vertex C (hypothesis). We must prove that $\delta(\triangle ABC) = 0°$.

By Statement 4, there exists a rectangle $\square DEFG$ such that $DG > AC$ and $FG > BC$. Choose a point B' on $\overline{GF}$ such that $GB' = CB$ and a point A' on $\overline{GD}$ such that $GA' = CA$. By SAS, $\triangle ABC \cong \triangle A'B'G$. In order to simplify notation, we will assume that $G = C$, $A' = A$, and $B' = B$.

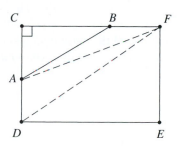

FIGURE 6.39: $(4 \Rightarrow 5)$ Embed $\triangle ABC$ in a rectangle and use additivity of defect

Since $\square CDEF$ is a rectangle, it is convex (Theorem 6.7.7) and has defect $0°$. Hence $\delta(\triangle DEF) = \delta(\triangle CDF) = 0°$ (Additivity of defect, Theorem 6.9.2). Subdividing again gives $0° = \delta(\triangle CDF) = \delta(\triangle ADF) + \delta(\triangle AFC)$ (Theorem 6.9.2) and so $\delta(\triangle AFC) = 0°$. Subdividing once more gives $0° = \delta(\triangle AFC) = \delta(\triangle AFB) + \delta(\triangle ABC)$ and so $\delta(\triangle ABC) = 0°$.

$(5 \Rightarrow 6)$ By Lemma 6.9.6 the triangle can be subdivided into two right triangles.

If each of the right triangles has defect $0°$, then Additivity of Defect implies that the original has defect $0°$ as well.

($6 \Rightarrow 1$) This implication is obvious.   □

The last theorem indicates that the question of whether or not there might exist defective triangles can be answered by determining whether or not there exists a rectangle. As a result, many geometers have thought about the problem of the existence of rectangles. There have historically been many different approaches to the problem. The straightforward approach is simply to assume that rectangles exist. (This does, after all, seem intuitively obvious to most people.) That is the approach followed by Clairaut.

Clairaut's Axiom. *There exists a rectangle.*

By Theorem 6.9.4, Clairaut's Axiom implies that every triangle has defect $0°$ and thus implies the Euclidean Parallel Postulate (Theorem 6.8.4). On the other hand, the Euclidean Parallel Postulate implies that every triangle has angle sum $180°$, so the Euclidean Parallel Postulate implies Clairaut's Axiom. Hence we see that Clairaut's Axiom is also equivalent to the Euclidean Parallel Postulate.

Corollary 6.9.7. *Clairaut's Axiom is equivalent to the Euclidean Parallel Postulate.*

Many other mathematicians worked to construct rectangles. There are two standard ways to begin the construction of a rectangle. The quadrilaterals in the constructions have been named after two mathematicians who worked on the problem. Notice that both kinds of quadrilaterals described below "should" be rectangles in the sense that it would be easy to prove that they are rectangles using familiar theorems of Euclidean geometry.

Definition 6.9.8. A *Saccheri quadrilateral* is a quadrilateral □$ABCD$ such that $\angle ABC$ and $\angle DAB$ are right angles and $\overline{AD} \cong \overline{BC}$. The segment $\overline{AB}$ is called the *base* of the Saccheri quadrilateral and the segment $\overline{CD}$ is called the *summit*. The two right angles $\angle ABC$ and $\angle DAB$ are called the *base angles* of the Saccheri quadrilateral and the angles $\angle CDA$ and $\angle BCD$ are called the *summit angles* of the Saccheri quadrilateral.

FIGURE 6.40: A Saccheri quadrilateral

It is easy to construct a Saccheri quadrilateral. Start with a line segment $\overline{AB}$ and erect perpendiculars at the endpoints. Choose points C and D, on the same side of $\overleftrightarrow{AB}$, such that $\overleftrightarrow{AD} \perp \overleftrightarrow{AB}$, $\overleftrightarrow{BC} \perp \overleftrightarrow{AB}$, and $\overline{AD} \cong \overline{BC}$.

Definition 6.9.9. A *Lambert quadrilateral* is a quadrilateral in which three of the angles are right angles.

FIGURE 6.41: A Lambert quadrilateral

Again it is easy to construct a Lambert quadrilateral. Simply start with a right angle with vertex at B and choose a point D in the interior of the right angle. Drop perpendiculars from D to the two sides of the angle and call the feet of the perpendiculars A and C.

We have followed common practice in naming the two special types of quadrilaterals after Giovanni Saccheri (1667–1733) and Johann Lambert (1728–1777), respectively. This assignment of names appears, however, to reflect a western bias that does not adequately recognize the contributions of Islamic mathematicians. The Islamic world contributed much to geometry between the time of Euclid and the western Renaissance. In their exploration of Euclid's Fifth Postulate, Islamic mathematicians studied the very quadrilaterals we now name in honor of Saccheri. It has been contended that it would be more fitting to name them after the Persian geometer and poet Omar Khayyam (Umar al-Khayyami, 1048–1131) than Saccheri.[2]

The next two theorems spell out the important properties of Saccheri and Lambert quadrilaterals (at least those that can be proved in neutral geometry).

Theorem 6.9.10 (Properties of Saccheri quadrilaterals). *If $\square ABCD$ is a Saccheri quadrilateral with base $\overline{AB}$, then*

1. *the diagonals $\overline{AC}$ and $\overline{BD}$ are congruent,*
2. *the summit angles $\angle BCD$ and $\angle ADC$ are congruent,*
3. *the segment joining the midpoint of $\overline{AB}$ to the midpoint of $\overline{CD}$ is perpendicular to both $\overline{AB}$ and $\overline{CD}$,*
4. *$\square ABCD$ is a parallelogram,*
5. *$\square ABCD$ is a convex quadrilateral, and*
6. *the summit angles $\angle BCD$ and $\angle ADC$ are either right or acute.*

Proof. Exercise 6.46.   □

---

[2]See [31], for example.

Theorem 6.9.11 (Properties of Lambert quadrilaterals). *If □ABCD is a Lambert quadrilateral with right angles at vertices A, B, and C, then*

1. *□ABCD is a parallelogram,*
2. *□ABCD is a convex quadrilateral,*
3. *∠ADC is either right or acute, and*
4. *BC ≤ AD.*

Proof. Exercise 6.49.   □

Both Saccheri and Lambert hoped to prove (in neutral geometry) that the quadrilaterals that bear their names are rectangles. Each of them succeeded only in ruling out the possibility that one of the angles is obtuse. Neither was able to eliminate the possibility that the unknown angles in their quadrilaterals might be acute. Thus their efforts to prove the Euclidean Parallel Postulate in this way failed. Despite that, their exploration of the properties of these quadrilaterals played an important part in the developments leading to the discovery of non-Euclidean geometry.

Saccheri and Lambert quadrilaterals continue to play a role in the study of parallelism. We will make extensive use of both kinds of quadrilaterals and their properties in later chapters when we investigate parallelism in more depth. Theorems 6.9.10 and 6.9.11 will be applied over and over again. Two sample applications are included in Exercises 6.47 and 6.48. The proof of the next theorem also illustrates what useful tools these quadrilaterals are. The result is known as Aristotle's Theorem because the philosopher Aristotle made use of this "fact" to argue that the universe is finite.

Theorem 6.9.12 (Aristotle's Theorem). *If ∠BAC is an acute angle and P and Q are two points on side $\overrightarrow{AB}$ such that $A * P * Q$, then $d(P, \overleftrightarrow{AC}) < d(Q, \overleftrightarrow{AC})$. Furthermore, for every positive number d_0 there exists a point R on $\overrightarrow{AB}$ such that $d(R, \overleftrightarrow{AC}) > d_0$.*

Proof. Exercise 6.51.   □

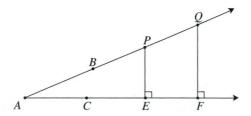

FIGURE 6.42: Aristotle's Theorem: $d(P, \ell)$ is an increasing function of AP and $d(P, \ell) \to \infty$ as $AP \to \infty$

6.10   THE UNIVERSAL HYPERBOLIC THEOREM

The Euclidean Parallel Postulate asserts that for every line ℓ and for every external point P there exists a unique parallel line through P, while the Hyperbolic Parallel Postulate asserts that for every line ℓ and for every external point P there are multiple parallel lines through P. Once one begins to think about this, an obvious question arises: Is it possible that for some lines and some external points there is a unique parallel while for others there are multiple parallels? The answer is that there are no mixed possibilities, that either there is a unique parallel in every situation or there are multiple parallels in every situation. The main theorem in the section asserts that if a model contains just one line and one external point where uniqueness of parallels fails, then uniqueness of parallels fails for every line and for every external point in that model. It is another all-or-nothing theorem, like the theorem regarding the existence of defective triangles.

Theorem 6.10.1 (The Universal Hyperbolic Theorem). *If there exists one line ℓ_0, an external point P_0, and at least two lines that pass through P_0 and are parallel to ℓ_0, then for every line ℓ and for every external point P there exist at least two lines that pass through P and are parallel to ℓ.*

Proof. Assume there exists a line ℓ_0, an external point P_0, and at least two lines that pass through P_0 and are parallel to ℓ_0 (hypothesis). This hypothesis implies that the Euclidean Parallel Postulate fails. Hence no rectangle can exist (Corollary 6.9.7).

Let ℓ be a line and let P an external point (hypothesis). We must prove that there exist at least two lines through P that are both parallel to ℓ. We begin by constructing one such line in the usual way. Drop a perpendicular to ℓ and call the foot of that perpendicular Q. Let m be the line through P that is perpendicular to $\overleftrightarrow{PQ}$. Choose a point R on ℓ that is different from Q and let t be the line through R that is perpendicular to ℓ. Drop a perpendicular from P to t and call the foot of that perpendicular S.

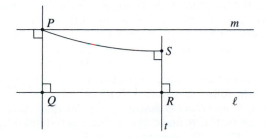

FIGURE 6.43: Proof of the Universal Hyperbolic Theorem

Now $\square PQRS$ is a Lambert quadrilateral. It cannot be a rectangle because, as noted in the previous paragraph, the hypotheses of the theorem imply that no rectangle can exist. Hence $\angle QPS$ is not a right angle and $\overleftrightarrow{PS} \neq m$. But $\overleftrightarrow{PS}$ is parallel to ℓ (Alternate Interior Angles Theorem, applied using the transversal t). This completes the proof. □

Notice that the hypothesis of the Universal Hyperbolic Theorem is just the negation of the Euclidean Parallel Postulate while the conclusion is the Hyperbolic Parallel Postulate. Therefore, the theorem can be restated in the following way.

Corollary 6.10.2. *The Hyperbolic Parallel Postulate is equivalent to the negation of the Euclidean Parallel Postulate.*

The following corollary is a fitting conclusion to the chapter.

Corollary 6.10.3. *In any model for neutral geometry either the Euclidean Parallel Postulate or the Hyperbolic Parallel Postulate will hold.*

In the next two chapters we will investigate the two possibilities separately.

SUGGESTED READING

Chapters 3 and 4 of *The Non-Euclidean Revolution*, [70].

EXERCISES

6.1. Find the first of Euclid's proofs in which he makes use of his Fifth Postulate.

6.2. Prove the Converse to the Isosceles Triangle Theorem (Theorem 6.2.2).

6.3. Prove the following companion to Theorem 6.2.3. *For every line ℓ and for every point P that lies on ℓ, there exists a unique line m such that P lies on m and $m \perp \ell$.*

6.4. Prove that it is possible to construct a congruent copy of a triangle on a given base (Theorem 6.2.4).

6.5. Let $\triangle ABC$ be the spherical triangle shown in Figure 6.5. Perform the construction in the proof of Theorem 6.3.2 on the sphere, starting with this triangle. Verify that the point F constructed is in the interior of $\angle ACD$ if and only if $\mu(\angle BAC) < 90°$.

6.6. Prove the Angle-Angle-Side Triangle Congruence Condition (Theorem 6.3.4).

6.7. Prove the Hypotenuse-Leg Theorem (Theorem 6.3.6).

6.8. Prove the Side-Side-Side Triangle Congruence Condition (Theorem 6.3.7).

6.9. Suppose $\triangle ABC$ and $\triangle DEF$ are two triangles such that $\angle BAC \cong \angle EDF$, $AC = DF$, and $CB = FE$ (the hypotheses of ASS). Prove that either $\angle ABC$ and $\angle DEF$ are congruent or are they are supplements.

6.10. Complete the proof of the Scalene Inequality (Theorem 6.4.1).

6.11. Prove the Triangle Inequality (Theorem 6.4.2).

6.12. Prove the following result: *A, B, and C are three points such that $AB + BC = AC$, then A, B, and C are collinear.* (It follows that the assumption in the definition of between that A, B, and C are collinear is redundant.)

6.13. Use the Hinge Theorem to prove SSS.

6.14. Prove that the shortest distance from a point to a line is measured along the perpendicular (Theorem 6.4.4).

6.15. Prove the Pointwise Characterization of Angle Bisectors (Theorem 6.4.6).

6.16. Prove the Pointwise Characterization of Perpendicular Bisectors (Theorem 6.4.7).

6.17. Prove the Corresponding Angles Theorem (Corollary 6.5.4).

6.18. Prove Corollary 6.5.5.

6.19. Prove Corollary 6.5.8.

6.20. Prove that the sum of the measures of any two interior angles in a triangle is less than 180° (Lemma 6.6.3).

6.21. Prove Lemma 6.6.4.

6.22. Prove Corollary 6.6.6.

6.23. Prove the Converse to Euclid's Fifth Postulate (Corollary 6.6.7).

6.24. Prove that the hypotenuse is always the longest side of a right triangle.

6.25. Prove that the angle sum of any convex quadrilateral is $\leqslant 360°$ (Theorem 6.7.5).

6.26. Prove that every parallelogram is convex (Theorem 6.7.7).

6.27. An alternate definition of convex quadrilateral. The opposite sides $\overline{AB}$ and $\overline{CD}$ of a quadrilateral $\square ABCD$ are said to be *semiparallel* if $\overline{AB} \cap \overleftrightarrow{CD} = \emptyset$ and $\overline{CD} \cap \overleftrightarrow{AB} = \emptyset$. Prove the following two results.
  (a) If one pair of opposite sides of $\square ABCD$ is semiparallel, then the other is.
  (b) $\square ABCD$ is convex if and only if both pairs of opposite sides are semiparallel.

6.28. A quadrilateral $\square ABCD$ is called a *trapezoid* if either $\overleftrightarrow{AB} \parallel \overleftrightarrow{CD}$ or $\overleftrightarrow{BC} \parallel \overleftrightarrow{AD}$. Prove that every trapezoid is convex.

6.29. Prove Theorem 6.7.8.

6.30. Prove that a quadrilateral is convex if the diagonals have a point in common (the remaining part of Theorem 6.7.9).

6.31. Prove Corollary 6.7.10.

6.32. A *rhombus* is a quadrilateral that has four congruent sides. Prove the following theorems about rhombi.
  (a) Every rhombus is convex.
  (b) The diagonals of a rhombus intersect at a point that is the midpoint of each diagonal.
  (c) The diagonals of a rhombus are perpendicular.
  (d) Every rhombus is a parallelogram.

6.33. Existence of rhombi. Squares do not necessarily exist in neutral geometry, but rhombi exist in profusion. Prove this by showing that if $\overline{AB}$ and $\overline{CD}$ are two segments that share a common midpoint and $\overleftrightarrow{AB} \perp \overleftrightarrow{CD}$, then $\square ACBD$ is a rhombus.

6.34. Suppose $\overline{AB}$ and $\overline{CD}$ are two segments that share a common midpoint. Prove that $\square ACBD$ is a parallelogram. Prove the following theorem. *If the diagonals of a convex quadrilateral intersect in a point that is the midpoint of each diagonal, then the quadrilateral is a parallelogram.*

6.35. Kites and darts. Let $\square ABCD$ be a quadrilateral such that $AB = AD$ and $CB = CD$. If such a quadrilateral is convex, it is called a *kite*; if it is not convex, it is called a *dart*. Draw a picture of both cases to discover the reason for the names. Prove that $\angle ABC \cong \angle ADC$ in each case.

6.36. Prove that Euclid's Postulate V implies the Euclidean Parallel Postulate (the second half of Theorem 6.8.2).

6.37. Prove that the Euclidean Parallel Postulate is equivalent to the first statement in Theorem 6.8.3.

6.38. Prove that the Euclidean Parallel Postulate is equivalent to the second statement in Theorem 6.8.3.

6.39. Prove that the Euclidean Parallel Postulate is equivalent to the third statement in Theorem 6.8.3.

6.40. Prove that the Euclidean Parallel Postulate is equivalent to the fourth statement in Theorem 6.8.3.

6.41. Prove that the Euclidean Parallel Postulate implies that the angle sum of any triangle is 180° (the second half of Theorem 6.8.4).

6.42. Prove Additivity of Defect for triangles (Part 1 of Theorem 6.9.2).

6.43. Prove Additivity of Defect for convex quadrilaterals (Part 2 of Theorem 6.9.2).

6.44. Prove the following converse to Lemma 6.9.6: *If* $\triangle ABC$ *is a triangle and the foot of the perpendicular from C to* $\overleftrightarrow{AB}$ *lies between A and B, then angles* $\angle CAB$ *and* $\angle CBA$ *are both acute.*

6.45. Prove the following result. *If* $\overline{AB}$ *is the longest side of* $\triangle ABC$, *then the foot of the perpendicular from C to* $\overleftrightarrow{AB}$ *lies in the segment* $\overline{AB}$. Must the foot be between A and B?

6.46. Prove that every Saccheri quadrilateral has the properties listed in Theorem 6.9.10.

6.47. Prove the following result: *If* ℓ *and* m *are two distinct lines and there exist points* P *and* Q *on* m *such that* $d(P, \ell) = d(Q, \ell)$, *then either* m *and* ℓ *intersect at the midpoint of* $\overline{PQ}$ *or* $m \parallel \ell$.

6.48. Prove the following result: *If* ℓ *and* m *are two lines and there exist three distinct points* P, Q, *and* R *on* m *such that* $d(P, \ell) = d(Q, \ell) = d(R, \ell)$, *then either* $m = \ell$ *or* $m \parallel \ell$.

6.49. Prove that every Lambert quadrilateral has the properties listed in Theorem 6.9.11.

6.50. Let $\angle BAC$ be an acute angle and let P and Q be two points on $\overrightarrow{AB}$ such that $A * P * Q$ and $AP = PQ$. Prove that $d(Q, \overleftrightarrow{AC}) \geqslant 2d(P, \overleftrightarrow{AC})$.

6.51. Prove Artistotle's Theorem (Theorem 6.9.12).

6.52. Which axiom in the UCSMP system (Appendix B) is equivalent to the Euclidean Parallel Postulate?

6.53. The UCSMP system of axioms includes a "Corresponding Angle Postulate" (see Appendix B). Which half of this axiom [part (a) or (b)?] is a redundant axiom in the sense that it could be proved as a theorem? Why would the authors of this textbook knowingly include a redundant axiom? Find another example of a redundant axiom in the UCSMP system.

6.54. Explain how Birkhoff's Axiom of Similarity implies the Euclidean Parallel Postulate. (The converse follows from theorems in the next chapter.)

CHAPTER 7

Euclidean Geometry

7.1 GEOMETRY WITH THE PARALLEL POSTULATE

In this chapter we finally add the Euclidean Parallel Postulate to our list of axioms and explore the theorems we can prove using it. For the entire chapter we will assume the six axioms of neutral geometry as well as the Euclidean Parallel Postulate; those seven axioms are the unstated hypotheses in every theorem in the chapter. Geometry based on the neutral axioms together with the Euclidean Parallel Postulate is called *Euclidean geometry*. It is the geometry you were taught in school, so you will finally see many theorems that look reassuringly familiar. Don't get too comfortable, however, because after a brief excursion into Euclidean geometry in this chapter we will back up and replace the Euclidean Parallel Postulate with the Hyperbolic Parallel Postulate in the next chapter. When we do that we will find that we can build a different but equally consistent geometry on the foundation of the six neutral axioms together with the Hyperbolic Parallel Postulate.

Even though the title of this chapter is "Euclidean Geometry," our treatment of the topic differs significantly from that of Euclid. The main reason for the difference is the fact that we are basing our treatment of geometry on the undefined terms *point* and *line* and we have not yet introduced some of the other constructs of geometry, such as circles and area. In this chapter and the next we continue to explore parallel lines, triangles, and quadrilaterals. The emphasis in the present chapter is placed on the theorems regarding such objects that are unique to Euclidean geometry. The next chapter will have a similar emphasis, but it will explore parallel lines and triangles in the context of hyperbolic geometry. In later chapters we will study area, circles, and transformations; those chapters will contain many more Euclidean theorems.

A major goal of this chapter is to provide a proof of the Pythagorean Theorem. The Pythagorean Theorem is the culmination of Book I of Euclid's *Elements*, and it is universally recognized as one of the most important theorems in geometry. One indication of the fundamental nature of the theorem is the fact that its proof can be approached from many different directions.[1]

Euclid, along with most high school textbooks, bases his first proof of the Pythagorean Theorem on the concept of area. A good reason for doing so is the fact that area-based proofs are simple and intuitive. In fact, the proof Euclid includes in Book I is brilliant and believed to be one of the few proofs in the *Elements* that is due to Euclid himself. We have postponed our discussion of area to a later chapter, however, so we do not follow that path just yet. Instead we choose to base our first proof of the Pythagorean Theorem on the theory of similar triangles; we will study area-based proofs of the theorem in Chapter 9. Both kinds of proofs are important and useful, so we want to study and understand them all. High school textbooks, such as [71], also include both types of proofs but present the area-based proofs first.

It is thought that the first Greek proofs of the Pythagorean Theorem were, like the proof in this chapter, based on the theory of similar triangles. However, the early Greek geometers made the assumption that any two lengths are *commensurable* (i.e., given any two lengths there is a third length such that the two given lengths are both integral multiples of the third). In modern terminology they were assuming that the ratio of two lengths is always a rational number. The discovery by the Pythagoreans that the lengths of the side and diagonal of a square are incommensurable (see Proposition 4.2.2) invalidated the earlier proofs of the similar triangles theorem. Even though the Greeks never doubted the truth of the theorems about similar triangles, adequate proofs of the general case were unavailable for many years. During that period it became standard to formulate the proof of the Pythagorean Theorem in terms of area.

It was Eudoxus of Cnidus (408 B.C.–355 B.C.) who found correct proofs of the Similar Triangles Theorem that covered both the commensurable and the incommensurable cases. At the time of Euclid, this theory was still relatively new and Euclid followed the custom of postponing it until late in the development of geometry. Euclid devotes an entire book of the *Elements* (Book V) to Eudoxus's theory of comparison of segments and uses that theory to prove the Similar Triangles Theorem in Book VI. Definitions 4 and 5 in Book V are usually attributed to Eudoxus. Definition 4 includes what we now call the Archimedean Property of Real Numbers while Definition 5 allows the comparison of incommensurable lengths. There is a sense in which the modern definition of irrational number is implicit in Euclid's Definition 5.

In our treatment we follow an arrangement of topics in which the first proof of the Pythagorean Theorem is based on the theory of ratios of lengths. The difficulty that confounded the early Greeks was how to handle what we now call irrational ratios. We face this difficulty in the proof of the Parallel Projection Theorem, below.

---

[1] For a discussion of the history of proofs of the Pythagorean Theorem, see [30, Volume 1, pages 350–369].

The case of a rational ratio is dealt with first and then the Comparison Theorem (Theorem 4.2.4) is used to generalize to the irrational case. Thus the modern theory of real numbers makes all the difficult work of Euclid's Book V unnecessary. While the terminology we use is different from that of Euclid, the idea behind the proof is still basically the same: First prove the rational case and then prove the irrational case by making appropriate comparisons with rational cases.

A section on trigonometry is included in this chapter. The purpose is to show how trigonometry is built on the foundation of Euclidean geometry. The point of view represented by trigonometry reflects the thinking of Alexandrian Greeks who came after Euclid. These later Greek geometers were much more concerned with practical applications than the classical Greeks had been. No trigonometry is to be found in Euclid's *Elements*.

The chapter ends with an optional exploration of further Euclidean geometry of the triangle. The results in the section are "modern" in the sense that most of them were discovered well after Euclid wrote his *Elements*. The exploration is based on the use of dynamic computer software. The software will be used to discover and examine the statements of a number of theorems. No proofs are included in the section, but proofs of many of the theorems are outlined in the exercises at the end of the chapter.

7.2   BASIC THEOREMS OF EUCLIDEAN GEOMETRY

The axioms that are assumed as unstated hypotheses in this chapter are the six neutral axioms plus the Euclidean Parallel Postulate. When we accept the Euclidean Parallel Postulate as an additional axiom, all the statements we proved equivalent to it in neutral geometry become theorems. So we have already proved the following eight theorems in Euclidean geometry.

Theorem 7.2.1 (Converse to Alternate Interior Angles Theorem). *If two parallel lines are cut by a transversal, then both pairs of alternate interior angles are congruent.*

Theorem 7.2.2 (Euclid's Postulate V). *If ℓ and ℓ' are two lines cut by a transversal t such that the sum of the measures of the two interior angles on one side of t is less than $180°$, then ℓ and ℓ' intersect on that side of t.*

Theorem 7.2.3 (Angle Sum Theorem). *For every triangle $\triangle ABC$, $\sigma(\triangle ABC) = 180°$.*

Theorem 7.2.4 (Wallis's Postulate). *If $\triangle ABC$ is a triangle and $\overline{DE}$ is a segment, then there exists a point F such that $\triangle ABC \sim \triangle DEF$.*

Theorem 7.2.5 (Proclus's Axiom). *If ℓ and ℓ' are parallel lines and $t \neq \ell$ is a line such that t intersects ℓ, then t also intersects ℓ'.*

Theorem 7.2.6. *If ℓ and ℓ' are parallel lines and t is a transversal such that $t \perp \ell$, then $t \perp \ell'$.*

Theorem 7.2.7. *If ℓ, m, n, and k are lines such that $k \parallel \ell$, $m \perp k$, and $n \perp \ell$, then either $m = n$ or $m \parallel n$.*

Theorem 7.2.8 (Transitivity of Parallelism). *If $\ell \parallel m$ and $m \parallel n$, then either $\ell = n$ or $\ell \parallel n$.*

There are many other results that can be proved using the theorems above. The one that will prove to be most useful in this chapter is the following theorem, which asserts that the opposite sides of a parallelogram are congruent. In the next chapter we will see that the opposite sides of a Lambert quadrilateral are not congruent in hyperbolic geometry, so the theorem is unique to Euclidean geometry.

Theorem 7.2.9. *The opposite sides of a parallelogram are congruent.*

Proof. Exercise 7.2.   □

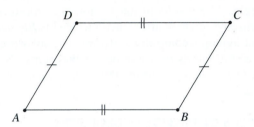

FIGURE 7.1: A Euclidean parallelogram

The last theorem in the section is also unique to Euclidean geometry. In hyperbolic geometry some parallelograms satisfy the conclusion and some do not. The converse to the theorem is true in neutral geometry, however (Exercise 6.34).

We say that two segments *bisect* each other if they intersect in a point that is the midpoint of each segment.

Theorem 7.2.10. *The diagonals of a parallelogram bisect each other.*

Proof. Exercise 7.4 (see Fig. 7.2).   □

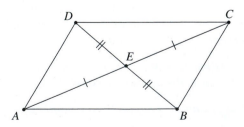

FIGURE 7.2: The diagonals of a parallelogram bisect each other

7.3   THE PARALLEL PROJECTION THEOREM

In this section we prove the Parallel Projection Theorem, a theorem that is a fundamental ingredient in many of the results of Euclidean geometry. In particular, it will be used in the next section to prove the Fundamental Theorem on Similar Triangles, and that theorem, in turn, will be used in the following section to prove the familiar Pythagorean Theorem. The general form of the Parallel Projection Theorem proved in this section does not appear in Euclid. However, the special case that is really needed in the proof of the Similar Triangles Theorem does appear as Proposition VI.2.

Theorem 7.3.1 (Parallel Projection Theorem). *Let ℓ, m, and n be distinct parallel lines. Let t be a transversal that cuts these lines at points A, B, and C, respectively, and let t' be a transversal that cuts the lines at points A', B', and C', respectively. Assume that $A * B * C$. Then*

$$\frac{AB}{AC} = \frac{A'B'}{A'C'}.$$

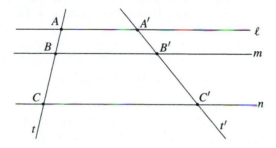

FIGURE 7.3: The Parallel Projection Theorem

The reason for the name of the theorem is that we think of the segments $\overline{AB}$ and $\overline{BC}$ as being projected onto the segments $\overline{A'B'}$ and $\overline{B'C'}$ in the direction of lines ℓ, m, and n. The theorem asserts that parallel projection preserves ratios of lengths. It should be obvious from Fig. 7.3 that parallel projection does not preserve the lengths themselves.

It is the transitivity of parallelism in Euclidean geometry that allows us to speak of three mutually parallel lines. The first sentence in the statement of the theorem would not make sense without transitivity of parallelism.

We will prove a special case first.

Lemma 7.3.2. *Let ℓ, m, and n be distinct parallel lines. Let t be a transversal that cuts these lines at points A, B, and C, respectively, and let t' be a transversal that cuts the lines at points A', B', and C', respectively. Assume that $A * B * C$. If $\overline{AB} \cong \overline{BC}$, then $\overline{A'B'} \cong \overline{B'C'}$.*

Proof. Let t'' be the line through A' such that $t'' \parallel t$ and let t''' be the line through B' that is parallel to t. (There exist unique lines with these properties by the Euclidean

Parallel Postulate.) Let B'' be the point at which t'' crosses m and let C''' be the point at which t''' crosses n. (These points exist by Theorem 7.2.5.)

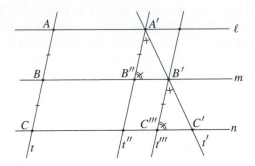

FIGURE 7.4: Proof of the special case $AB = BC$

If $A' = A$, then $B'' = B$ and $\overline{AB} = \overline{A'B''}$. Similarly, if $B' = B$, then $C''' = C$ and $\overline{BC} = \overline{B'C'''}$. If $A' \neq A$ and $B'' \neq B$, then $\overline{A'B''} \cong \overline{AB}$ and $\overline{B'C'''} \cong \overline{BC}$ by Theorem 7.2.9. Combining this with the hypothesis $\overline{AB} \cong \overline{BC}$ gives $\overline{A'B''} \cong \overline{B'C'''}$ in every case. By Theorem 7.2.8, $t'' \parallel t'''$ or $t'' = t'''$. Hence the Converse to the Alternate Interior Angles Theorem[2] implies that $\angle B''A'B' \cong \angle C'''B'C'$ and $\angle A'B''B' \cong \angle B'C'''C'$. Thus $\triangle B''A'B' \cong \triangle C'''B'C'$ (ASA). It follows that $\overline{A'B'} \cong \overline{B'C'}$ (definition of congruent triangles) and the proof is complete. $\square$

Proof of the Parallel Projection Theorem. We first consider the special case in which AB/AC is a rational number; let us say $AB/AC = p/q$, where p and q are positive integers. Use the Ruler Postulate to choose points $A_0, A_1, \ldots, A_q$ on t such that $A_0 = A$, $A_q = C$, and for each i, $A_iA_{i+1} = AC/q$. Note that $A_p = B$. For each $i, 1 \leq i \leq q$, there exists a line ℓ_i such that A_i lies on ℓ_i and $\ell_i \parallel \ell$. Let $A_0' = A'$ and let $A_i', i \geq 1$, be the point at which ℓ_i crosses t' (Theorem 7.2.5). By Lemma 7.3.2, $A_i'A_{i+1}' = A'C'/q$ for each i. Since $\ell_p = m$ and $\ell_q = n$, we must have that $A_p' = B'$ and $A_q' = C'$. It follows that

$$\frac{A'B'}{A'C'} = \frac{A_0'A_p'}{A_0'A_q'} = \frac{A_0A_p}{A_0A_q} = \frac{p}{q}$$

and so the proof of the rational case is complete.

Now suppose that $AB/AC = x$ (a possibly irrational real number) and $A'B'/A'C' = y$. We will use the Comparison Theorem (Theorem 4.2.4) and the previous paragraph to prove that $x = y$. Let r be a rational number such that $0 < r < x$. Choose a point D on t such that $AD/AC = r$. Let m' be the line such that D lies on m' and $m' \parallel m$, and let D' be the point at which m' meets t'. By the previous paragraph, $A'D'/A'C' = r$. Since ℓ, m, and m' are parallel, $A' * D' * B'$

---

[2]We are actually applying the Converse to the Corresponding Angles Theorem; see Exercise 7.1.

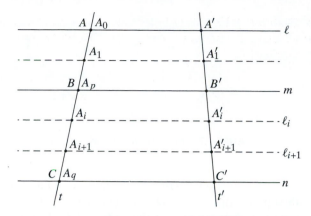

FIGURE 7.5: Parallel projection in case $\dfrac{AB}{AC}$ is rational

and therefore $r = A'D'/A'C' < A'B'/A'C' = y$. A similar argument shows that if r is a rational number such that $0 < r < y$, then $r < x$. Hence $x = y$ by the Comparison Theorem (Theorem 4.2.4).   □

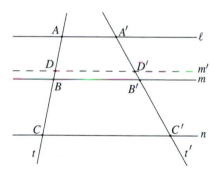

FIGURE 7.6: Parallel projection in case $\dfrac{AB}{AC}$ is irrational

7.4   SIMILAR TRIANGLES

We already know from Wallis's Postulate that noncongruent similar triangles exist in Euclidean geometry. We now prove the familiar theorem regarding lengths of sides of similar triangles. It can be paraphrased this way: Ratios of corresponding sides of similar triangles are equal. The theorem is called the Fundamental Theorem on Similar Triangles or simply the Similar Triangles Theorem.

Theorem 7.4.1 (Fundamental Theorem on Similar Triangles). *If $\triangle ABC$ and $\triangle DEF$ are two triangles such that $\triangle ABC \sim \triangle DEF$, then*

$$\frac{AB}{AC} = \frac{DE}{DF}.$$

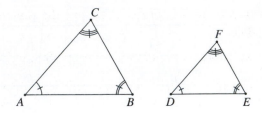

FIGURE 7.7: Similar Triangles Theorem: $AB/AC = DE/DF$

Proof. If $AB = DE$, then $\triangle ABC \cong \triangle DEF$ (ASA) and the conclusion is evident. So we will suppose that $AB \neq DE$. Either $AB > DE$ or $AB < DE$ (trichotomy). Change notation, if necessary, so that $AB > DE$. Choose a point B' on $\overline{AB}$ such that $AB' = DE$. Let m be the line through B' such that m is parallel to $\ell = \overleftrightarrow{BC}$ and let C' be the point at which m intersects $\overline{AC}$. (C' exists by Pasch's Axiom.) Then $\triangle AB'C' \cong \triangle DEF$ (Converse to Alternate Interior Angles Theorem and ASA). Let n be the line through A that is parallel to ℓ and m. Applying the Parallel Projection Theorem to lines $\ell, m,$ and n gives $AB'/AB = AC'/AC$ and so $DE/AB = DF/AC$. Cross multiplying gives $DE/DF = AB/AC$ as desired. □

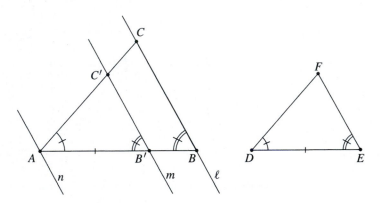

FIGURE 7.8: Proof of the Similar Triangles Theorem

The "ratios of corresponding sides are equal" formulation of the Fundamental Theorem on Similar Triangles is the most useful for most applications, but another way to view the conclusion is that all the side lengths of the similar triangle are multiplied by the same ratio.

Corollary 7.4.2. *If $\triangle ABC$ and $\triangle DEF$ are two triangles such that $\triangle ABC \sim \triangle DEF$, then there is a positive number r such that*

$$DE = r \cdot AB, \ DF = r \cdot AC, \ and \ EF = r \cdot BC.$$

Proof. Exercise 7.7. □

The number r in the corollary is called the *common ratio* of the sides of the similar triangles.

The next theorem provides a Side-Angle-Side criterion for similarity.

Theorem 7.4.3 (SAS Similarity Criterion). *If $\triangle ABC$ and $\triangle DEF$ are two triangles such that $\angle CAB \cong \angle FDE$ and $AB/AC = DE/DF$, then $\triangle ABC \sim \triangle DEF$.*

Proof. Exercise 7.8.   □

Remark. Birkhoff [7] took the the Side-Angle-Side Similarity Criterion as an axiom. He used this one axiom to substitute for both the Euclidean Parallel Postulate and Side-Angle-Side. That is one of the ways in which Birkhoff was able to get by with such a remarkably small number of postulates.

The following converse to the Similar Triangles Theorem can also be viewed as a Side-Side-Side criterion for similarity.

Theorem 7.4.4 (Converse to Similar Triangles Theorem). *If $\triangle ABC$ and $\triangle DEF$ are two triangles such that $AB/DE = AC/DF = BC/EF$, then $\triangle ABC \sim \triangle DEF$.*

Proof. Exercise 7.9.   □

7.5 THE PYTHAGOREAN THEOREM

In this section we will use the Similar Triangles Theorem to prove the Pythagorean Theorem. The notation that is used in the statement of the Pythagorean Theorem is so standardized that we should adopt it as well.

Notation. Let $\triangle ABC$ be a triangle. We will denote $\angle CAB$ by $\angle A$, $\angle ABC$ by $\angle B$, and $\angle ACB$ by $\angle C$. We use lowercase letters to denote the lengths of the sides of the triangle: $a = BC, b = AC$, and $c = AB$.

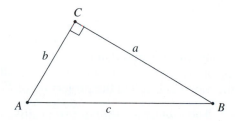

FIGURE 7.9: Notation for the Pythagorean Theorem

Theorem 7.5.1 (Pythagorean Theorem). *If $\triangle ABC$ is a right triangle with right angle at vertex C, then $a^2 + b^2 = c^2$.*

Euclid proved the Pythagorean Theorem as the culmination of Book I of the *Elements*. He would not, however, recognize our statement of the theorem. We have followed the modern practice of stating the conclusion as an algebraic equation. Euclid would have thought of the theorem as asserting that the sum of the areas of the two squares built on the legs of a right triangle is equal to the area of the square on the hypotenuse. We will study the theorem from Euclid's point of view in a later chapter.

Proof of the Pythagorean Theorem. Drop a perpendicular from C to $\overleftrightarrow{AB}$ and call the foot of that perpendicular D. By Lemma 6.9.6, D is in the interior of $\overline{AB}$. Now $\mu(\angle A) + \mu(\angle B) = 90°$ and $\mu(\angle A) + \mu(\angle ACD) = 90°$ (Angle Sum Theorem). It follows that $\angle B \cong \angle ACD$. In a similar way we see that $\angle A \cong \angle DCB$. Hence there are three similar triangles:

$$\triangle ABC \sim \triangle CBD \sim \triangle ACD.$$

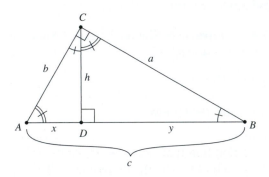

FIGURE 7.10: Proof of the Pythagorean Theorem

To simplify notation, let $x = AD$, $y = BD$, and $h = CD$. Applying the Similar Triangles Theorem twice gives $x/b = b/c$ and $y/a = a/c$. Hence $b^2 = cx$ and $a^2 = cy$. Adding those two equations gives $a^2 + b^2 = c(x + y)$. But $x + y = c$, so the proof is complete.   □

Terminology. The various segments in the proof of the Pythagorean Theorem have names. The segment $\overline{CD}$ is an *altitude* of the triangle and its length, $h = CD$, is called the *height* of the triangle. The segment $\overline{AD}$ is called the *projection* of $\overline{AC}$ onto $\overline{AB}$. In the same way, $\overline{BD}$ is called the projection of $\overline{BC}$ onto $\overline{AB}$.

Definition 7.5.2. The *geometric mean* of two positive numbers x and y is defined to be $\sqrt{xy}$.

Theorem 7.5.3. *The height of a right triangle is the geometric mean of the lengths of the projections of the legs.*

Proof. Exercise 7.11.   □

Theorem 7.5.4. *The length of one leg of a right triangle is the geometric mean of the length of the hypotenuse and the length of the projection of that leg onto the hypotenuse.*

Proof. Exercise 7.12.   □

Remark. The terminology *geometric mean* is used because of area considerations. A rectangle of length x and width y has the same area as a square whose side has length $\sqrt{xy}$. In that geometric sense the length $\sqrt{xy}$ is the average of x and y. In later chapters we will explore the geometric relationship between the $x \times y$ rectangle and the $\sqrt{xy} \times \sqrt{xy}$ square in more depth. We will also see how to construct the geometric mean of two lengths using compass and straightedge.

The converse to the Pythagorean Theorem is the final theorem in Book I of the *Elements*. In our statement we continue to use the notation $a = BC, b = AC$, and $c = AB$.

Theorem 7.5.5 (Converse to the Pythagorean Theorem). *If $\triangle ABC$ is a triangle such that $a^2 + b^2 = c^2$, then $\angle BCA$ is a right angle.*

Proof. Exercise 7.13.   □

7.6 TRIGONOMETRY

In this section we digress to discuss the essentials of trigonometry. The purpose of the section is simply to make it clear that the Fundamental Theorem on Similar Triangles and the Pythagorean Theorem form the foundation on which trigonometry is based. Let us begin with the definition of the trigonometric functions.

Definition 7.6.1. Let θ be an acute angle with vertex A. Then θ consists of two rays with common endpoint A. Pick a point B on one of the rays and drop a perpendicular to the other ray. Call the foot of that perpendicular C. Define the sine and cosine functions by

$$\sin \theta = \frac{BC}{AB} \quad \text{and} \quad \cos \theta = \frac{AC}{AB}.$$

If θ is an obtuse angle, let θ' denote its supplement. Define

$$\sin \theta = \sin \theta' \quad \text{and} \quad \cos \theta = -\cos \theta'.$$

If θ has measure $0°$, define

$$\sin \theta = 0 \quad \text{and} \quad \cos \theta = 1.$$

If θ is a right angle, define

$$\sin \theta = 1 \quad \text{and} \quad \cos \theta = 0.$$

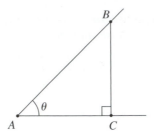

FIGURE 7.11: $\sin\theta = BC/AB$, $\cos\theta = AC/AB$

In the definition of the trigonometric functions we had to make two choices: which of the two rays to use first and then which point B to choose on that ray. Whichever choices we make, the result is a triangle $\triangle ABC$ which includes angle θ at A and a right angle at C. Since the angle sum of the triangle is $180°$, the measure of third angle is determined. Hence any two triangles constructed in this way will be similar. The Fundamental Theorem on Similar Triangles implies that the trigonometric functions are well defined in the sense that the numerical values of the sine and cosine functions are the same regardless of which choices are made.

Sine and cosine are functions. In each case the domain is the set of angles. It is clear, however, that congruent angles have the same sine and cosine, so we can think of the domain of each function as consisting of angle measures. Remember that we have only defined angles whose measures lie in the interval $[0, 180)$. Therefore, the domain of each function is the half-open interval $[0, 180)$. With this domain, the range of the sine function is $[0, 1]$ and the range of the cosine function is the interval $(-1, 1]$.

The Pythagorean Theorem states that the lengths in the triangle above will satisfy the equation $(AC)^2 + (BC)^2 = (AB)^2$. Dividing both sides of that equation by $(AB)^2$ gives the famous Pythagorean Identity.

Theorem 7.6.2 (Pythagorean Identity). *For any angle θ,*

$$\sin^2\theta + \cos^2\theta = 1.$$

The trigonometric functions satisfy many other identities. The two most important theorems of trigonometry are the Law of Sines and the Law of Cosines, which relate the trigonometric functions of the angles of a triangle to the lengths of the sides of a triangle. These two theorems allow one to "solve" a triangle; that is, to determine the remaining parts of a triangle once three of the six parts are known (as long as at least one of the known parts is a side). This is what makes trigonometry so useful for surveying and navigating. The Law of Cosines is a generalization of the Pythagorean Theorem to triangles that are not necessarily right-angled.

Notation. For the next two theorems we adopt the notation that was used in the statement and proof of the Pythagorean Theorem (i.e., $\triangle ABC$ is a triangle, $a = BC$, $b = AC$, and $c = AB$). It is not assumed that $\angle C$ is a right angle.

Theorem 7.6.3 (Law of Sines). *If △ABC is any triangle, then*

$$\frac{a}{\sin \angle A} = \frac{b}{\sin \angle B} = \frac{c}{\sin \angle C}.$$

Proof. Exercise 7.14.   □

Theorem 7.6.4 (Law of Cosines). *If △ABC is any triangle, then*

$$c^2 = a^2 + b^2 - 2ab \cos \angle C.$$

Proof. Exercise 7.15.   □

7.7 EXPLORING THE EUCLIDEAN GEOMETRY OF THE TRIANGLE

Euclidean geometry did not end with Euclid. Over the years since Euclid wrote his *Elements* many people have worked to extend Euclidean geometry and have proved an incredibly large number of fascinating and surprising results. The nineteenth century, in particular, saw a period of intense activity in Euclidean geometry, and many of the most famous theorems of the subject were discovered during that time.

This book does not attempt a comprehensive coverage of the subject of modern (or advanced) Euclidean geometry, but will provide a glimpse of the subject. The exploration begins in this section and it is continued in Section 10.7, after circles have been introduced and studied. It is hoped that enough has been included here to give some sense of what the modern subject is like and to inspire the reader to explore more of it. Those who want to study Euclidean geometry in more depth will find numerous sources available. For example, the books by Eves [26] and Dodge [20] give good, traditional coverage of the subject. The recent book by Posamentier [59] integrates Geometer's Sketchpad into the study of advanced Euclidean geometry.

We will explore some modern Euclidean geometry using dynamic geometry software.[3] This section is quite different from those that come before in that the emphasis is on exploration and discovery and not on proof. Proofs of all but one of the theorems we examine are developed in the exercises and in later chapters, but the section itself simply explores the theorems without any attempt to prove them. The computer software makes an ideal tool for such exploration and we will use it to discover a number of surprising and intricate results about triangles.

The terminology used will be that associated with Geometer's Sketchpad, although the other computer programs mentioned in the Technology section of the Preface could just as well be used. Whatever software you choose, you should be able to perform the following basic operations:

- Plot points.
- Draw segments, rays, and lines determined by those points.
- Mark points of intersection.
- Construct perpendiculars.

---

[3]Read the comments on "Technology" in the Preface if you have not already done so.

- Locate midpoints.
- Find angle bisectors.
- Measure lengths and angles.
- Label objects.
- Create Tools and Scripts (Sketchpad) or Macros (Cabri).

We begin by studying several points associated with a triangle. The first four of these points were known to the ancients and so their study cannot really be called "modern" Euclidean geometry. But they provide a natural starting place for our exploration. The fifth one was discovered by Pierre de Fermat in the seventeenth century. These points are all called *centers* for the triangle because each of them has some claim to being at the center of the triangle. Each of the first four triangle centers is specified as the point of intersection of three lines. The first surprise in each case is the (nonobvious) fact that the three lines have a point in common.

Definition 7.7.1. Three lines are *concurrent* if there is a point P such that P lies on all three of the lines. The point P is called the *point of concurrency*. Three segments are *concurrent* if they have an interior point in common.

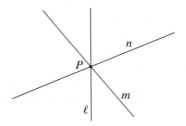

FIGURE 7.12: Concurrent lines

It is clear from Theorem 5.3.7 that distinct lines can have at most one point of concurrency.

Now it is time to begin the exploration. Open your computer software and familiarize yourself with the basic commands listed above. Construct a triangle and label the vertices A, B, and C. The best way to do this in Sketchpad is to use the "Point" tool to plot the three vertices and then construct the three segments connecting the vertices. Locate the midpoints of the three sides of the triangle and label them as indicated in Fig. 7.13.

Construct segments that connect the vertices to the midpoints of the opposite sides. A segment joining a vertex of a triangle to the midpoint of the opposite side is called a *median* for the triangle. Move the vertices of the triangle around in the plane and watch how the medians change. What you should observe is that no matter how you distort the triangle, the three medians are always concurrent.

The fact that the three medians are concurrent is our first theorem; it will be stated shortly, but first we want to explore some additional relationships. The point of concurrency of the three medians is called the *centroid* of the triangle. It is the

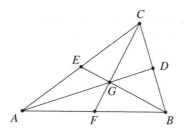

FIGURE 7.13: The three medians and the centroid of △ABC

same centroid that you studied in calculus. Label the centroid G in your diagram. (*Note*: You can only mark the point of intersection of two segments. Just use any two of the three medians. You will always observe that the third median also goes through the point you have marked.)

Use the "Measurement" tool to measure AG and GD, and then calculate AG/GD. Make an observation about the ratio. Now measure BG, GE, CG, and GF, and then calculate BG/GE and CG/GF. Leave the calculations displayed on the screen while you move the vertices of the triangle. You should see that the centroid always divides the medians in a ratio of exactly 2:1. This exploration has led us to the following theorem.

Theorem 7.7.2 (Median Concurrence Theorem). *The three medians of any triangle are concurrent; that is, if* △ABC *is a triangle and D, E, and F are the midpoints of the sides opposite A, B, and C, respectively, then $\overline{AD}$, $\overline{BE}$, and $\overline{CF}$ all intersect in a common point G. Moreover, $AG = 2GD$, $BG = 2GE$, and $CG = 2GF$.*

Two proofs of the Median Concurrence Theorem are outlined in the exercises. The proof in Exercise 7.17 applies the Parallel Projection Theorem while the proof in Exercise 7.19 uses the Fundamental Theorem of Similar Triangles. The concurrency part of the theorem is true in neutral geometry, but the proof is much more complicated in that setting. The usual strategy is to prove the theorem in some specific model for hyperbolic geometry.

We will want to make use of the centroid of a triangle in our further explorations, so you should make a tool that will find the centroid of a triangle without the necessity of going through all the steps of the construction (Exercise T7.1). If possible, you should save all the tools you create in Sketchpad's "Tool Folder" so that they are available to you when you need them later. As you work through the continued exploration below, you should define tools that find the other triangle centers as well. Be sure to make good use of colors to help distinguish the various objects you create.

An *altitude* for a triangle is a line through one vertex that is perpendicular to the line determined by the other two vertices. Hide the medians for your triangle, but keep the centroid visible. Construct the three altitudes for your triangle. Note that they too are concurrent. The point of concurrency is called the *orthocenter* of the triangle. Mark the orthocenter of your triangle and label it H. Again move one vertex and watch what happens to H.

Exercise 7.23 outlines a proof that the three altitudes of any Euclidean triangle are concurrent. We will see in Chapter 10 that the assertion that the three altitudes of any triangle are concurrent is equivalent in neutral geometry to the Euclidean Parallel Postulate (Exercise 10.9).

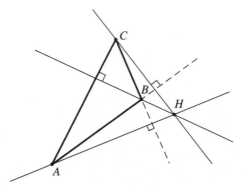

FIGURE 7.14: The three altitudes and the orthocenter of $\triangle ABC$

A point P is said to be *inside* $\triangle ABC$ if P is in the interior of each of the three interior angles of $\triangle ABC$, and P is said to be *outside* $\triangle ABC$ if P is neither inside nor on $\triangle ABC$. The centroid always stayed inside the triangle, but the orthocenter can be outside the triangle or even on the triangle. Determine by experimentation the shape of triangles for which the orthocenter is outside the triangle. Find a shape for the triangle so that the orthocenter of the triangle is equal to one of the vertices of the triangle. Observe what happens to the orthocenter when one vertex crosses over the line determined by the other two vertices. Hide the altitudes, but keep H visible.

Now construct the three perpendicular bisectors for the sides of your triangle. Once again these perpendicular bisectors should be concurrent. Mark the point of concurrency and label it O. This point is called the *circumcenter* of the triangle for reasons that will become clear when we study circles associated with triangles in Chapter 10.

Move one vertex of the triangle around and observe how the circumcenter changes. The three perpendicular bisectors should continue to be concurrent. Note what happens when one vertex crosses over the line determined by the other two vertices. Find example triangles which show that the circumcenter may be inside, on, or outside the triangle, depending on the shape of the triangle. We will see in Chapter 10 that the assertion that the perpendicular bisectors of the sides of a triangle are concurrent is equivalent in neutral geometry to the Euclidean Parallel Postulate (Theorem 10.3.3). Measure the distances OA, OB, and OC and make an observation about them. Now hide the perpendicular bisectors and the circumcircle, but keep the triangle $\triangle ABC$ and the point O visible.

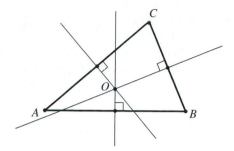

FIGURE 7.15: The three perpendicular bisectors and the circumcenter of $\triangle ABC$

At this point in the exploration you should have a triangle $\triangle ABC$ with three additional points, the centroid G, the orthocenter H, and the circumcenter O, displayed. If not, reconstruct each of the three points G, H, and O. Hide all the lines that were used in the construction of the three points. Now move the vertices of the triangle around in the plane and make an observation about how the three points are located relative to each other. Measure the distances HG and GO and calculate HG/GO. Observe what happens to these three numbers as the vertices of the triangle are moved around in the plane.

Cabri Geometry has a built-in tool that can be used to check for collinearity. In Sketchpad you will have to improvise. If things are working correctly, you have discovered the following theorem.

Theorem 7.7.3 (Euler Line Theorem). *The orthocenter H, the circumcenter O, and the centroid G of any triangle are collinear. Furthermore, G is between H and O (unless the triangle is equilateral, in which case the three points coincide) and HG = 2GO.*

Recall that a triangle is *equilateral* if all three sides have equal lengths. The line through H, O, and G is called the *Euler line* of the triangle. The line is named in honor of Leonard Euler, who discovered Theorem 7.7.3 in the eighteenth century. A proof of the theorem is outlined in Exercise 7.28.

A triangle has many other "centers" besides the three we have looked at. Two additional triangle centers will be explored in Exercises T7.4 and T7.5.

Associated with the centers we have constructed there are several triangles that are of special interest. Construct a new triangle $\triangle ABC$. Label the midpoints of the sides D, E, and F as in Fig. 7.16. The triangle $\triangle DEF$ is called the *medial triangle* for $\triangle ABC$. Use the tools you have made to find the centroids of $\triangle ABC$ and $\triangle DEF$ and compare them. Find the circumcenter of $\triangle ABC$ and the orthocenter of $\triangle DEF$ and compare the two. Make observations about the relationships between these points. If necessary, look at Exercise 7.24 where you will find statements of the two theorems being hinted at here. The medial triangle is quite useful in the proof of the Median Concurrence Theorem.

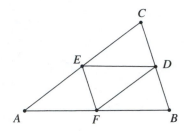

FIGURE 7.16: △DEF is the medial triangle for △ABC

In order to save words, let us start referring to the line through a side of a triangle as a *sideline* of the triangle. Thus the lines $\overleftrightarrow{AB}$, $\overleftrightarrow{BC}$, and $\overleftrightarrow{AC}$ are the sidelines for triangle △ABC.

Just as there was a triangle associated with the centroid of a triangle, there is also a triangle associated with the orthocenter. Let A', B', and C', respectively, be the feet of the perpendiculars from A, B, and C to opposite sidelines of △ABC. The triangle △$A'B'C'$ is called the *orthic triangle* for △ABC.

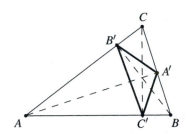

FIGURE 7.17: △$A'B'C'$ is the orthic triangle for △ABC

Construct the orthic triangle. Move the vertices of your triangle around and observe what happens to the associated orthic triangle. It is not necessarily contained in the original triangle. (See Fig. 7.18.) In fact it need not be a triangle at all, but can degenerate into a line segment. Try to make two vertices of the orthic triangle coincide with one vertex of the original triangle. Try to determine what must be true of △ABC in order for the orthic triangle to be contained in the original triangle.

The constructions of the medial and orthic triangles are special cases of the following more general construction, which you should carry out using your dynamic software. Start with a triangle △ABC. Draw the three sidelines of the triangle. Pick a point P that does not lie on any of those lines. Let L be the point at which $\overleftrightarrow{AP}$ intersects $\overleftrightarrow{BC}$, let M be the point at which $\overleftrightarrow{BP}$ intersects $\overleftrightarrow{AC}$, and let N be the point at which $\overleftrightarrow{CP}$ intersects $\overleftrightarrow{AB}$. Construct the triangle △LMN. Such a triangle is called a *Cevian triangle* for the original triangle △ABC.

The medial triangle is the special case of a Cevian triangle in which P is the centroid and the orthic triangle is the special case in which P is the orthocenter of

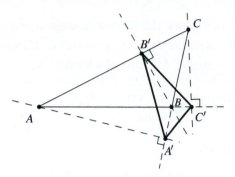

FIGURE 7.18: Another location for the orthic triangle

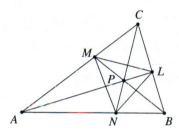

FIGURE 7.19: $\triangle LMN$ is the Cevian triangle associated with $\triangle ABC$ and P

$\triangle ABC$. Move the points A, B, C, and P around and observe how the Cevian triangle changes. To keep things relatively simple for now, it is probably best to choose P inside $\triangle ABC$. It would also be a good idea to hide the lines $\overleftrightarrow{AP}$, $\overleftrightarrow{BP}$, and $\overleftrightarrow{CP}$ so that the diagram does not become too complicated.

Cevian triangles are named after the seventeenth-century Italian mathematician Giovanni Ceva, who discovered a remarkable fact about them. He discovered that there is a relationship among the distances determined by the points on the sides of the original triangle. To see it, measure the distances AN, NB, BL, LC, CM, and MA and then calculate

$$d = \frac{AN}{NB} \cdot \frac{BL}{LC} \cdot \frac{CM}{MA}.$$

Move the points A, B, C, and P around and observe how d changes. It doesn't! This combination of distances is always equal to 1. Not only that, but the fact that d is equal to 1 characterizes triangles that are constructed by the Cevian process.

Any line through a vertex of a triangle is called a *Cevian line* (or simply a Cevian). A Cevian is specified by naming the vertex it passes through along with the point at which it intersects the line through the opposite side. Thus the assertion that $\overleftrightarrow{AL}$ is a Cevian for $\triangle ABC$ is understood to imply that L is a point on $\overleftrightarrow{BC}$. A Cevian line is *proper* if passes through only one vertex of the triangle.

We will now work the Cevian triangle construction backward. Start with a triangle $\triangle ABC$ and choose arbitrary points L, M, and N on the sidelines $\overleftrightarrow{BC}$, $\overleftrightarrow{AC}$, and $\overleftrightarrow{AB}$, respectively. Measure the lengths AN, NB, BL, LC, CM, and MA and then calculate $d = (AN \cdot BL \cdot CM)/(NB \cdot LC \cdot MA)$ again. You probably do not see a pattern this time. Try to adjust the points so that d is exactly equal to 1. Draw the lines $\overleftrightarrow{AL}$, $\overleftrightarrow{BM}$, and $\overleftrightarrow{CN}$ and make an observation about when the lines are concurrent. The statement of the following theorem assumes the notation of the last two paragraphs.

Theorem 7.7.4 (Ceva's Theorem). *Let $\triangle ABC$ be a triangle. The proper Cevian lines* $\overleftrightarrow{AL}$, $\overleftrightarrow{BM}$, *and* $\overleftrightarrow{CN}$ *are concurrent (or mutually parallel) if and only if*

$$\frac{AN}{NB} \cdot \frac{BL}{LC} \cdot \frac{CM}{MA} = 1.$$

There is another aspect of Ceva's Theorem that should be explored. When P, the point of concurrency, is inside $\triangle ABC$, the points L, M, and N all lie on the sides of the triangle. When P moves outside $\triangle ABC$, some of the other points move outside the triangle as well. Try to find a location for P such that exactly one of the points L, M, N is outside $\triangle ABC$. Then try to have exactly two and exactly three of L, M, N outside the triangle. What you will soon discover is that the only possibilities are that either zero or two of the points L, M, and N can lie outside $\triangle ABC$.

The way the fact that some of the points may be outside $\triangle ABC$ is incorporated into the theorem is by reinterpreting the ratios AN/NB, BL/LC, and CM/MA, as what are called *sensed ratios*. In general, if N is a point on $\overleftrightarrow{AB}$, we will give the quotient AN/NB a special meaning. The absolute value $|AN/NB|$ is always the ratio of the two distances. We will consider AN/NB to be positive if N is between A and B and negative if it is not. With this interpretation of the ratios, Ceva's Theorem contains the additional information that either zero or two of the points L, M, N are outside the triangle $\triangle ABC$ since the only way the product of the three numbers can be $+1$ is if either zero or two of them are negative.

Most of the results in the section so far have been based on the fact that certain triples of lines are concurrent; the one exception is the Euler Line Theorem, which asserts that three points are collinear. Concurrency and collinearity are important examples of *duality*, which is the principle that there should be a symmetry in the way points and lines behave relative to incidence. (Three points are collinear if they are all incident with one line and three lines are concurrent if they are all incident with one point.) We will now pursue that idea further and look for a theorem that is dual to Ceva's Theorem.

Construct a triangle $\triangle ABC$ and extend the lines through the sides. Choose points L, M, and N on the lines $\overleftrightarrow{BC}$, $\overleftrightarrow{AC}$, and $\overleftrightarrow{AB}$, respectively. These points are called *Menelaus points* for the triangle. A Menelaus point for $\triangle ABC$ is *proper* if it is not equal to any of the vertices of the triangle. Calculate the product

$$d = \frac{AN}{NB} \cdot \frac{BL}{LC} \cdot \frac{CM}{MA}$$

as before. Try to move the points around to make d as close to 1 as possible and make an observation. Now pick an arbitrary line ℓ and choose L, M, and N to be the points at which ℓ crosses the lines $\overleftrightarrow{BC}$, $\overleftrightarrow{AC}$, and $\overleftrightarrow{AB}$, respectively. Make an observation about the value of the d with the points in this configuration. Determine exactly how many of the three points L, M, and N can lie on $\triangle ABC$. Determine the value of d taking account of the sense of the ratios.

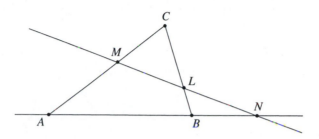

FIGURE 7.20: L, M, and N are collinear

The theorem you have discovered is due to Menelaus of Alexandria who lived at around A.D. 100. The theorem was not well known until it was rediscovered by Ceva in 1678.

Theorem 7.7.5 (Theorem of Menelaus). *Let $\triangle ABC$ be a triangle. Three proper Menelaus points L, M, and N on the lines $\overleftrightarrow{BC}$, $\overleftrightarrow{AC}$, and $\overleftrightarrow{AB}$, respectively, are collinear if and only if*

$$\frac{AN}{NB} \cdot \frac{BL}{LC} \cdot \frac{CM}{MA} = -1.$$

Even though some of the results we have explored in this section were first discovered thousands of years after Euclid, most of them are still hundreds of years old and therefore do not seem very modern to us. We should end our exploration with a theorem that is definitely modern in flavor and is a fundamental departure from anything Euclid could conceivably have done.

As is discussed elsewhere in this book, the ancient Greeks were never able to construct the trisector for a general angle, but using real number measurements of angles this is easy. The dynamic software we are using has measurement built into it, so we should take advantage of that. The first step in this investigation is to make a tool that trisects angles. To do that, make an angle $\angle BAC$. Use the measurement menu to find $\mu(\angle BAC)$ and then calculate $\mu(\angle BAC)/3$. Rotate one side of the angle by this amount and then rotate again by the same amount. This will construct the two rays that trisect $\angle BAC$.

Now take an arbitrary triangle $\triangle ABC$. Construct the two angle trisectors for each of the three interior angles of $\triangle ABC$. These six rays intersect in a total of twelve points in the interior of $\triangle ABC$. Label the point at which the rays through B and C that are closest to $\overline{BC}$ intersect as A'. Similarly, label the intersection of the two trisectors closest to $\overline{AC}$ as B' and label the intersection of the two trisectors closest to $\overline{AB}$ as C'. The triangle $\triangle A'B'C'$ is called the *Morley triangle* for the triangle $\triangle ABC$. (See Fig. 7.21.)

Hide all the angle trisectors and concentrate on the Morley triangle $\triangle A'B'C'$. Measure all the sides and the angles of $\triangle A'B'C'$. Make observations about the Morley triangle. Move the vertices of the original triangle $\triangle ABC$ around and see if your observations continue to hold.

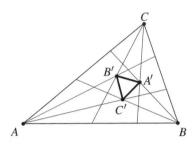

FIGURE 7.21: $\triangle A'B'C'$ is the Morley triangle for $\triangle ABC$

The following theorem was discovered by the American mathematician Frank Morley in the early twentieth century.

Theorem 7.7.6. *If $\triangle ABC$ is any triangle, then the associated Morley triangle is equilateral.*

SUGGESTED READING

Chapter V of *Mathematics in Western Culture*, [40].

EXERCISES

7.1. State and prove the Converse to the Corresponding Angles Theorem.

7.2. Prove that the opposite sides of a parallelogram are congruent (Theorem 7.2.9).

7.3. Prove that the opposite angles of a parallelogram are congruent; i.e., if $\square ABCD$ is a parallelogram, then $\angle DAB \cong \angle BCD$ and $\angle ABC \cong \angle CDA$.

7.4. Prove that the diagonals of a parallelogram bisect each other (Theorem 7.2.10).

7.5. Suppose $\square ABCD$ is a quadrilateral such that $\overleftrightarrow{AB} \parallel \overleftrightarrow{CD}$ and $\overline{AB} \cong \overline{CD}$. Prove that $\square ABCD$ is a parallelogram.

7.6. Suppose $\square ABCD$ is a quadrilateral such that $\overleftrightarrow{AB} \parallel \overleftrightarrow{CD}$ and $\overline{AB} \cong \overline{AD}$. Prove that $\overrightarrow{DB}$ is the bisector of $\angle ADC$.

7.7. Prove Corollary 7.4.2.

7.8. Prove the SAS Similarity Criterion (Theorem 7.4.3).

7.9. Prove the Converse to the Similar Triangles Theorem (Theorem 7.4.4).

7.10. Prove the following Angle-Side-Angle criterion for similarity: If $\triangle ABC$ and $\triangle DEF$ are two triangles such that $\angle CAB \cong \angle FDE$, $\angle ABC \cong \angle DEF$, and $DE = r \cdot AB$, then $\triangle ABC \sim \triangle DEF$ with common ratio r.

7.11. Prove Theorem 7.5.3.

7.12. Prove Theorem 7.5.4.

7.13. Prove the converse to the Pythagorean Theorem (Theorem 7.5.5).

7.14. Prove the Law of Sines (Theorem 7.6.3).

7.15. Prove the Law of Cosines (Theorem 7.6.4).

7.16. Use the Law of Cosines to prove the converse to the Pythagorean Theorem (Theorem 7.5.5).

7.17. This exercise outlines a proof of the Median Concurrence Theorem. Let $\triangle ABC$ be a triangle, and let D, E, and F be the midpoints of the sides opposite A, B, and C, respectively. Let $\ell = \overleftrightarrow{AD}$ and let m and n be the lines parallel to ℓ that contain the points E and F, respectively. Let H be the point at which m crosses $\overline{BC}$, let I be the point at which n crosses $\overline{BC}$, and let J be the point at which n crosses $\overline{BE}$.

    (a) Use the Crossbar Theorem to prove that any two of the medians must intersect.

    (b) Use the Parallel Projection Theorem to prove that $CH = HD$ and $DI = IB$. Conclude that $CH = HD = DI = IB$.

    (c) Let G be the point at which the medians $\overline{AD}$ and $\overline{BE}$ intersect. Prove that $BJ = JG = GE$. Conclude that $BG = 2GE$.

    (d) Use the fact that part (c) is valid for the point of intersection of any two medians to conclude that all three medians intersect at G.

7.18. This exercise shows that the medial triangle divides the original triangle into four congruent triangles, each of which is similar to the original with common ratio 1/2. Let $\triangle ABC$ be a triangle, and let D, E, and F be the midpoints of the sides opposite A, B, and C, respectively.

    (a) Prove that $\triangle EDC \sim \triangle ABC$ with common ratio 2.

    (b) Prove that $\triangle EDC \cong \triangle AFE \cong \triangle FBD \cong \triangle DEF$.

7.19. This exercise outlines a second proof of the Median Concurrence Theorem. Let $\triangle ABC$ be a triangle, and let D, E, and F be the midpoints of the sides opposite A, B, and C, respectively.

    (a) Use the Crossbar Theorem to prove that $\overline{AD}$ and $\overline{BE}$ intersect in a point G.

    (b) Use Exercise 7.18 to prove that $\overleftrightarrow{DE} \parallel \overleftrightarrow{AB}$ and conclude that $\triangle DEG \sim \triangle ABG$.

    (c) Use Exercise 7.18 to show that $AB = 2ED$.

    (d) Use parts (b) and (c) of this exercise and the Fundamental Theorem on Similar Triangles to conclude that $AG = 2GD$ and $BG = 2GE$.

    (e) Use the fact that part (d) is valid for the point of intersection of any two medians to conclude that all three medians intersect at G.

7.20. Let $\square ABCD$ be an arbitrary quadrilateral and let L be the midpoint of $\overline{AB}$, M be the midpoint of $\overline{BC}$, N the midpoint of $\overline{CD}$, and O the midpoint of $\overline{DA}$. Prove that $\square LMNO$ is a parallelogram.

7.21. Prove that the perpendicular bisectors of the sides of any triangle are concurrent and that the point of concurrency is equidistant from the vertices of the triangle.

7.22. Given a triangle $\triangle ABC$, draw a line through each vertex that is parallel to the opposite side. The three lines intersect to form a new triangle $\triangle A'B'C'$. (See Fig. 7.22.) Prove that $\triangle ABC$ is the medial triangle for $\triangle A'B'C'$. Prove that the perpendicular bisector of a side of $\triangle A'B'C'$ is an altitude for $\triangle ABC$.

7.23. Use Exercises 7.22 and 7.21 to prove that the three altitudes of any triangle are concurrent.

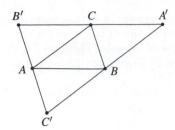

FIGURE 7.22: $\triangle ABC$ is the medial triangle for $\triangle A'B'C'$

7.24. Let $\triangle ABC$ be a triangle and let $\triangle DEF$ be the associated medial triangle.
  (a) Prove that $\triangle ABC$ and $\triangle DEF$ have the same centroid.
  (b) Prove that the circumcenter of $\triangle ABC$ is the orthocenter of $\triangle DEF$.
7.25. Prove that a triangle is isosceles if and only if two medians are congruent.
7.26. Prove that sides of the orthic triangle cut off three triangles from $\triangle ABC$ that are all similar to $\triangle ABC$. Specifically, if $\triangle A'B'C'$ is the orthic triangle for $\triangle ABC$, then $\triangle ABC \sim \triangle AB'C' \sim \triangle A'BC' \sim \triangle A'B'C$.
7.27. Prove that a triangle is equilateral if and only if its centroid and circumcenter coincide.
7.28. Fill in the details in the following proof of the Euler Line Theorem (Theorem 7.7.3). By the previous exercise, it may be assumed that $G \neq O$ (explain). Choose a point H' on $\overrightarrow{OG}$ such that G is between O and H' and $GH' = 2OG$. (See Fig. 7.23.) The objective is to show that $H' = H$. It suffices to show that H' is on the altitude through C (explain). By the SAS Similarity Criterion, $\triangle GOF \sim \triangle GH'C$. Therefore, $\overleftrightarrow{CH'} \parallel \overleftrightarrow{OF}$ and thus $\overleftrightarrow{CH'} \perp \overleftrightarrow{AB}$.

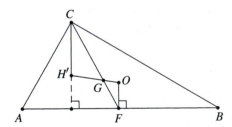

FIGURE 7.23: Construct the point H' and then prove that $H' = H$

7.29. Use the construction in the preceding exercise to give a new proof of the concurrency of the altitudes of a triangle.
7.30. Let A and B be two distinct points.
  (a) Prove that for each real number $x \neq -1$ there exists a unique point X on $\overleftrightarrow{AB}$ such that $AX/XB = x$. (The fraction AX/XB is to be interpreted as a sensed ratio.)
  (b) Prove that there is no point X on $\overleftrightarrow{AB}$ for which $AX/XB = -1$.

(c) Draw a graph that illustrates how AX/XB varies as the point X moves along the line $\overleftrightarrow{AB}$.

7.31. This exercise outlines a proof of the Theorem of Menelaus. Let L, M, and N be proper Menelaus point for triangle $\triangle ABC$.

(a) Assume, first, that L, M, and N are collinear. Use theorems from Chapter 5 to prove that either exactly two of the Menelaus points are on $\triangle ABC$ or none are. Assume that exactly two of the Menelaus points lie on the triangle (see Figure 7.20). Drop perpendiculars from A, B, and C to the line $\overleftrightarrow{LM}$ and call the feet R, S, and T, respectively. Let $r = AR$, $s = BS$, and $t = CT$. Use similar triangles to express each of the sensed ratios AN/NB, BL/LC, and CM/MA in terms of r, s, and t and then use algebra to derive the Menelaus formula. Use a similar argument to prove the Menelaus formula in case all three Menelaus points are outside the triangle.

(b) Now assume that the Menelaus formula holds. Let $\ell = \overleftrightarrow{LM}$. Use Exercise 7.30(b) to prove that ℓ cannot be parallel to $\overleftrightarrow{AB}$. Let N' be the point at which ℓ meets $\overleftrightarrow{AB}$. Use Exercise 7.30(a) to prove that $N' = N$.

7.32. This exercise contains an outline of a proof that the Theorem of Menelaus implies Ceva's Theorem. Suppose $\triangle ABC$ is a triangle and $\overleftrightarrow{AL}$, $\overleftrightarrow{BM}$, and $\overleftrightarrow{CN}$ are three proper Cevian lines.

(a) Assume, first, that the three Cevian lines are concurrent at a point P. Apply the Theorem of Menelaus to each of the triangles $\triangle ABL$ and $\triangle ALC$. Combine the results to derive the formula in Ceva's Theorem.

(b) Now assume that the formula in Ceva's Theorem holds. If the three Cevian lines are not mutually parallel, then it may be assumed that $\overleftrightarrow{AL}$ and $\overleftrightarrow{BM}$ intersect. Let P be the point at which $\overleftrightarrow{AL}$ and $\overleftrightarrow{BM}$ intersect and let N' be the point at which $\overleftrightarrow{CP}$ intersects $\overleftrightarrow{AB}$. Use Exercise 7.30(a) to prove that $N' = N$.

7.33. Use Ceva's Theorem to prove the Median Concurrence Theorem.

7.34. Use Ceva's Theorem to prove that the three altitudes of a triangle concur.

TECHNOLOGY EXERCISES

T7.1. Make a tool (macro) that finds the centroid of a triangle. Your tool should accept three vertices of a triangle as the "givens" and produce the triangle and the centroid G as "results."

T7.2. Make a tool (macro) that finds the orthocenter of a triangle. Your tool should accept three vertices of a triangle as givens and produce the triangle and the centroid H as results.

T7.3. Make a tool (macro) that finds the circumcenter of a triangle. Your tool should accept three vertices of a triangle as givens and produce the triangle and the circumcenter O as results.

T7.4. Construct a triangle $\triangle ABC$ and the three angle bisectors of the interior angles of $\triangle ABC$. Observe that the three angle bisectors are concurrent. The point of concurrency is called the *incenter* for $\triangle ABC$. The reason for the name will be explained in Chapter 10, where it will be shown that the angle bisectors are concurrent in neutral geometry (Theorem 10.3.8). Make a tool (macro) that finds the incenter of a triangle. Your tool should accept three vertices of a triangle as givens and produce the triangle and the incenter *I* as results. Drop perpendiculars from the incenter to each of the sides of the triangle. Measure the distance from the incenter to each of the sides. What do you observe?

T7.5. Construct a triangle $\triangle ABC$ and a point *P* inside the triangle. Calculate the sum of distances $PA + PB + PC$. Move the point *P* around and try to find the point *P* at which the sum of the distances is minimized. Give a geometric description of the location of this point that does not involve distance. This point is called the *Fermat point* for the triangle.

T7.6. Make a tool (macro) that constructs the medial triangle for a triangle $\triangle ABC$.

T7.7. Make a tool (macro) that constructs the orthic triangle for a triangle $\triangle ABC$.

T7.8. Let $\triangle ABC$ be a triangle and let *P* be a point that is not on any of the sidelines of $\triangle ABC$. Drop perpendiculars from *P* to the sidelines of the triangle; let A'' be the foot of the perpendicular to $\overleftrightarrow{BC}$, let B'' be the foot of the perpendicular to $\overleftrightarrow{AC}$, and let C'' be the perpendicular from *P* to $\overleftrightarrow{AB}$. The triangle $\triangle A''B''C''$ is called a *pedal triangle* for $\triangle ABC$. Make a tool that constructs the pedal triangle for a given triangle $\triangle ABC$ and point *P*.

T7.9. Make a tool (macro) that constructs the Cevian triangle determined by triangle $\triangle ABC$ and point *P*.

T7.10. Two triangles $\triangle ABC$ and $\triangle A'B'C'$ are *copolar* if the three lines $\overleftrightarrow{AA'}$, $\overleftrightarrow{BB'}$, and $\overleftrightarrow{CC'}$ are concurrent. Draw diagrams illustrating copolar triangles. Two triangles $\triangle ABC$ and $\triangle A'B'C'$ are *coaxial* if the three points *L*, *M*, and *N* of intersection of the lines through the corresponding sides are collinear. (To be specific, *L* is the point at which $\overleftrightarrow{BC}$ intersects $\overleftrightarrow{B'C'}$, *M* is the point at which $\overleftrightarrow{AC}$ intersects $\overleftrightarrow{A'C'}$, and *N* is the point at which $\overleftrightarrow{AB}$ intersects $\overleftrightarrow{A'B'}$.) Draw some examples of coaxial triangles. Draw a diagram that illustrates the following theorem: *Two triangles $\triangle ABC$ and $\triangle A'B'C'$ are copolar if and only if they are coaxial.* (Assume that no two of the lines determined by the points *A*, *B*, *C*, A', B', and C' are parallel.) This theorem was first proved by Gérard Desargues (1591–1661) and is therefore known as Desargues's Theorem.

T7.11. Make a tool (macro) that trisects angles. Your tool should accept three points *A*, *B*, *C* as givens and should produce the two rays in the interior of $\angle BAC$ that trisect $\angle BAC$ as results. (See comments in the section about how to do this.)

T7.12. Make a tool (macro) that constructs the Morley triangle determined by $\triangle ABC$.

T7.13. Construct a triangle $\triangle ABC$. On each side of $\triangle ABC$ construct an external equilateral triangle. Let *U* be the centroid of the triangle on the side opposite *A*, let *V* be the centroid of the triangle on the side opposite *B* and let *W* be the centroid of the triangle on the side opposite *C*.

  (a) Verify that the three lines $\overleftrightarrow{AU}$, $\overleftrightarrow{BV}$, and $\overleftrightarrow{CW}$ are concurrent. The point of concurrency is called the *Napoleon point* of the triangle.

  (b) Under what conditions is the Napoleon point inside the triangle? Under what conditions is the Napoleon point outside the triangle?

  (c) Is it possible for the Napoleon point to equal one of the vertices?

  (d) Verify that the triangle $\triangle UVW$ is equilateral. This theorem is known as Napoleon's Theorem. It is named after the French emperor Napoleon Bonaparte (1769–1821) even though it seems unlikely that Napoleon himself actually discovered it.

CHAPTER 8

Hyperbolic Geometry

In this chapter we study *hyperbolic geometry*, the geometry that is based on the six postulates of neutral geometry together with the Hyperbolic Parallel Postulate. The geometry in this chapter might more accurately be called axiomatic hyperbolic geometry because we prove many theorems based on the axioms of hyperbolic geometry, but do not address the question of whether or not there is a model for hyperbolic geometry. That problem will be faced in a later chapter.

Following the pattern of the previous chapter, we will focus on theorems regarding parallel lines, triangles, and quadrilaterals, with the major part of the chapter spent studying different kinds of parallel lines. We will see that there are two essentially different kinds of parallel lines in hyperbolic geometry. Parallel lines of the first kind tend to diverge from one another while parallel lines of the second kind approach each other asymptotically. Study of these asymptotic parallels will, for the first time in this course, require the extensive use of ideas related to continuity that have their roots in calculus and analysis. The theorems regarding limiting parallelism have interesting implications regarding defects of hyperbolic triangles.

Before delving into the technical results of hyperbolic geometry we take a quick look at the origins of the subject.

8.1   THE DISCOVERY OF HYPERBOLIC GEOMETRY

Most of the theorems in this chapter were first discovered by the Jesuit priest Giovanni Girolamo Saccheri (1667–1733). Like many mathematicians of earlier times, Saccheri wanted to improve on Euclid by demonstrating that Euclid's Fifth Postulate could be eliminated as an axiom. While others followed the direct approach of trying to prove the Fifth Postulate based on simpler axioms, Saccheri took the indirect approach of assuming the negation of Euclid's Fifth Postulate and trying to derive a contradiction. He found that he could use this assumption to prove many interesting but strange and apparently unnatural theorems. He never did reach the logical contradiction he expected to find, but he did eventually conclude that the theorems he was able to deduce from his hypothesis were so strange that they could not possibly be correct. In 1733 (the year of his death) he published his work in a book [62] that he entitled "Euclid Vindicated and Freed of all Blemish."

What Saccheri had actually discovered was a whole new geometry in which the Euclidean Parallel Postulate is false. This geometry is called either non-Euclidean or hyperbolic geometry.[1] Even though Saccheri worked out the basic theorems of hyperbolic geometry, he never fully appreciated the significance of what he had discovered. As a result, he is not usually credited with the discovery of non-Euclidean geometry; that honor is reserved for the nineteenth-century mathematicians János Bolyai (1802–1860), Nikolai Ivanovich Lobachevsky (1792–1856), and Carl Friedrich Gauss (1777–1855). They were the ones who first recognized that it was logically possible to accept the negation of Euclid's Fifth Postulate as an axiom and to develop a geometry based on that assumption that is as logically consistent as Euclidean geometry.

The difference between what Saccheri did and what Bolyai and others did is largely psychological. Saccheri still viewed Euclidean geometry as the mathematical description of the real world and could not conceive of the possibility of some other geometry. By the nineteenth century, understanding of mathematics had developed to the point at which it was possible for people to think of geometry as simply a subject in which we make certain assumptions and work out their logical consequences. Once the subject is viewed that way, it becomes natural to replace Euclid's postulate with another postulate and explore the logical consequences of that assumption. Thus the possibility of non-Euclidean geometry was somehow "in the air" and it was almost inevitable that many people would independently "discover" it at about the same time.

The main accomplishment of the discoverers of hyperbolic geometry was to show that it is possible to develop an extensive and interesting geometry in which the Euclidean Parallel Postulate is false. Even though they did not prove that the new geometry is consistent in the technical sense in which we understand the term today, they did prove enough theorems and develop the geometry sufficiently that it became clear that it is as intellectually rich and logically consistent as Euclidean geometry. Bolyai, Lobachevsky, Lambert, and others also worked out formulas for hyperbolic distance and proved many technical results concerning such topics as hyperbolic trigonometry. Later Georg Friedrich Bernhard Riemann (1826–1866) developed a

---

[1]The term *hyperbolic* comes from the Greek *hyperbole* or excess. That is a good description of the geometry because the number of parallel lines exceeds what one would expect based on experience with Euclidean geometry. The term *non-Euclidean* is less descriptive since it can also refer to spherical or elliptic geometries. The name *hyperbolic* was first applied to this geometry by Felix Klein around 1900; before that hyperbolic geometry was known as Lobachevskian geometry. In Chapter 13 there is a hyperboloid model for hyperbolic geometry that also helps explain the name.

more comprehensive theory of geometry that encompasses both Euclidean and non-Euclidean geometries (see Chapter 14). Still later Eugenio Beltrami (1835–1900), Henri Poincaré (1854–1912), and others constructed detailed models for hyperbolic geometry. Their models are constructed within Euclidean geometry and show that if Euclidean geometry is consistent, then hyperbolic geometry is as well. We will describe those models in Chapter 13.

It should also be mentioned that during the ninth through twelfth centuries, Islamic mathematicians were studying Euclid and, in particular, were investigating the Parallel Postulate. Many of the theorems of hyperbolic geometry as we now understand it are implicit in their work. For example, Umar al-Khayyami introduced an axiom that he considered to be simpler than Euclid's Fifth Postulate and used it to give a proof of Euclid's postulate. In that proof he made use of the quadrilaterals that we now call Saccheri quadrilaterals. Later Nasir al-Din al-Tusi tried to derive a contradiction from the assumption that the summit angles in such a quadrilateral are acute. His work was published in Rome in 1594 and may well have been the starting point for Saccheri's work.

8.2   BASIC THEOREMS OF HYPERBOLIC GEOMETRY

The axioms that are assumed as unstated hypotheses in this chapter are the six neutral axioms plus the Hyperbolic Parallel Postulate. Since the Hyperbolic Parallel Postulate is equivalent to the negation of the Euclidean Parallel Postulate (Corollary to the Universal Hyperbolic Theorem), we can conclude that any statement that is equivalent to the Euclidean Parallel Postulate is false in hyperbolic geometry. Therefore, the next three theorems are immediate.

Theorem 8.2.1. *For every triangle $\triangle ABC$, $\sigma(\triangle ABC) < 180°$.*

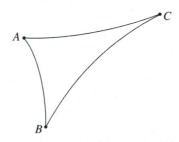

FIGURE 8.1: A hyperbolic triangle

Theorem 8.2.2. *For every convex quadrilateral $\square ABCD$, $\sigma(\square ABCD) < 360°$.*

Theorem 8.2.3. *There does not exist a rectangle.*

The defect of a triangle is defined by $\delta(\triangle ABC) = 180° - \sigma(\triangle ABC)$, where $\sigma(\triangle ABC)$ is the angle sum of $\triangle ABC$. It is clear from this definition that the defect of any triangle is less than $180°$. Hence Theorem 8.2.1 has the following corollary.

Corollary 8.2.4. *For every triangle* $\triangle ABC$, $0° < \delta(\triangle ABC) < 180°$.

One surprise is that different triangles can have different defects—see Exercise 8.1. In fact, later in the chapter we will study the defect of hyperbolic triangles in more depth and prove that for every number d in the interval $(0, 180)$ there is a triangle of defect $d°$.

Theorem 8.2.2 has immediate consequences for Saccheri quadrilaterals and Lambert quadrilaterals.

Corollary 8.2.5. *The summit angles in a Saccheri quadrilateral are acute.*

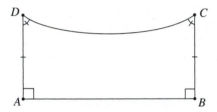

FIGURE 8.2: A Saccheri quadrilateral in hyperbolic geometry

Corollary 8.2.6. *The fourth angle in a Lambert quadrilateral is acute.*

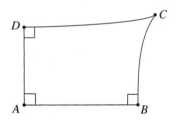

FIGURE 8.3: A Lambert quadrilateral in hyperbolic geometry

Digression on diagrams. Some of the diagrams in this chapter will look strange at first. It must be remembered that the purpose of diagrams is just to help us understand the relationships between the various geometric objects being discussed. They are only *schematic* diagrams, meant to aid our understanding, and should not be taken literally. That has been the case with the diagrams all along because we are studying geometry axiomatically and have not been committed to any particular model of geometry, but that point may have gotten lost by now. The diagrams we draw throughout the text should be viewed as one possible interpretation of the terms being studied. An important lesson of the early part of the course was that we should not base our proofs on relationships that are apparent from the diagram but rather we should base them entirely on the assumptions made explicit in the axioms or the theorems that were proved using the axioms.

We are accustomed to interpreting the diagrams we draw through Euclidean eyes, and it is difficult to imagine other possibilities. In this chapter we want to draw diagrams that help us understand hyperbolic relationships, so some Euclidean conventions will necessarily be violated. For example, we will often draw some of the lines in a way that makes them appear to be curved. One purpose of drawing the diagrams this way is to make it possible to portray the size of the angles accurately. Thus we can visualize a triangle whose angle sum is less than 180° or a quadrilateral whose angle sum is less than 360° (see the first three figures in the chapter). A second purpose of drawing the lines with curves in them is to help us visualize how it might be possible for more than one line through an external point to be parallel to a given line (see Fig. 8.4, for example).

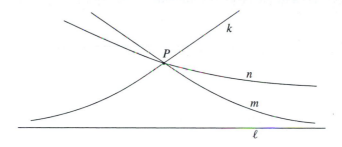

FIGURE 8.4: The lines m, n, and k are all parallel to ℓ

Some of our Euclidean preconceptions are contradicted by the diagrams, but all the axioms of neutral geometry are still assumed to hold. Thus, for example, there is only one line passing through two points. We may draw that line as curved in order to represent other relationships more faithfully, but there is still only one line through the two points. It would be a mistake to imagine that there is one curved line and one straight one. There is a tradeoff that must be made in the diagrams: in order to accurately represent some properties, we must distort others.

Even though we have not defined the term *straight*, the lines in hyperbolic geometry are to be thought of as straight. They look curved to us only because we are viewing them from outside hyperbolic space through Euclidean eyes. The diagrams make it clear that the study of hyperbolic geometry seriously challenges our conventional ideas of what it means for a line to be straight.

In later chapters we will explore models for hyperbolic geometry and also try to understand the relationship between the geometries we study in this course and the geometry of the physical space in which we live. At that time we will discuss straightness and try to clarify some of these matters. Meanwhile the diagrams should be viewed simply as schematic representations of the relationships we are attempting to understand.   [End of digression.]

Let us now return to our development of hyperbolic geometry. The last corollary above asserts that the fourth angle in a Lambert quadrilateral is acute. Another way in which a Lambert quadrilateral deviates from the Euclidean norm is in the lengths of the sides.

Theorem 8.2.7. *In a Lambert quadrilateral, the length of a side between two right angles is strictly less than the length of the opposite side.*

Proof. Let □$ABCD$ be a Lambert quadrilateral. Assume that the angles at vertices A, B, and C are right angles. We must show that $BC < AD$. We already know that $BC \leqslant AD$ (Theorem 6.9.11), so we need only show that $BC = AD$ is impossible.

If $BC = AD$, then □$ABCD$ is a Saccheri quadrilateral. By Theorem 6.9.10, $\angle BAD \cong \angle CDA$. This means that all four angles are right angles and □$ABCD$ is a rectangle. But that contradicts Corollary 8.2.5. Hence we can conclude that $BC \neq AD$ and the proof is complete. $\square$

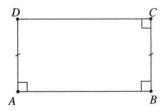

FIGURE 8.5: If $BC = AD$, then □$ABCD$ is a rectangle

Definition 8.2.8. Let □$ABCD$ be a Saccheri quadrilateral with base $\overline{AB}$. The segment joining the midpoint of $\overline{AB}$ to the midpoint of $\overline{CD}$ is called the *altitude* of the quadrilateral. The length of the altitude is called the *height* of the quadrilateral.

Recall that, in neutral geometry, the altitude of a Saccheri quadrilateral is perpendicular to both the base and the summit (Theorem 6.9.10). In hyperbolic geometry the length of the altitude is less than the length of the sides.

Corollary 8.2.9. *In a Saccheri quadrilateral, the length of the altitude is less than the length of a side.*

Proof. Exercise 8.2. $\square$

Corollary 8.2.10. *In a Saccheri quadrilateral, the length of the summit is greater than the length of the base.*

Proof. Exercise 8.3. $\square$

Since Wallis's Postulate is equivalent to the Euclidean Parallel Postulate, we already know that noncongruent similar triangles are harder to construct in

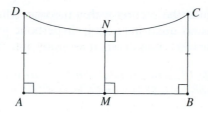

FIGURE 8.6: MN is the altitude of the Saccheri quadrilateral $\square ABCD$

hyperbolic geometry than they are in Euclidean geometry. The next theorem shows that such triangles do not exist in hyperbolic geometry; it asserts that the only similar triangles are congruent triangles. As a result, Angle-Angle-Angle is a valid triangle congruence condition in hyperbolic geometry. This is one of the most surprising theorems in axiomatic hyperbolic geometry.

Theorem 8.2.11 (AAA). *If* $\triangle ABC$ *is similar to* $\triangle DEF$, *then* $\triangle ABC$ *is congruent to* $\triangle DEF$.

Proof. Let $\triangle ABC$ and $\triangle DEF$ be two triangles such that $\triangle ABC \sim \triangle DEF$ (hypothesis). We must show that $\triangle ABC \cong \triangle DEF$. If any one side of $\triangle ABC$ is congruent to the corresponding side of $\triangle DEF$, then $\triangle ABC \cong \triangle DEF$ (ASA).

Suppose $AB \neq DE$, $BC \neq EF$, and $AC \neq DF$ (RAA hypothesis). Since there are three comparisons and each can go only one of two ways, at least two must turn out the same. In other words, either there are two sides of $\triangle ABC$ that are both longer than the corresponding sides of $\triangle DEF$ or there are two sides of $\triangle DEF$ that are both longer than the corresponding sides of $\triangle ABC$. Without loss of generality, we may assume that $AB > DE$ and $AC > DF$.

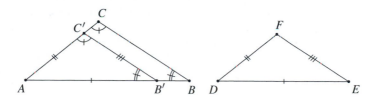

FIGURE 8.7: Proof of Angle-Angle-Angle

Choose a point B' on $\overline{AB}$ such that $AB' = DE$ and choose a point C' on $\overline{AC}$ such that $AC' = DF$. By Lemma 6.7.8, $\square BCC'B'$ is convex. In addition, $\triangle AB'C' \cong \triangle DEF$ (SAS), so $\angle AB'C' \cong \angle ABC$ and $\angle AC'B' \cong \angle ACB$. Since $\angle BB'C'$ is the supplement of $\angle AB'C'$ and $\angle CC'B'$ is the supplement of $\angle AC'B'$, it follows that $\sigma(\square BCC'B') = 360°$. But this contradicts Theorem 8.2.2, so we reject the RAA hypothesis and conclude that $\triangle ABC \cong \triangle DEF$. □

We conclude the section with a theorem which shows that it is also difficult to rescale Saccheri quadrilaterals in hyperbolic geometry. This theorem will be important in the next chapter when we study area in hyperbolic geometry.

Theorem 8.2.12. *If $\square ABCD$ and $\square A'B'C'D'$ are two Saccheri quadrilaterals such that $\delta(\square ABCD) = \delta(\square A'B'C'D')$ and $\overline{CD} \cong \overline{C'D'}$, then $\square ABCD \cong \square A'B'C'D'$.*

Proof. Let $\square ABCD$ and $\square A'B'C'D'$ be two Saccheri quadrilaterals such that $\delta(\square ABCD) = \delta(\square A'B'C'D')$ and $\overline{CD} \cong \overline{C'D'}$ (hypothesis). Since $\delta(\square ABCD) = \delta(\square A'B'C'D')$, it follows that $\sigma(\square ABCD) = \sigma(\square A'B'C'D')$. The base angles in a Saccheri quadrilateral are right angles and the two summit angles in any Saccheri quadrilateral are congruent, so $\angle ADC \cong \angle A'D'C' \cong \angle DCB \cong \angle D'C'B'$.

Choose a point E on $\overrightarrow{DA}$ and a point F on $\overrightarrow{CB}$ such that $\overline{DE} \cong \overline{D'A'} \cong \overline{CF}$. The proof will be completed in two steps: First we will shown that $\square EFCD \cong \square A'B'C'D'$ and then we will show that $E = A$ and $F = B$.

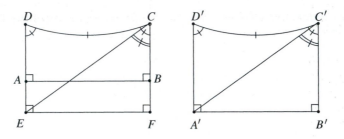

FIGURE 8.8: Proof of Theorem 8.2.12

By SAS, $\triangle EDC \cong \triangle A'D'C'$. It follows by subtraction that $\angle ECF \cong \angle A'C'B'$. Hence $\triangle ECF \cong \triangle A'C'B'$ (SAS again). It follows that $\angle EFC$ is a right angle. A similar proof shows that $\angle FED$ is a right angle. Therefore, $\square EFCD \cong \square A'B'C'D'$.

Suppose $E \neq A$ (RAA hypothesis). Then $\overleftrightarrow{AB} \parallel \overleftrightarrow{EF}$ (Alternate Interior Angles Theorem), so $F \neq B$ and $\square ABFE$ is a rectangle. Since no rectangle exists, we have a contradiction. Thus we reject the RAA hypothesis and conclude that $A = E$ and $B = F$. $\qquad \square$

8.3 COMMON PERPENDICULARS

Our life-long experience with Euclidean geometry has taught us to think of parallel lines as lines that stay the same distance apart. In this section we see that that view of parallel lines is not valid in hyperbolic geometry. Recall that the distance from an external point P to a line ℓ is defined to be the distance $d(P, \ell)$ from the point to the foot of the perpendicular from P to ℓ (Theorem 6.4.4). The next theorem asserts that there cannot be three distinct points on one line that are equidistant from a second line.

Theorem 8.3.1. *If ℓ is a line, P is an external point, and m is a line such that P lies on m, then there exists at most one point Q such that $Q \neq P$, Q lies on m, and $d(Q, \ell) = d(P, \ell)$.*

Proof. Let ℓ and m be two lines (hypothesis). Suppose there exist three distinct points P, Q, and R on m such that $d(P, \ell) = d(Q, \ell) = d(R, \ell)$ (RAA hypothesis). Let P', Q', and R', respectively, denote the feet of the perpendiculars from P, Q, and R to ℓ.

None of the three points P, Q, or R lies on ℓ since $d(P, \ell) > 0$. Therefore, at least two of the three points must lie on the same side of ℓ (Plane Separation Postulate). Let us say that P and Q lie on the same side of ℓ. Then $\square PP'Q'Q$ is a Saccheri quadrilateral. Hence $\ell \parallel m$ (Theorem 6.9.10, Part 4) and all three of the points P, Q, and R lie on the same side of ℓ.[2]

Without loss of generality, we may assume that $P * Q * R$. Then $\square PP'Q'Q$ and $\square QQ'R'R$ are both Saccheri quadrilaterals. Thus angles $\angle PQQ'$ and $\angle RQQ'$ are both acute (Theorem 8.2.5). But this contradicts the fact that angles $\angle PQQ'$ and $\angle RQQ'$ are supplements. Therefore, we must reject the RAA hypothesis and conclude that it is impossible for three points on m to be equidistant from ℓ. $\square$

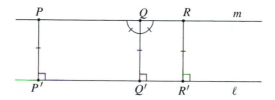

FIGURE 8.9: Three equidistant points in Theorem 8.3.1

Suppose ℓ and m are two parallel lines. By Theorem 8.3.1, there are at most two points on ℓ at any given distance from m. We now explore the consequences of having two points on ℓ that are equidistant from m.

Definition 8.3.2. Lines ℓ and m *admit a common perpendicular* if there exists a line n such that $n \perp m$ and $n \perp \ell$. If ℓ and m admit a common perpendicular, then the line n intersects ℓ at a point P and intersects m at a point Q; the line n is called the *common perpendicular* (line) while the segment $\overline{PQ}$ is called a *common perpendicular segment.*

Theorem 8.3.3. *If ℓ and m are parallel lines and there exist two points on m that are equidistant from ℓ, then ℓ and m admit a common perpendicular.*

Proof. Exercise 8.4. $\square$

---

[2] The proof up to this point is valid in neutral geometry—see Exercise 6.48. The final paragraph of the proof requires the Hyperbolic Parallel Postulate.

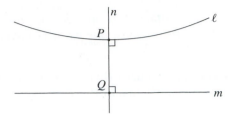

FIGURE 8.10: Common perpendicular

Theorem 8.3.4. *If lines ℓ and m admit a common perpendicular, then that common perpendicular is unique.*

Proof. Exercise 8.5. □

The existence of a common perpendicular is related to the question of whether or not the conclusion of the Converse of the Alternate Interior Angles Theorem holds for a particular transversal. The next theorem indicates that the parallel lines and the transversal would have to be quite special before the alternate interior angles would be congruent. It can happen in hyperbolic geometry that the alternate interior angles are congruent, but it is relatively rare.

Theorem 8.3.5. *Let ℓ and m be parallel lines cut by a transversal t. Alternate interior angles formed by ℓ and m with transversal t are congruent if and only if ℓ and m admit a common perpendicular and t passes through the midpoint of the common perpendicular segment.*

Proof. Let ℓ and m be parallel lines cut by a transversal t. Let R be the point at which t crosses ℓ and let S be the point at which t crosses m.

Assume first that both pairs of alternate interior angles formed by ℓ and m with transversal t are congruent (hypothesis). We must show that ℓ and m admit a common perpendicular and that t passes through the midpoint of the common perpendicular segment. If the alternate interior angles are right angles, then t is the common perpendicular and the proof is complete. Hence we may assume that the alternate interior angles are not right angles.

Let M be the midpoint of $\overline{RS}$. Drop a perpendicular from M to ℓ and call the foot P. Drop a perpendicular from M to m and call the foot Q. Notice that the fact that the alternating interior angles are congruent combines with the Exterior Angle Theorem to show that P and Q are on opposite sides of t. (If the congruent alternate interior angles are acute, then they will both be interior angles for the right triangles $\triangle SQM$ and $\triangle RPM$ and if they are obtuse then they will both be exterior angles for the same two triangles.) Therefore $\triangle SQM \cong \triangle RPM$ (hypothesis and AAS) and so $\angle SMQ \cong \angle RMP$. Hence $\overrightarrow{MP}$ and $\overrightarrow{MQ}$ are opposite rays (see Exercise 5.36) and segment $\overline{PQ}$ is a common perpendicular segment for ℓ and m.

The proof of the converse is left as an exercise (Exercise 8.6). □

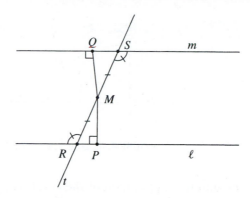

FIGURE 8.11: Proof of Theorem 8.3.5

8.4 LIMITING PARALLEL RAYS AND ASYMPTOTICALLY PARALLEL LINES

Until now we have always constructed parallel lines by dropping a perpendicular from an external point and then constructing a perpendicular to the perpendicular. That double perpendicular construction, which is ubiquitous in our treatments of parallelism in neutral and Euclidean geometries, gives us both the parallel line and the common perpendicular at the same time. It is clear from the results in the preceding section that the common perpendicular between two parallel lines is a key to understanding at least some parallel lines in hyperbolic geometry as well. This raises the question of whether or not every pair of parallel lines admits a common perpendicular. In this section we will investigate a second construction of parallel lines. We will see that the new construction gives us a genuinely different kind of parallel line in hyperbolic geometry.

Construction. Let ℓ be a line and let P be an external point. Drop a perpendicular from P to ℓ and call the foot of that perpendicular A. Let $B \neq A$ be a point on ℓ. (The purpose of B is to specify a particular side of $\overleftrightarrow{PA}$.) For each real number r with $0 < r \leq 90$ there exists a point D_r, on the same side of $\overleftrightarrow{PA}$ as B, such that $\mu(\angle APD_r) = r°$ (Protractor Postulate, Part 3). Define

$$K = \{r \mid \overrightarrow{PD_r} \cap \overrightarrow{AB} \neq \varnothing\}.$$

Notice that K is a set of real numbers, that K is nonempty ($\mu(\angle APB) \in K$, for example), and that K is bounded above (by 90, for example). Therefore, K has a least upper bound, which we will denote by r_0 (the Least Upper Bound Postulate, Axiom 4.2.7). Since 90 is an upper bound for K, it is clear that $r_0 \leq 90$. It is also clear that $90 \notin K$ since the perpendicular at P is parallel to ℓ.

Definition 8.4.1. The set K is called the *intersecting set* for P and $\overrightarrow{AB}$. The number r_0 is called the *critical number* for P and $\overrightarrow{AB}$.

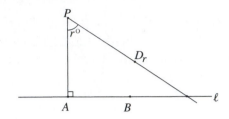

FIGURE 8.12: Construction of K

Here is an intuitive description of the construction: If D is a point such that $\mu(\angle APD)$ is small, then $\overrightarrow{PD}$ will intersect $\overrightarrow{AB}$. As the measure of $\angle APD$ increases, there will be a first ray that does not intersect $\overrightarrow{AB}$ and r_0 is the measure of the angle made by that ray. Any ray that makes an angle of larger measure will not intersect $\overrightarrow{AB}$. The following theorem makes the intuitive description precise.

Theorem 8.4.2. *If* $0 < r < r_0$, *then* $r \in K$. *If* $r_0 \leqslant r \leqslant 90$, *then* $r \notin K$.

Proof. Let r be given, $0 < r \leqslant 90$. Assume first that $r < r_0$ (hypothesis). There must exist a number $s \in K$ such that $r < s$ (otherwise r would be an upper bound for K that is smaller than r_0). Because $s \in K$, $\overrightarrow{PD_s}$ intersects $\overrightarrow{AB}$ at a point T. Since $r < s$, D_r is in the interior of $\angle APD_s$. Hence $\overrightarrow{PD_r}$ must intersect $\overline{AT}$ (Crossbar Theorem) and so $r \in K$.

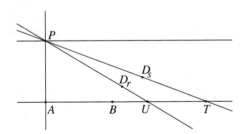

FIGURE 8.13: Proof of Theorem 8.4.2

Now assume that $r \geqslant r_0$ (hypothesis). Suppose $r \in K$ (RAA hypothesis). Then $\overrightarrow{PD_r}$ intersects $\overrightarrow{AB}$ at a point U. Choose a point T such that $A * U * T$ and define $s = \mu(\angle APT)$. Because $\overrightarrow{PD_s} \cap \overrightarrow{AB} \neq \varnothing$, $s \in K$. Furthermore, $s > r$ (Protractor Postulate) and so $s > r_0$. This contradicts the fact that r_0 is an upper bound for K. Therefore we reject the RAA hypothesis and conclude that $r \notin K$. □

We have used a deep property of the real numbers (the Least Upper Bound Postulate) to prove that there is a first ray from P that does not intersect $\overrightarrow{AB}$. As the second part of the proof of Theorem 8.4.2 shows, there is no last ray that intersects $\overrightarrow{AB}$.

Definition 8.4.3. Suppose P, A, and B are as in the definition of the intersecting set and that r_0 is the critical number for P and $\overrightarrow{AB}$. Let D be a point on the same side of $\overleftrightarrow{PA}$ as B such that $\mu(\angle APD) = r_0$. The angle $\angle APD$ is called the *angle of parallelism* for P and $\overrightarrow{AB}$.

Since there are two essentially different ways in which to choose the point B that determines the ray $\overrightarrow{AB}$ (one on each side of $\overleftrightarrow{PA}$), there are actually two angles of parallelism associated with P and ℓ, one determined by P and $\overrightarrow{AB}$ and the other determined by P and the opposite ray $\overrightarrow{AB'}$. As one would expect from looking at Fig. 8.14, these two angles of parallelism are congruent. In fact the next theorem shows that the measure of the angle of parallelism depends only on the distance from P to ℓ. As a result, we will usually refer to the angle of parallelism determined by a line ℓ and an external point P and not distinguish between the two congruent angles determined by the opposite rays $\overrightarrow{AB}$ and $\overrightarrow{AB'}$ on ℓ.

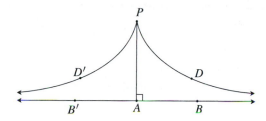

FIGURE 8.14: Two angles of parallelism for P and ℓ

Theorem 8.4.4. *The critical number depends only on $d(P, \ell)$.*

Proof. Let ℓ be a line, let P be an external point, let A be the foot of the perpendicular from P to ℓ, and let B be a point on ℓ that is different from A. Let P', ℓ', A', and B' be another setup such that that $PA = P'A'$. We must show that the critical number for P and $\overrightarrow{AB}$ is equal to the critical number for P' and $A'B'$. We will accomplish that by showing that the intersecting sets are the same.

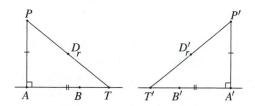

FIGURE 8.15: Proof of Theorem 8.4.4

Call the intersecting sets K and K'. Suppose $r \in K$. Then $\overrightarrow{PD_r}$ intersects $\overrightarrow{AB}$ at a point T. Choose a point T' on $\overrightarrow{A'B'}$ such that $A'T' = AT$. Then $\triangle PAT \cong \triangle P'A'T'$ (SAS) and so $r \in K'$. In a similar way it follows that if $r \in K'$, then $r \in K$. Therefore, $K = K'$ and the proof is complete. $\square$

Since the critical number depends only on $d(P, \ell)$, we can think of it as a function of real numbers.

Definition 8.4.5. Given a positive real number x, locate a point P and a line ℓ such that $d(P, \ell) = x$. Then define $\kappa(x)$ to be the critical number associated with P and ℓ; that is, choose a point $B \neq A$ on ℓ and define $\kappa(x) = \mu(\angle APD)$, where $\angle APD$ is the angle of parallelism for P and $\overrightarrow{AB}$. By Theorem 8.4.4, κ is a well defined function of x and does not depend on the particular choice of P, ℓ, or B. The function $\kappa : (0, \infty) \to (0, 90]$ is called the *critical function*.

The critical function will be studied in more depth in the next section. For now we need just the following property, which asserts that the angle of parallelism between P and ℓ does not increase as P moves away from ℓ.

Theorem 8.4.6. $\kappa : (0, \infty) \to (0, 90]$ *is a decreasing function; that is, $a < b$ implies* $\kappa(a) \geqslant \kappa(b)$.

Proof. Let a and b be two positive numbers such that $a < b$ (hypothesis). Choose P, A, B, and D to be four points such that $PA = a$ and $\angle APD$ is the angle of parallelism for P and $\overrightarrow{AB}$. Then choose Q to be a point on $\overrightarrow{AP}$ such that $QA = b$. We must show that the measure of the angle of parallelism at Q is no greater than the angle of parallelism at P.

Define $r = \mu(\angle APD)$ and choose a point E on the same side of $\overleftrightarrow{AP}$ as B such that $\mu(\angle AQE) = r$. Now $\overleftrightarrow{QE} \parallel \overrightarrow{PD}$ (Corresponding Angles Theorem), so $\overrightarrow{QE}$ does not intersect $\overrightarrow{AB}$. (All the points of $\overleftrightarrow{QE}$ are on one side of $\overleftrightarrow{PD}$ while all the points of $\overrightarrow{AB}$ are on the other.) Since r is not in the intersecting set for Q and $\overrightarrow{AB}$, r cannot be less than the critical number for Q and $\overrightarrow{AB}$ (Theorem 8.4.2). Therefore, r, which is the measure of the angle of parallelism at P, is greater than or equal to the measure of the angle of parallelism at Q. $\square$

The definition of the angle of parallelism and the three theorems proved so far in this section depend only on theorems from neutral geometry. Thus the angle of parallelism could have been defined in Chapter 6, where it would play a role analogous to that of the defect of a triangle. Just like defect, the angle of parallelism can be used to distinguish Euclidean geometry from hyperbolic geometry. In particular, every angle of parallelism is a right angle in Euclidean geometry and every angle of parallelism is acute in hyperbolic geometry. We have already proved neutral theorems of that sort for defect of triangles and therefore have chosen not to do the same for the angle of parallelism. Instead we will study the angle of parallelism in hyperbolic geometry exclusively.

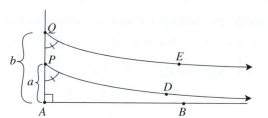

FIGURE 8.16: The measure of the angle of parallelism is nonincreasing

In hyperbolic geometry the angle of parallelism is a very interesting and highly nontrivial concept. As we will see, it is the key to understanding the second type of parallel line. Most of the remainder of this chapter will be devoted to a study of its properties. The next theorem tells us that the angle of parallelism always has measure strictly less than 90° in hyperbolic geometry. It is the first genuinely hyperbolic result in the section.

Theorem 8.4.7. *Every angle of parallelism is acute and every critical number is less than 90.*

Proof. Let ℓ be a line and let P be an external point. Drop a perpendicular from P to ℓ and call the foot A. We will show that $\kappa(PA) < 90$.

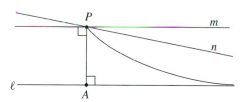

FIGURE 8.17: There are at least two parallel lines, so one makes an acute angle

Let m be the line such that P lies on m and $m \perp \overleftrightarrow{PA}$. As we have seen many times before, $m \parallel \ell$. By the Hyperbolic Parallel Postulate, there must be at least one line n such that n is distinct from m, P lies on n, and $n \parallel \ell$. Since n is not perpendicular to $\overleftrightarrow{PA}$, the angle between n and $\overrightarrow{PA}$ must measure less than 90° on one side or the other. The measure of the angle between n and $\overrightarrow{PA}$ is not in the intersecting set, so the critical number cannot be larger than this measure. Thus the critical number is less than 90 and angle of parallelism is acute.   □

Definition 8.4.8. Suppose P, A, B, and D are as in the definition of the angle of parallelism. The ray $\overrightarrow{PD}$ is called a *limiting parallel ray* for $\overrightarrow{AB}$. We will use the notation $\overrightarrow{PD} \mid \overrightarrow{AB}$ to indicate that $\overrightarrow{PD}$ is a limiting parallel ray for $\overrightarrow{AB}$.

The following theorem shows that if $\overrightarrow{PD}$ is a limiting parallel ray for $\overrightarrow{AB}$, then any ray contained in $\overrightarrow{PD}$ is a limiting parallel ray for a subray of $\overrightarrow{AB}$.

Theorem 8.4.9. *Suppose $\overrightarrow{PD} \mid \overrightarrow{AB}$, $Q \in \overrightarrow{PD}$ and C is the foot of the perpendicular from Q to $\overleftrightarrow{AB}$. If B' is a point on $\overrightarrow{AB}$ such that $A * C * B'$ and D' is a point on $\overrightarrow{PD}$ such that $P * Q * D'$, then $\overrightarrow{QD'} \mid \overrightarrow{CB'}$.*

Proof. Exercise 8.7.   □

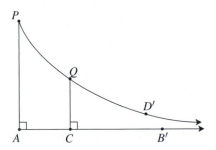

FIGURE 8.18: A subray of a limiting parallel ray is also a limiting parallel ray, Theorem 8.4.9

The rays $\overrightarrow{PD}$ and $\overrightarrow{AB}$ can be extended to lines $\overleftrightarrow{PD}$ and $\overleftrightarrow{AB}$. The next theorem asserts that these extensions are parallel lines.

Theorem 8.4.10. *If $\overrightarrow{PD} \mid \overrightarrow{AB}$, then $\overleftrightarrow{PD} \parallel \overleftrightarrow{AB}$.*

Proof. Exercise 8.8.   □

Theorem 8.4.10 justifies the following definition.

Definition 8.4.11. If $\overrightarrow{PD} \mid \overrightarrow{AB}$, then the lines $\overleftrightarrow{PD}$ and $\overleftrightarrow{AB}$ are called *asymptotically parallel* lines. More specifically, we will say that lines ℓ and m are *asymptotically parallel in the direction $\overrightarrow{AB}$*, where A and B are points on ℓ, if there exist points P and D on m such that $\overrightarrow{PD} \mid \overrightarrow{AB}$.

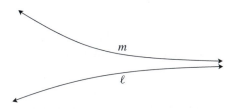

FIGURE 8.19: Asymptotically parallel lines

The notation | is similar to ‖ because limiting parallel rays are part of parallel lines. The fact that there is only one vertical bar is meant to suggest that the rays are just barely parallel: There is no room for any other parallels between them in the sense that any closer rays with the same endpoints will intersect.

Notice that the relationship of limiting parallelism is not exactly a symmetric relationship for hyperbolic rays (i.e., $\overrightarrow{PD} \mid \overrightarrow{AB}$ does not imply that $\overrightarrow{AB} \mid \overrightarrow{PD}$). The reason for this is that the construction starts with a perpendicular dropped from P to $\overleftrightarrow{AB}$, so $\angle PAB$ is a right angle while $\angle APD$ is not. However, the relationship of asymptotic parallelism is a symmetric relationship for lines.

Theorem 8.4.12 (Symmetry of Limiting Parallelism). *Suppose $\overrightarrow{PD} \mid \overrightarrow{AB}$ and Q is the foot of the perpendicular from A to $\overrightarrow{PD}$. If D' is a point on $\overrightarrow{PD}$ such that $P * Q * D'$, then $\overrightarrow{AB} \mid \overrightarrow{QD'}$.*

Proof. Exercise 8.9. □

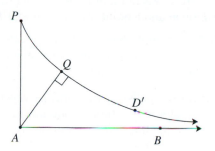

FIGURE 8.20: Symmetry of limiting parallelism

Corollary 8.4.13 (Symmetry of Asymptotic Parallelism). *If ℓ is asymptotically parallel to m, then m is asymptotically parallel to ℓ.*

As promised, we have defined a new kind of parallelism in hyperbolic geometry. The next theorem shows that asymptotically parallel lines are genuinely different from parallel lines that admit a common perpendicular.

Theorem 8.4.14 (Classification of Parallels, Part 1). *If ℓ and m are asymptotically parallel lines, then ℓ and m do not admit a common perpendicular. If ℓ and m admit a common perpendicular, then ℓ and m are not asymptotically parallel.*

Proof. The second statement is just the contrapositive of the first. We will prove the second.

Let ℓ and m be two lines that admit a common perpendicular t, and let us say that t crosses ℓ and m at points S and R, respectively. Observe, first, that no subray of m with endpoint R can be a limiting parallel ray because the angle of parallelism must be acute (Theorem 8.4.7).

Let P be a point on m that is different from R and drop a perpendicular from P to point A on ℓ. Let D be a point on m such that $R * P * D$. We will show that neither $\overrightarrow{PD}$ nor $\overrightarrow{PR}$ is a limiting parallel ray for a subray of ℓ. Since P was an arbitrary point on m, this will complete the proof.

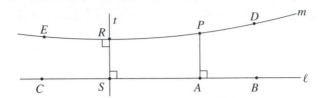

FIGURE 8.21: Theorem 8.4.14

Choose a point B on ℓ such that B and D lie on the same side of $\overleftrightarrow{PA}$. Since $\square SAPR$ is a Lambert quadrilateral, $\angle APR$ is acute (Corollary 8.2.6) and so $\angle APD$ is obtuse. Therefore $\overrightarrow{PD}$ is not a limiting parallel ray for $\overrightarrow{AB}$ (Theorem 8.4.7). Now choose a point E on m and a point C on ℓ such that $E * R * D$ and $C * S * B$. The ray $\overrightarrow{PE}$ cannot be a limiting parallel ray for $\overrightarrow{AC}$ because, if it were, $\overrightarrow{RE}$ would be a limiting parallel ray for $\overrightarrow{SC}$ (Theorem 8.4.9). We have already observed that that is not the case, so the proof is complete. $\square$

The next theorem spells out one concrete way in which the two kinds of parallel lines differ. In case the lines admit a common perpendicular, they get further and further apart; in case the lines are asymptotically parallel they get closer and closer together.

Theorem 8.4.15. *Suppose $\ell \parallel m$. Let P, Q, and R be points on m such that $P * Q * R$ and let A, B and C be the feet of the perpendiculars from P, Q, and R to ℓ.*

1. *If $\overleftrightarrow{PA} \perp m$, then $PA < QB < RC$.*

2. *If $\overrightarrow{PQ} \mid \overrightarrow{AB}$, then $PA > QB > RC$.*

Proof. Exercises 8.10 and 8.11. $\square$

We can improve on Theorem 8.4.15. Not only does the distance steadily increase in case the lines admit a common perpendicular and decrease in case the lines are asymptotically parallel, but the distance between the lines approaches infinity in the first case and approaches zero in the second case. This justifies the use of the term *asymptotic* for the second kind of parallel and suggests that parallel lines that admit a common perpendicular should really be called "divergently parallel." We have the tools in place to prove the theorem about divergently parallel lines now. The theorem about asymptotic parallels will be proved in the next section.

Theorem 8.4.16. *If ℓ and m are parallel lines that admit a common perpendicular, then for every positive number d_0 there exists a point P on m such that $d(P, \ell) > d_0$. Furthermore, P may be chosen to lie on either side of the common perpendicular.*

Proof. Assume ℓ and m admit a common perpendicular and let the positive number d_0 be given. Let us say that the common perpendicular cuts m and ℓ at points R and S, respectively. Let $\angle SRD$ be the angle of parallelism for R and ℓ on the specified side of $\overleftrightarrow{RS}$. Choose a point Q that lies on m and is on the same side of $\overleftrightarrow{RS}$ as D. By Aristotle's Theorem (Theorem 6.9.12) there exists a point P on $\overrightarrow{RQ}$ such that $d(P, \overleftrightarrow{RD}) > d_0$.

Let F be the foot of the perpendicular from P to $\overleftrightarrow{RD}$ and let E be the foot of the perpendicular from P to ℓ. Because D is in the interior of $\angle SRP$ (Theorem 5.7.12), P and S are on opposite sides of $\overleftrightarrow{RD}$. Therefore, P and E are on opposite sides of $\overleftrightarrow{RD}$ and so $\overline{PE}$ intersects $\overleftrightarrow{RD}$ at a point G. Now $PG > PF$ by the Scalene Inequality (Theorem 6.4.1) and $PE > PG$ since G is between P and E, so $PE > d_0$. But $PE = d(P, \ell)$, so the proof of the theorem is complete. $\square$

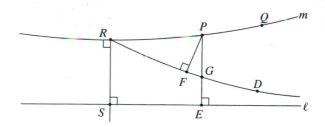

FIGURE 8.22: The proof of Theorem 8.4.16

We end the section with statements of two theorems that shed additional light on the nature of parallelism in hyperbolic geometry. The proofs of the two theorems are omitted because they are beyond the scope of these notes and because the results themselves are not needed for what follows. It is good to be familiar with the statements, however, because the theorems put the results of the section in their proper context.

The first of these theorems asserts that the relationship of asymptotic parallelism in hyperbolic geometry is, in a sense, a transitive relationship. This is surprising since we proved that transitivity of ordinary parallelism is equivalent to the Euclidean Parallel Postulate. A nice exposition of the proof of Theorem 8.4.17 may be found on pages 380 and 381 of the book by Moise [51].

Theorem 8.4.17 (Transitivity of Limiting Parallelism). *If ℓ is asymptotically parallel to m in the direction $\overrightarrow{AB}$ and ℓ is asymptotically parallel to n in the direction $\overrightarrow{AB}$, then either $m = n$ or m is asymptotically parallel to n.*

We have proved that there are two kinds of parallelism in hyperbolic geometry. The obvious question is this: If there are two different kinds of parallelism, might there be others? The last theorem in the section tells us that the answer is no, there is no other kind of parallelism. It is important to state the theorem because it answers the obvious question and completes the picture of parallelism in hyperbolic geometry. An outline of the proof is found on page 199 of Greenberg's book [27]. The idea, due to Hilbert, is to show that if two lines are parallel but not asymptotically parallel, then one of the lines must contain two points that are equidistant from the other line.

Theorem 8.4.18 (Classification of Parallels, Part 2). *If $\ell \parallel m$, then either ℓ and m admit a common perpendicular or ℓ and m are asymptotically parallel.*

Summary

Let us summarize the results of this section. Start with a line ℓ and an external point P. There are two limiting parallel rays for P and ℓ so there are two distinct lines, m and m', that contain P and are asymptotically parallel to ℓ. Points on those two lines get closer and closer to ℓ as the points get further from P in one direction and get further from ℓ in the other direction. Every line between m and m' is also parallel to ℓ, but any such line will admit a common perpendicular with ℓ. On any such line n there will be one point that is closest to ℓ and other points of n will get further from ℓ as they move in either direction from that unique closest point. (See Fig. 8.23.) Most parallels through P admit common perpendiculars in the sense that there are just two asymptotically parallel lines through P but there are infinitely many other parallel lines.

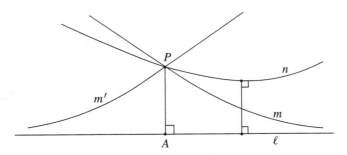

FIGURE 8.23: The parallels to ℓ at P

Given a point P and two lines m and m' such that P lies on both m and m', the set of lines consisting of m, m', and all lines between m and m' is called a *pencil of lines*. In that terminology, we have seen that the set of lines that are parallel the line ℓ and contain the point P form a pencil of lines. The first and last lines in the pencil are asymptotically parallel to ℓ and all the others admit common perpendiculars.

Asymptotic parallelism in hyperbolic geometry is in some ways like parallelism in Euclidean geometry. For example, given a line $\ell = \overleftrightarrow{AB}$ and an external point P, there is exactly one line m such that P lies on m and m is asymptotically parallel to ℓ in the direction $\overrightarrow{AB}$. Another example of a way in which they are alike is that both kinds of parallelism are transitive.

8.5 PROPERTIES OF THE CRITICAL FUNCTION

We now take a closer look at the critical function $\kappa : (0, \infty) \rightarrow (0, 90)$ and study its properties as a function. The starting point is Theorem 8.4.6, which says that κ is a decreasing (or nonincreasing) function. We will prove the stronger result that κ is strictly decreasing and we will also compute the limit of $\kappa(x)$ as x approaches the two endpoints of its domain, zero and infinity. It is worth noting that the proof that κ is decreasing is valid in neutral geometry, but the fact that κ is strictly decreasing is a uniquely hyperbolic result.

Theorem 8.5.1. $\kappa : (0, \infty) \rightarrow (0, 90)$ *is a strictly decreasing function; that is, $a < b$ implies $\kappa(a) > \kappa(b)$.*

Proof. Let a and b be given, $0 < a < b$. Choose a ray $\overrightarrow{AB}$ and points P and Q on a perpendicular ray such that $d(P, \overleftrightarrow{AB}) = a$ and $d(Q, \overleftrightarrow{AB}) = b$ as indicated in Fig. 8.24. Choose points D and E on the same side of $\overleftrightarrow{AQ}$ as B so that $\overrightarrow{PD} \mid \overrightarrow{AB}$ and $\angle AQE \cong \angle APD$.

Since $\overleftrightarrow{PQ}$ is a transversal for lines $\overleftrightarrow{PD}$ and $\overleftrightarrow{QE}$ with congruent alternate interior angles, lines $\overleftrightarrow{PD}$ and $\overleftrightarrow{QE}$ admit a common perpendicular (Theorem 8.3.5). Thus there exists a point R on $\overrightarrow{QE}$ such that $d(R, \overleftrightarrow{PD}) > b$ (Theorem 8.4.16). Let F be the foot of the perpendicular from R to $\overleftrightarrow{PD}$ and let H be the foot of the perpendicular from R to $\overleftrightarrow{AB}$. Since R and H are on opposite sides of $\overleftrightarrow{PD}$, $\overline{RH}$ intersects $\overleftrightarrow{PD}$ at a point G.

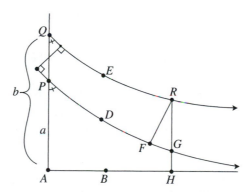

FIGURE 8.24: The lines admit a common perpendicular

By exactly the same argument as in the proof of Theorem 8.4.16, $RH > RF$ and thus $d(R, \overleftrightarrow{AB}) > d(Q, \overleftrightarrow{AB})$. Therefore, $\overrightarrow{QE}$ cannot be a limiting parallel ray for $\overrightarrow{AB}$ (Theorem 8.4.15) and $\kappa(b) \neq \kappa(a)$. Since we already know that $\kappa(b) \leq \kappa(a)$ (Theorem 8.4.6), this completes the proof. $\square$

As an application of Theorem 8.5.1, we prove that the distance between asymptotically parallel lines approaches zero. We already know that it decreases (Theorem 8.4.15); we need to prove that it becomes arbitrarily small.

Theorem 8.5.2. *If ℓ and m are asymptotically parallel lines, then for every positive number ϵ there exists a point T on m such that $d(T, \ell) < \epsilon$.*

Proof. Suppose ℓ and m are asymptotically parallel and that $\epsilon > 0$ is given. Then there exist points P and D on m and A and B on ℓ such that $\overleftrightarrow{PA} \perp \ell$ and $\overrightarrow{PD} \mid \overrightarrow{AB}$. Choose a point Q between P and A such that $QA < \epsilon$. The proof of the theorem will be completed by showing that there is a point T on m such that $d(T, \ell) < d(Q, \ell)$.

Let n be the line such that Q lies on n and $n \perp \overleftrightarrow{PA}$. Since $PQ < PA$, $\kappa(PQ) > \kappa(PA)$ (Theorem 8.5.1). Thus $\mu(\angle QPD)$ is in the intersecting set for P and n and $\overrightarrow{PD}$ must intersect n in a point R. Let F be the foot of the perpendicular from R to ℓ. We may assume that $P * R * D$ and $A * F * B$. (If not, choose different points B and D for which this is the case.)

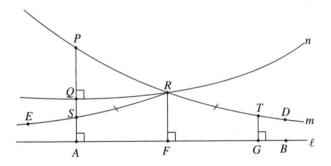

FIGURE 8.25: The proof of Theorem 8.5.2

Choose a point E, on the opposite side of $\overleftrightarrow{RF}$ from B and D, such that $\angle FRE$ is the angle of parallelism for R and $\overrightarrow{FA}$. By Exercise 8.18, $\overrightarrow{RE}$ must intersect $\overline{QA}$ in a point S. Since S is between Q and A, $d(S, \ell) < d(Q, \ell) < \epsilon$. Choose a point T on $\overrightarrow{RD}$ such that $RT = RS$. It follows from Exercise 8.19 that $d(T, \ell) = d(S, \ell)$, so the proof is complete. $\square$

We now compute the limit of κ at each end of its domain.

Theorem 8.5.3. $\lim\limits_{x \to \infty} \kappa(x) = 0$.

Proof. The fact that that κ is a decreasing function means that it is enough to prove the following: For every $\epsilon > 0$ there exists an x such that $\kappa(x) \leq \epsilon$.

Suppose there exists a positive number ϵ such that $\kappa(x) > \epsilon$ for every x (RAA hypothesis). We will derive a contradiction by showing that the RAA hypothesis allows us to construct a triangle whose defect is greater than $180°$. The construction is one of the repeated constructions that uses the Archimedean Property of Real Numbers.

The construction begins with a right angle. Call the vertex P_0. Choose a point P_1 on one side of the angle such that $P_0 P_1 = 1$ and construct a ray that makes an angle of measure $\epsilon°$ at P_1. Since $\kappa(1) > \epsilon$, that ray must intersect the second side of the right angle at a point B_1. (Refer to Fig. 8.26.) Drop a perpendicular from P_0 to $\overleftrightarrow{P_1 B_1}$ and call the foot Q_1.

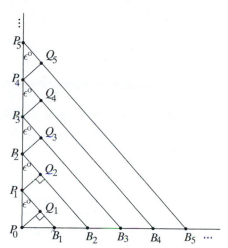

FIGURE 8.26: Proof of Theorem 8.5.3

Choose a point P_2 on $\overrightarrow{P_0 P_1}$ such that $P_0 P_2 = 2$ and construct an angle of measure $\epsilon°$ at P_2. Since $\kappa(2) > \epsilon$, that ray must intersect the second side of the right angle at a point B_2. Drop a perpendicular from P_1 to $\overleftrightarrow{P_2 B_2}$ and call the foot Q_2. Continue that construction inductively to produce Figure 8.26.

Notice that $\triangle P_i Q_i P_{i-1} \cong \triangle P_1 Q_1 P_0$ for every i (AAS) and so $\delta(\triangle P_i Q_i P_{i-1}) = \delta(\triangle P_1 Q_1 P_0)$ for every i. Let us denote $\delta(\triangle P_1 Q_1 P_0)$ by δ_0. The Archimedean Property of Real Numbers implies that there exists an n such that $n\delta_0 > 180$.

It follows from Additivity of Defect (see Exercise 8.12) that

$$\delta(\triangle P_n B_n P_0) > \delta(\triangle P_1 Q_1 P_0) + \delta(\triangle P_2 Q_2 P_1) + \cdots + \delta(\triangle P_n Q_n P_{n-1}).$$

Hence $\delta(\triangle P_n B_n P_0) > n\delta_0° > 180°$. This contradicts the fact that all triangles have defect less than $180°$ (definition of defect of a triangle) and so the proof is complete. $\square$

Theorem 8.5.4. $\lim\limits_{x \to 0^+} \kappa(x) = 90$.

Proof. Again the fact that we have already proved that κ is decreasing makes this proof a little easier. Since $\kappa(x) < 90$ for every x and κ is decreasing, it suffices to show that for every $\epsilon > 0$ there is a positive number d such that $\kappa(d) > 90 - \epsilon$.

Let $\epsilon > 0$ be given. Construct an angle of measure $(90 - \epsilon)°$ and call the vertex P. Choose a point B on one side of the angle and drop a perpendicular from B to the other side of the angle. Call the foot of the perpendicular A. Notice that $A \neq P$ since $\mu(\angle APB) < 90°$. The triangle $\triangle APB$ is the desired triangle.

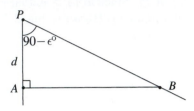

FIGURE 8.27: Proof of Theorem 8.5.4

Let $d = AP$. Then $\kappa(d)$ must be greater than $90 - \epsilon$ because the existence of $\triangle APB$ shows that $90 - \epsilon$ is in the intersecting set for P and $\overrightarrow{AB}$. This completes the proof. □

The properties of κ that have been verified in this section are the ones that we will need in the next section when we study the defects of triangles. There is, however, one more property that should at least be mentioned: κ is onto. Once we know that κ is onto, Lemma 5.7.26 can be used to show that κ is continuous. The proofs of Theorem 8.5.5 and Corollary 8.5.6 will be omitted because the results are not actually needed in what follows. Proofs are outlined in Exercises 8.20 through 8.22.

Theorem 8.5.5. *κ is onto; that is, for every number $y \in (0, 90)$ there exists a number $x \in (0, \infty)$ such that $\kappa(x) = y$.*

Proof. Exercise 8.20. □

Corollary 8.5.6. *$\kappa : (0, \infty) \rightarrow (0, 90)$ is continuous.*

Proof. Exercises 8.21 and 8.22. □

Summary

The theorems in this section give us a fairly good idea of how the angle of parallelism behaves. When the point P is close to ℓ, the angle of parallelism is approximately a right angle. As P moves away from ℓ, the measure of the angle of parallelism gradually decreases and it approaches zero as $d(P, \ell)$ approaches ∞. The graph of κ will look roughly like that shown in Fig. 8.28.

Another way to summarize the results of this section is to say that if P is close to ℓ then the pencil of lines through P that are parallel to ℓ is very thin. As P moves away from ℓ, the pencil of parallel lines gets fatter and eventually includes nearly all the lines through P.

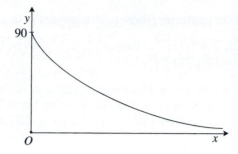

FIGURE 8.28: A graph of the critical function

8.6 THE DEFECT OF A TRIANGLE

The theorems in the previous section have many surprising consequences concerning defects of triangles. Such facts about the defect are collected in this section. The results are interesting and important in their own right; they will also be used in the next chapter when we investigate the relationship between defect and area. The main result is that the defect of hyperbolic triangles is variable. In fact, we will see that for every d in the open interval $(0, 180)$ there is a triangle whose defect is $d°$. The first step in that direction is to prove that the defect can be close to $180°$. Since the defect is defined to be $180°$ minus the angle sum, a triangle whose defect is close to $180°$ has largest possible defect and very small angle sum.

Theorem 8.6.1. *For every $\epsilon > 0$ there exists an isosceles triangle $\triangle ABC$ such that $\sigma(\triangle ABC) < \epsilon°$ and $\delta(\triangle ABC) > (180 - \epsilon)°$.*

Proof. Let $\epsilon > 0$ be given. Choose a large enough so that $\kappa(a) < \epsilon/4$ (Theorem 8.5.3). Let $\triangle PAB$ be a triangle with right angle at vertex A and both legs of length a. Since the ray $\overrightarrow{PB}$ intersects $\overrightarrow{AB}$, the angle $\angle APB$ must measure less than $\kappa(a) = (\epsilon/4)°$. The triangle is isosceles, so the measure of $\angle ABP$ must also be less than $(\epsilon/4)°$. Placing two copies of this triangle next to each other as indicated in Fig. 8.29 results in one triangle $\triangle CBP$ whose angle sum is less than $\epsilon°$. $\qquad\square$

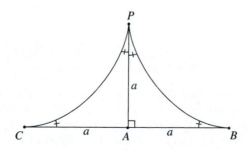

FIGURE 8.29: A very defective triangle

At the other extreme, there exist triangles with arbitrarily small defect.

Theorem 8.6.2. *For every $\epsilon > 0$ there is a right triangle $\triangle APB$ such that $\sigma(\triangle APB) > (180 - \epsilon)°$ and $\delta(\triangle APB) < \epsilon°$.*

Proof. Let $\epsilon > 0$ be given. Proceeding exactly as in the proof of Theorem 8.5.4, we can construct a triangle such that one interior angle is a right angle and a second interior angle has measure $(90 - \epsilon)°$. (See Fig. 8.27.) This triangle has the desired properties.   □

While it is interesting to know that triangles with small defect exist, it is more useful to have conditions under which the defect will be small. The next theorem says that small triangles have small defect. This means that small hyperbolic triangles are nearly indistinguishable from small Euclidean triangles.

Theorem 8.6.3. *For every positive number ϵ there is a positive number d such that if $\triangle ABC$ is a triangle in which every side has length less than d, then $\delta(\triangle ABC) < \epsilon$.*

Proof. Let $\epsilon > 0$ be given. By Theorem 8.6.2 there exists a right triangle $\triangle EFG$ such that $\delta(\triangle EFG) < \epsilon/2$. Let us say that the right angle is at vertex E. Choose $d > 0$ to be the length of the shortest side of $\triangle EFG$.

We must now prove that the d we have chosen satisfies the conclusion of the theorem. Let $\triangle ABC$ be any triangle each of whose sides has length less than d. Without loss of generality, we may assume that the interior angles at vertices A and B are acute. As indicated in Fig. 8.30, we can split $\triangle ABC$ into two right triangles, $\triangle ADC$ and $\triangle BCD$.

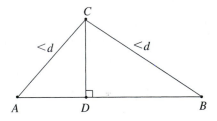

FIGURE 8.30: Subdivide the small triangle into two smaller right triangles

Notice that each of these latter triangles has sides of length less than d. Both $\overline{AD}$ and $\overline{DB}$ must be shorter than d because the sum of their lengths is AB. By the Scalene Inequality (Theorem 6.4.1), $CD < AC$ and so $CD < d$ as well.

Now observe that each of the triangles $\triangle ADC$ and $\triangle BCD$ is congruent to a triangle that is contained in $\triangle EFG$. In order to see this, choose a point A' between E and F such that $EA' = DA$. Then choose a point C' between E and G such that $EC' = DC$. By SAS, $\triangle ADC \cong \triangle A'EC'$. By Additivity of Defect, this implies that $\delta(\triangle ADC) < \delta(\triangle EFG) < \epsilon/2$. In a similar way we see that $\delta(\triangle BCD) < \epsilon/2$ and so $\delta(\triangle ABC) < \epsilon$. This completes the proof.   □

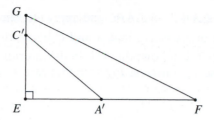

FIGURE 8.31: Embed each small right triangle in the large right triangle

In order for a triangle to have small defect it is not necessary for every side of the triangle to be short. In fact, the triangle constructed in the proof of Theorem 8.6.2 (see Figure 8.27) will have one short side while the other two sides could be quite long. The real theorem is that thin triangles have small defect. The next theorem is a major step toward making that statement precise. The result will be useful later when we prove that the defect is continuous. The continuity of the defect will be an important ingredient in the proof (in the next chapter) that defect is proportional to area.

Theorem 8.6.4. *For every pair of points A and B and for every positive number ϵ, there exists a number $d > 0$ such that if C is any point not on $\overleftrightarrow{AB}$ with $AC < d$, then $\delta(\triangle ABC) < \epsilon$.*

Proof. Choose a point P such that $P * A * B$. Construct an angle of measure $(90 - \epsilon)^\circ$ such that $\overrightarrow{PB}$ is one side of the angle (Protractor Postulate). Drop a perpendicular from B to the other side of the angle and let Q denote the foot of that perpendicular. (See Fig. 8.32.) It is clear that $\sigma(\triangle PQB) > (180 - \epsilon)^\circ$, so $\delta(\triangle PQB) < \epsilon^\circ$. Define d to be the smaller of the two distances $d(A, \overleftrightarrow{PQ})$ and $d(A, \overleftrightarrow{QB})$.

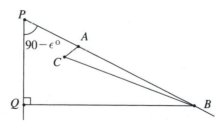

FIGURE 8.32: Construction of $\triangle PQB$ in the proof of Theorem 8.6.4

In order to complete the proof we must demonstrate that the number d defined in the previous paragraph has the property specified in the statement of the theorem. Let C be a point not on $\overleftrightarrow{AB}$ with $AC < d$. We may assume that C and Q are on the same side of $\overleftrightarrow{AB}$. (If C is on the opposite side, choose a point C' on this side of

$\overleftrightarrow{AB}$ such that $\triangle ABC' \cong \triangle ABC$ and apply the remainder of the proof to $\triangle ABC'$.)
The definition of d ensures that A and C lie on the same side of $\overleftrightarrow{PQ}$ and on the same side of $\overleftrightarrow{QB}$. It follows that C is in the interior of each of the interior angles of $\triangle QPB$. Choose a point D on $\overline{PQ}$ such that $D * C * B$ (Crossbar Theorem).

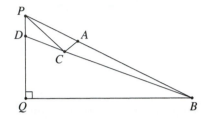

FIGURE 8.33: Application of Additivity of Defect in the proof of Theorem 8.6.4

Several applications of Additivity of Defect give

$$\delta(\triangle PQB) = \delta(\triangle QBD) + \delta(\triangle ABC) + \delta(\triangle ACP) + \delta(\triangle PCD)$$

(see Fig. 8.33). Therefore, $\delta(\triangle ABC) < \delta(\triangle PQB) < \epsilon$ and the proof is complete. $\square$

We now prove that the defect varies continuously. Since the domain of the defect function consists of the collection of triangles, we must specify what we mean by continuity in this context. The following construction makes that precise.

Construction. Let $\triangle ABC$ be a triangle and let $c = AB$. For each number $x \in [0, c]$ there is a unique point P_x on $\overline{AB}$ such that $AP_x = x$ (Ruler Postulate). For $x > 0$, define $f(x) = \delta(\triangle AP_xC)$. Note that $P_0 = A$ and $P_c = B$, so $f(c) = \delta(\triangle ABC)$. Define $f(0) = 0$.

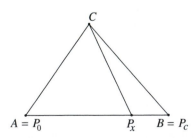

FIGURE 8.34: Continuity Construction

Theorem 8.6.5 (Continuity of Defect). *The function $f : [0, c] \to \mathbb{R}$ defined above is a continuous function.*

Proof. We assume the notation of the construction that precedes the statement of the theorem. Here is what we must prove: For every positive number ϵ and for every number $x \in [0, c]$, there exists a positive number d such that $|f(x) - f(y)| < \epsilon$ whenever $0 < |x - y| < d$.

Let $\epsilon > 0$ and x be given. Apply Theorem 8.6.4 with $A = P_x$ and $B = C$. The output is a number $d > 0$ with the property that if $P_x P_y < d$, then $\delta(\triangle CP_x P_y) < d$. The proof will be completed by demonstrating that this number d has the properties specified in the previous paragraph.

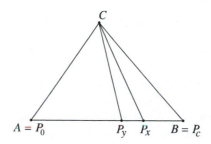

FIGURE 8.35: Proof of the Continuity of Defect

Suppose y is a number in $[0, c]$ such that $|x - y| < d$. Then $P_x P_y < d$ and so $\delta(\triangle CP_x P_y) < \epsilon$ (choice of d). Now $|f(x) - f(y)| = |\delta(\triangle AP_xC) - \delta(\triangle AP_yC)| = \delta(\triangle CP_x P_y)$ (Additivity of Defect). Hence $|f(x) - f(y)| < \epsilon$.   ☐

It should be noted that the proof of the continuity of defect does not rely on the continuity of the critical function.

The final theorem of the section combines the previous results to show that the defect function is onto.

Theorem 8.6.6. *For every* $y \in (0, 180)$ *there exists a triangle* $\triangle ADC$ *such that* $\delta(\triangle ADC) = y°$.

Proof. Let $y \in (0, 180)$ be given. Since $y < 180$, Theorem 8.6.2 implies that there exists a triangle $\triangle ABC$ such that $\delta(\triangle ABC) > y°$. By Theorem 8.6.5 and the Intermediate Value Theorem, there exists a point D on $\overline{AB}$ such that $\delta(\triangle ADC) = y°$.   ☐

8.7 IS THE REAL WORLD HYPERBOLIC?

The study of a geometry that is different from what we were taught in school presents a challenge to some of our preconceptions. We have all been brought up on Euclidean geometry and so we tend to assume that the world is Euclidean. We believe that Euclidean geometry is the science of space, that it tells us the facts about spatial relationships in the real world. Some of the theorems of hyperbolic geometry contradict those of Euclidean geometry and so it is not possible for both geometries to be true descriptions of the world. Which one is correct? Does hyperbolic geometry have any practical use the way Euclidean geometry does? These and other questions naturally come to mind when we first encounter hyperbolic geometry. In this section we engage in some speculation about what a non-Euclidean world would be like.

Before we can apply the theorems of geometry to the real world, we must agree on physical interpretations for the undefined terms. For purposes of this discussion we will interpret *point* to be a location in space and *line* to be the path followed by a ray of light. This interpretation of line has the advantage that, whatever a line is, it will definitely look straight to us.

The fact is that our prejudice in favor of Euclidean geometry as the true description of the world is probably confirmed by the theorems of this chapter, some of which appear to contradict our experience. For example, the Angle-Angle-Angle triangle congruence condition implies that you cannot change the size of an object without changing its shape. But we do this all the time when we take photographs of things and print them up at whatever scale we choose. That would be impossible in a hyperbolic world. We should be cautious, however, about concluding too much from our apparent ability to rescale things. Our experience all takes place on a rather small scale compared to the overall size of the universe, so we do not really have any experience with what happens when we try to make a drastic increase in size.

A major area of application of Euclidean geometry is the use of trigonometry in navigation and surveying. The entire enterprise is based on the assumption that every triangle has angle sum 180°. So a result from hyperbolic geometry that would have practical implications is the theorem about angle sums of triangles. In fact, angle sums offer a way to determine experimentally whether our world is Euclidean or hyperbolic: We could carefully measure the angles of a physical triangle and check to see whether or not the measures add up to 180°. This very experiment is said to have been carried out by Carl Friedrich Gauss in the eighteenth century. Gauss was employed by the government to conduct a geodesic survey of the State of Hanover. While doing that he attempted to measure the angles of a triangle that had its vertices at the peaks of three mountains. The experiment was inconclusive; the angle sum turned out to be so close to 180° that it was impossible to determine whether the difference was due to inaccuracies in the measurements or was real.

Any such attempt to demonstrate experimentally that the world is Euclidean is bound to fail. Physical measurements are never exact and so we would never get an angle sum of precisely 180°. Even if the world were hyperbolic, Theorem 8.6.3 tells us that a triangle might have to be quite large before its defect would be measurable. The theorem does not give any hint as to just how large the triangle would have to be, but it is reasonable to guess that its size might have to be significant on an astronomical scale. Hence there is very little chance of our being able to measure the defect of a triangle, even if defective triangles exist. So it is safe to continue to use Euclidean geometry for most practical purposes in everyday life because, on small scales, the world is approximately Euclidean regardless of what it turns out to be on a large scale.

Another practical implication of hyperbolic geometry involves distance. In hyperbolic geometry there is an absolute standard of measurement for distances. One way to define such an absolute standard for distance is to choose a length x_0 such that $\kappa(x_0) = 45$. (There is such a length by the continuity of the critical function and the Intermediate Value Theorem.) A different method that could be used to define an absolute standard of measurement is to choose a triangle $\triangle ABC$ such that every angle in $\triangle ABC$ is a 30° angle. (See Exercise 8.16.) The length of one side is uniquely determined (AAA), so the length of the side can be used as an absolute unit of measurement. As a result it would not be necessary for the National Bureau of Standards to maintain a standard for length the way it does now. A unit of length would be built into nature and it could be used as the absolute standard of reference.

This last point is not quite as strange as it first sounds. Something much like that is already the case for angle measure. It is not necessary for any government agency to keep a standard degree or standard radian in storage somewhere. Rather, anyone who needs to can construct those standards of measure to whatever degree of accuracy is required. There is an absolute standard for 360° built into space and fractions of 360° can be constructed to whatever degree of accuracy is needed. Euclidean geometry has only a standard of measure for angles built into it while hyperbolic geometry comes with an absolute standard for both angle measure and length built in. From that point of view it is Euclidean geometry that is deficient.

In Chapter 14 we will attempt to give answers to the questions raised in the first paragraph of this section. Here is a preview of the answer: The simple answer to the question in the title of the section is that the geometry of the real world is not hyperbolic. But we can't just say that it is Euclidean either. The geometry of a small part of space is very closely approximated by Euclidean geometry. At an intermediate scale, the geometry changes from one part of space to another. In some places space is nearly Euclidean, while at other locations it is more hyperbolic or spherical At the same time there is one overall geometry that describes things on a large scale. Surprisingly, the exact nature of that overall, large-scale geometry of the physical universe has not yet been determined. Whether it is Euclidean, hyperbolic, or elliptic is an open question.

SUGGESTED READING

1. Part III, pages 93–150, of *Euclid's Window*, [50].
2. Chapter XXVI of *Mathematics in Western Culture*, [40].
3. Chapters 5 and 6 of *The Non-Euclidean Revolution*, [70].

EXERCISES

8.1. Prove the following theorem in neutral geometry: *If there exists a constant c and model for neutral geometry such that the defect of every triangle in that model is c, then c = 0°.* Use this result to prove that triangles in hyperbolic geometry cannot all have the same defect.

8.2. Prove Corollary 8.2.9.

8.3. Prove Corollary 8.2.10.

8.4. Prove Theorem 8.3.3.

8.5. Prove Theorem 8.3.4.

8.6. Complete the proof of Theorem 8.3.5.

8.7. Prove Theorem 8.4.9.

8.8. Prove Theorem 8.4.10.

8.9. Prove Theorem 8.4.12.

8.10. Prove Theorem 8.4.15, Part 1.

8.11. Prove Theorem 8.4.15, Part 2.

8.12. Use Additivity of Defect to prove the claim

$$\delta(\triangle P_n B_n P_0) > \delta(\triangle P_1 Q_1 P_0) + \delta(\triangle P_2 Q_2 P_1) + \cdots + \delta(\triangle P_n Q_n P_{n-1})$$

in the proof of Theorem 8.5.3.

8.13. Let $\triangle ABC$ be a triangle and let D, E, and F be the midpoints of the sides $\overline{BC}$, $\overline{AC}$, and $\overline{AB}$, respectively.

    (a) Prove that $\triangle EDC$ is not similar to $\triangle ABC$.

    (b) Prove that the congruences $\overline{AF} \cong \overline{ED}$, $\overline{AE} \cong \overline{FD}$, and $\overline{BD} \cong \overline{EF}$ cannot all hold.

    (c) Compare the results above with the Euclidean theorems about medial triangles in Exercise 7.18.

8.14. It follows from Theorem 5.7.10 (in neutral geometry) that if a point is in the interior of a segment joining one side of an angle to the other side of the angle, then it is in the interior of the angle.

    (a) Prove that the converse is false in hyperbolic geometry (i.e., there exists an angle $\angle BAC$ and a point P in the interior of $\angle BAC$ such that P does not lie on any segment joining a point of $\overrightarrow{AB}$ to a point of $\overrightarrow{AC}$).

    (b) Prove that the converse is true in Euclidean geometry (i.e., for every angle $\angle BAC$ and for every point P in the interior of $\angle BAC$ there exist points B' on $\overrightarrow{AB}$ and C' on $\overrightarrow{AC}$ such that P is in the interior of $\overline{B'C'}$).

8.15. Prove that for every $\epsilon > 0$ there exists a rhombus whose angle sum is less than $\epsilon°$.

8.16. A triangle is *equilateral* if the three sides are congruent. This exercise will show that the angle sum of an equilateral decreases to zero as the triangle grows larger.

    (a) Fix two distinct points P and A. Prove that there are points B and C such that $PA = PB = PC$ and $\mu(\angle APB) = \mu(\angle BPC) = \mu(\angle CPA) = 120°$.

    (b) Prove that $\triangle ABC$ is equilateral.

    (c) Choose points A', B', and C' on the rays $\overrightarrow{PA}$, $\overrightarrow{PB}$, and $\overrightarrow{PC}$, respectively, such that $PA' = PB' = PC' > PA$. Prove that $\sigma(\triangle A'B'C') < \sigma(\triangle ABC)$.

    (d) Prove that $A'B' > PA'$ so that the side length of $\triangle A'B'C'$ approaches ∞ as PA' approaches ∞.

    (e) Prove that
$$\lim_{PA' \to \infty} \sigma(\triangle A'B'C') = 0.$$

    (f) Prove that
$$\lim_{PA' \to 0} \sigma(\triangle A'B'C') = 180.$$

    (g) Prove that for every equilateral triangle $\triangle DEF$ there exist points A', B', C' on rays $\overrightarrow{PA}$, $\overrightarrow{PB}$, and $\overrightarrow{PC}$, respectively, such that $\triangle A'B'C' \cong \triangle DEF$.

8.17. Let a distance $D_0 > 0$ and a defect δ_0, $0 < \delta_0 < 180$, be given. Construct an isosceles triangle $\triangle ABC$ such that each side of $\triangle ABC$ has length greater than D_0 and $\delta(\triangle ABC) < \delta_0$. (Contrast this with the previous exercise and with Theorem 8.6.3; the defect of a triangle can be small even though every side of the triangle is large.)

8.18. Suppose ℓ is a line, R is an external point for ℓ, and n is a line such that R lies on n and $n \parallel \ell$. Drop a perpendicular from R to ℓ and call the foot F. Pick a point $Q \neq R$ on n and drop a perpendicular from Q to a point A on ℓ. (Refer to Figure 8.25.) Prove that the limiting parallel ray for R and $\overrightarrow{FA}$ must intersect $\overline{QA}$.

8.19. Suppose ℓ is a line, R is an external point, and $\overrightarrow{RD}$ and $\overrightarrow{RE}$ are the two limiting parallel rays for ℓ at R. Let T be a point on $\overrightarrow{RD}$ and S a point on $\overrightarrow{RE}$ such that $RS = RT$. (Refer to Figure 8.25.) Prove that $d(S, \ell) = d(T, \ell)$.

8.20. Fill in the details in the following outline of the proof of Theorem 8.5.5. Let $\angle BAC$ be given. The objective is to find a point P on $\overrightarrow{AC}$ and a point Q such that $\overrightarrow{AB} \mid \overrightarrow{PQ}$. For each $x \geq 0$ there exists a point P_x on $\overrightarrow{AC}$ such that $AP_x = x$. Let ℓ_x be the line such that P_x lies on ℓ_x and $\ell_x \perp \overleftrightarrow{AC}$. Define $S = \{x \geq 0 \mid \ell_x \cap \overrightarrow{AB} \neq \varnothing\}$. Claim: $S \neq \varnothing$ and S has an upper bound. To prove that S has an upper bound, use the fact that there is a positive number b such that $\kappa(b) < \mu(\angle BAC)$ and therefore $\ell_b \cap \overrightarrow{AB} = \varnothing$. Let a be the least upper bound of S, let $P = P_a$, and let Q be a point on ℓ_a such that Q lies on the same side of $\overleftrightarrow{AC}$ as B.

The proof is completed by showing that $\overrightarrow{AB} \mid \overrightarrow{PQ}$. Suppose this is not the case. It is easy to see that $\overrightarrow{AB}$ cannot intersect $\overrightarrow{PQ}$, so it must be that the measure of the angle of parallelism for A and $\overrightarrow{PQ}$ is smaller than $\mu(\angle PAB)$. Choose D such that $\overrightarrow{AD} \mid \overrightarrow{PQ}$. Use Theorem 8.4.12 and the fact that $d(P, \overleftrightarrow{AB}) > d(P, \overleftrightarrow{AD})$ to show that there exists a ray $\overrightarrow{PR}$ between $\overrightarrow{PQ}$ and $\overrightarrow{PA}$ such that $\overrightarrow{PR} \cap \overrightarrow{AB} = \varnothing$. Drop a perpendicular from R to $\overleftrightarrow{AC}$ and call the foot F. Note that $AF < AP = a$. Prove that $\ell_x \cap \overrightarrow{AB} = \varnothing$ for every x between AF and a. This contradicts the choice of a. □

8.21. Prove the following generalization of Lemma 5.7.26 to open intervals. *If $f : (a, b) \to (c, d)$ is strictly decreasing and onto, then f is continuous.*

8.22. Use the previous exercise and Theorem 8.5.5 to prove Corollary 8.5.6.

CHAPTER 9

Area

In this chapter we finally introduce the last undefined term: area. Like distance and angle measure, area is a measurement, which means that area is a function that assigns a real number to a geometric object. The basic axioms for area are valid in neutral geometry, but area behaves quite differently in Euclidean and hyperbolic geometries. As a result we will study area separately in the two geometries. In Euclidean geometry we will see that area can be used to give much simpler proofs of such theorems as the Pythagorean Theorem. We will also prove in both geometries that two regions of the same area can be cut up into congruent pieces. The study of that "dissection" problem leads to other interesting mathematics; in particular, it leads naturally to a study of the relationship between area and defect in hyperbolic geometry.

Euclid uses area right from the beginning of his treatment of geometry. He simply says that two triangles are equal when he means that they have the same area. In the next section we introduce the concept of area. This allows us, in the section after that, to fill in some of the theorems from Book I that were omitted in our earlier treatment. In particular, we follow Euclid in giving a proof of the Pythagorean Theorem that is based on the use of area. The rest of the chapter is devoted to a study of dissection theory. While this subject is not treated explicitly in Euclid, it is very closely related to the area constructions in Books I and II.

We assume that the existence of the area function is given to us as an axiom. Of course it is not necessary to treat area axiomatically; it is possible to define area and to construct an area function in both Euclidean and hyperbolic geometries without assuming any special axioms. There are, however, several reasons why we prefer to treat area axiomatically.

1. The axiomatic approach allows us to see that area, distance, and angle measure are in many ways alike.

2. The axiomatic approach allows us to give a unified treatment of area that applies to both Euclidean and hyperbolic geometries. If the area function is constructed, it must be constructed in quite different ways in the two geometries; the axiomatic approach allows us to emphasize the common features.

3. The axiomatic approach allows us to avoid certain technical difficulties that arise in showing that the area functions in the two geometries are well defined.

4. The axiomatic approach is consistent with that of most high school textbooks. In particular, both [64] and [71] treat area axiomatically.

The possibility of taking two different approaches is not unique to area. Distance and angle measure are also functions that can either be constructed or treated axiomatically.

In this chapter we only consider areas of polygonal regions. The limiting process of calculus must be used to generalize to other regions. The only such generalization we attempt in this book is to define area for circles in the next chapter. It is worthwhile to note that the kind of process that allows us to define and compute areas of more general regions is exactly the same as the kind of generalization that is used for lengths of curves. In this text we define lengths of straight segments. An obvious generalization extends the notion of length to polygonal paths. A limiting process is then needed to extend the definition to lengths for more general paths.

9.1 THE NEUTRAL AREA POSTULATE

Before we can state the axiom for area, we must carefully specify the kinds of objects that are measured by the area function.

Definition 9.1.1. Let $\triangle ABC$ be a triangle. The *interior of* $\triangle ABC$, denoted $\text{Int}(\triangle ABC)$, is the intersection of the interiors of the three interior angles $\angle ABC$, $\angle BCA$, and $\angle CAB$.

The interior of a triangle is a convex set since it is defined as the intersection of convex sets. It includes the points that we intuitively think of as being inside the triangle but does not include the points on the triangle itself.

Definition 9.1.2. Let $\triangle ABC$ be a triangle. The *associated triangular region* is the subset T of the plane consisting of all points that lie on the triangle or in its interior; that is,

$$T = \triangle ABC \cup \text{Int}(\triangle ABC).$$

Another way to say this is that the triangular region is the set consisting of the vertices, points on the edges, and points in the interior of the triangle. The vertices and edges of the triangle are also called the vertices and edges of the triangular region.

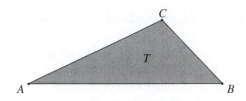

FIGURE 9.1: A triangular region

It is important to understand that the triangle and the associated triangular region are not the same things. A triangular region is the set of points that includes not only the points on the triangle itself but also the points that lie inside the triangle. We can draw on our intuitive experience with dimension and area to understand the difference: the triangle is one dimensional and has zero area while the triangular region is two dimensional and has positive area.

There is a special notation, $\blacktriangle ABC$, that we will occasionally use for the triangular region associated with the triangle $\triangle ABC$. This notation is helpful when it is necessary to make the distinction clear. We will use that notation sparingly, however. The distinction between triangle and triangular region must be made as we begin our study of area, but it soon becomes burdensome and awkward to continue stressing it. After the danger of confusion has passed, we will often drop the phrase "the triangular region associated with" and just refer to the triangle. It should be clear from the context whether we really mean the triangle or the triangular region.

Definition 9.1.3. A *polygonal region* is a subset R of the plane that can be written as the union of a finite number of triangular regions in such a way that if two of the triangular regions intersect, then the intersection is contained in an edge of each.

Notice that every triangular region is a polygonal region but not the other way round. Polygonal regions are not necessarily convex sets.

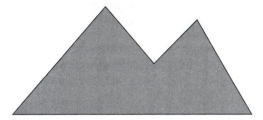

FIGURE 9.2: A polygonal region

Definition 9.1.4. A *triangulation* of a polygonal region is a decomposition of that region into triangles in such a way that the conditions in the definition of triangular region are satisfied. Specifically, a triangulation of a polygonal region R is a collection of triangular regions $T_1, T_2, \dots, T_n$ such that $R = T_1 \cup T_2 \cup \cdots \cup T_n$ and for each

$i \neq j$ either $T_i \cap T_j$ is empty or $T_i \cap T_j$ is a subset of an edge of T_i and a subset of an edge of T_j.

One polygonal region can have many different triangulations.

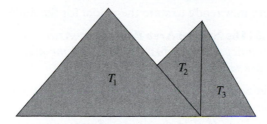

FIGURE 9.3: One triangulation of the region

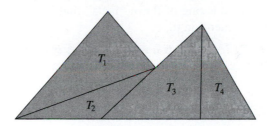

FIGURE 9.4: A different triangulation of the same region

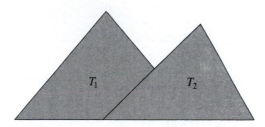

FIGURE 9.5: Not a triangulation (T_1 and T_2 overlap)

Remark. The definition of triangulation given here is slightly nonstandard. In other contexts it is often required that two triangular regions that intersect do so in exactly a vertex or an edge of each. The weaker condition assumed here, that the intersection should be a subset of an edge of each, is easier to achieve and is all we need for what we want to do. The stronger condition can always be attained by adding extra vertices and subdividing the given triangles into smaller triangles.

Two triangular regions intersect in a subset of an edge of each if and only if their interiors are disjoint.

Definition 9.1.5. Two polygonal regions R_1 and R_2 are *nonoverlapping* if $R_1 \cap R_2$ consists only of subsets of edges of each. Specifically, if T_1 is one of the triangular regions in R_1 and T_2 is one of the triangular regions in R_2, then Int $T_1 \cap T_2 = \emptyset$ and $T_1 \cap$ Int $T_2 = \emptyset$.

We are now ready to state the axiom for the undefined term area.

Axiom 9.1.6 (The Neutral Area Postulate). *Associated with each polygonal region R there is a nonnegative number $\alpha(R)$, called the area of R, such that the following conditions are satisfied.*

1. *(Congruence) If two triangles are congruent, then their associated triangular regions have equal areas.*
2. *(Additivity) If R is the union of two nonoverlapping polygonal regions R_1 and R_2, then $\alpha(R) = \alpha(R_1) + \alpha(R_2)$.*

The following technical result is often needed in connection with the Neutral Area Postulate. It is one of those "facts" that is so obvious from the diagram that we often do not mention it.

Theorem 9.1.7. *If $\triangle ABC$ is a triangle and E is a point on the interior of $\overline{AC}$, then $\blacktriangle ABC = \blacktriangle ABE \cup \blacktriangle EBC$. Furthermore, $\blacktriangle ABE$ and $\blacktriangle EBC$ are nonoverlapping regions.*

Proof. Exercise 9.2. □

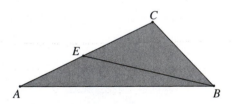

FIGURE 9.6: A triangular region split into two nonoverlapping triangular regions (Theorem 9.1.7)

9.2 AREA IN EUCLIDEAN GEOMETRY

The Neutral Area Postulate is not strong enough to uniquely determine areas of regions. In order to do that we must specify at least some areas. The natural place to start is with a rectangle and to define its area to be the length times the width. This will work in Euclidean geometry but won't do much good in hyperbolic geometry, where there are no rectangles. In this section we add a postulate specifying the area of a rectangle and explore its consequences for Euclidean geometry. In later sections of the chapter we will look at area in hyperbolic geometry. For the remainder of this section we assume the Euclidean Parallel Postulate.

We want to identify a polygonal region associated with a rectangle $\square ABCD$. By Theorem 6.7.7, $\square ABCD$ is convex and so the diagonals $\overline{AC}$ and $\overline{BD}$ intersect in a point (Theorem 6.7.9).

Definition 9.2.1. Let $\square ABCD$ be a rectangle and let E be the point at which its diagonals intersect. The *rectangular region* associated with $\square ABCD$ is the polygonal region R that is the union of the triangular regions corresponding to the four triangles $\triangle ABE$, $\triangle BCE$, $\triangle CDE$, and $\triangle ADE$. The *length* of R is AB and the *width* is BC.

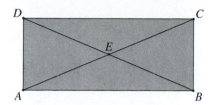

FIGURE 9.7: A rectangular region

Which side determines the length and which determines the width depends on how the vertices are labeled.

Axiom 9.2.2 (The Euclidean Area Postulate). *If R is a rectangular region, then $\alpha(R) = length(R) \times width(R)$.*

Once the area of a rectangle is specified it is easy to determine the area of a triangle. We begin with a right triangle.

Theorem 9.2.3. *If T is the triangular region corresponding to the right triangle $\triangle ABC$ with right angle at C, then $\alpha(T) = (1/2)AC \times BC$.*

Proof. Exercise 9.4. □

We want to generalize Theorem 9.2.3 to triangles that are not necessarily right.

Definition 9.2.4. Let T be a triangular region corresponding to $\triangle ABC$. The *base* of T is $\overline{AB}$. Drop a perpendicular from C to $\overleftrightarrow{AB}$ and call the foot of that perpendicular D. The *height* of T is the length of $\overline{CD}$.

Note that a given triangular region has three different bases since any one of the sides of the corresponding triangle can serve as the base. The triangular region can also have three different heights, depending on how the triangle is oriented.

Theorem 9.2.5. *The area of a triangular region is one-half the length of the base times the height; that is,*

$$\alpha(T) = (1/2) \text{ base } (T) \times \text{ height } (T).$$

Proof. Exercise 9.5. □

The Neutral Area Postulate assigns a unique value to the area of the triangle. Hence the Neutral Area Postulate and the Euclidean Area Postulate together imply that we will obtain the same value for $(1/2)$ base $(T) \times$ height (T) regardless of which side we use as the base of our triangle. This can also be proved directly from the Similar Triangles Theorem—see Exercise 9.6.

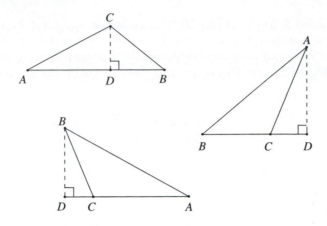

FIGURE 9.8: Three different heights for the same triangle

Terminology. In order to avoid unnecessary wordiness we will usually refer to the "area of a triangle" when what we really mean is the area of the corresponding triangular region. We take similar liberty with "area of a quadrilateral," "area of a square," "area of a rectangle," and so on.

The next theorem shows that area (in Euclidean geometry) varies as the square of length.

Theorem 9.2.6. *If two triangles are similar, then the ratio of their areas is the square of the ratio of the lengths of any two corresponding sides; i.e., if $\triangle ABC \sim \triangle DEF$ and $DE = r \cdot AB$, then $\alpha(\triangle DEF) = r^2 \cdot \alpha(\triangle ABC)$.*

Proof. Exercise 9.7. □

The use of area can greatly simplify some of the proofs in Euclidean geometry. To illustrate this point we present several proofs of the Pythagorean Theorem. It should be made clear, however, that the use of the area concept in the Pythagorean Theorem is not just a matter of convenience for the sake of the proof. As far as Euclid was concerned, the theorem is all about areas. Euclid would not have recognized our earlier statement of the Pythagorean Theorem (Theorem 7.5.1) in which the conclusion was stated as an algebraic equation.

Definition 9.2.7. A *square* is a quadrilateral that is both a rectangle and a rhombus. Given a segment $\overline{AB}$, a *square on* $\overline{AB}$ is a square $\square ABED$ that has $\overline{AB}$ as one side.

Since the vertices D and E can be chosen on either side of $\overleftrightarrow{AB}$, there are two squares on a given segment $\overline{AB}$. It is obvious that these two squares have the same area, so there is usually no need to distinguish between them. Given a segment $\overline{AB}$, it is easy to construct the square on $\overline{AB}$: simply choose one side of $\overleftrightarrow{AB}$, erect perpendiculars on that side at each endpoint of the segment (Protractor

Postulate) and then measure off segments $\overline{BE}$ and $\overline{AD}$ of the appropriate length (Ruler Postulate).

The following version of the Pythagorean Theorem is the penultimate Proposition in Book I of Euclid's *Elements*. Euclid concludes Book I with the converse to this theorem.

Theorem 9.2.8 (Euclid's version of the Pythagorean Theorem). *The area of the square on the hypotenuse of a right triangle is equal to the sum of the areas of the squares on the legs.*

As mentioned above, we will present several proofs of the theorem, all of them based on the concept of area. We can do no better than to start with Euclid's proof.

Euclid's proof of Theorem 9.2.8. Let $\triangle ABC$ be a triangle with right angle at C. We need to construct some points first, before we can describe the proof. Choose points D and E, on the opposite side of $\overleftrightarrow{AB}$ from C, such that $\square ABED$ is a square. Choose points F and G, on the opposite side of $\overleftrightarrow{BC}$ from A, such that $\square BFGC$ is a square. Finally choose points H and I, on the opposite side of $\overleftrightarrow{AC}$ from B, such that $\square ACHI$ is a square. We must show that $\alpha(\square ABED) = \alpha(\square BFGC) + \alpha(\square ACHI)$.

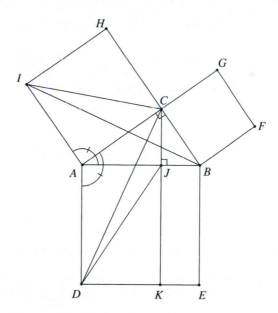

FIGURE 9.9: Euclid's proof of the Pythagorean Theorem

Drop a perpendicular from C to $\overleftrightarrow{AB}$ and call the foot J. Let K be the point where $\overleftrightarrow{CJ}$ intersects $\overline{DE}$. The next two paragraphs will show that $\alpha(\square ACHI) = \alpha(\square AJKD)$ and $\alpha(\square BFGC) = \alpha(\square KEBJ)$. It is clear from the Euclidean Area

Postulate that $\alpha(\square ABED) = \alpha(\square AJKD) + \alpha(\square KEBJ)$, so this will complete the proof of the theorem.

Since $\angle ACB$ is a right angle, H lies on $\overleftrightarrow{BC}$ and $\overleftrightarrow{BC} \parallel \overrightarrow{IA}$. Hence $\alpha(\triangle IAC) = \alpha(\triangle IAB)$. (Both have base $\overline{IA}$ and height AC, so Theorem 9.2.5 applies.) Now $\angle IAB \cong \angle CAD$ (because each is the sum of a right angle and $\angle BAC$). Hence $\triangle IAB \cong \triangle CAD$ (SAS) and so $\alpha(\triangle IAB) = \alpha(\triangle CAD)$ (Neutral Area Postulate). Finally, $\alpha(\triangle CAD) = \alpha(\triangle JAD)$ because the two triangles share base $\overline{AD}$ and both have height AJ. Combining the statements above gives $\alpha(\triangle IAC) = \alpha(\triangle JAD)$. But $\alpha(\triangle IAC) = (1/2)\alpha(\square ACHI)$ and $\alpha(\triangle ADJ) = (1/2)\alpha(\square ADKJ)$, so $\alpha(\square ACHI) = \alpha(\square ADKJ)$ as required.

The proof that $\alpha(\square BFGC) = \alpha(\square KEBJ)$ is similar. $\square$

Euclid's proof just begs to be made into a movie and many people have done that. There are various applets on the World Wide Web that animate the proof. A search of the web will probably turn up several. It is instructive to watch such a proof animation because it clearly shows the motion that takes $\triangle ACI$ to $\triangle AJD$. This motion consists of a shear followed by a rotation about A and then another shear. Each of these motions preserves the area of the triangle.

Our second proof of Theorem 9.2.8 is usually attributed to the 12th century Indian mathematician Bhaskara [30, Vol. 1, page 355].

Second proof of Theorem 9.2.8. Let $\triangle ABC$ be a right triangle with right angle at vertex C. As usual, a denotes the length of side $\overline{BC}$, b denotes the length of side $\overline{AC}$, and c denotes the length of side $\overline{AB}$. Label the vertices in such a way that $a \leqslant b$. (If this is not the case, simply interchange the labels on vertices A and B.)

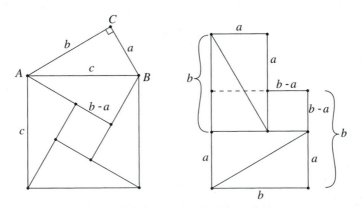

FIGURE 9.10: Bhaskara's proof of the Pythagorean Theorem

As illustrated in Fig. 9.10, the area of the square on $\overline{AB}$ is equal to 4 times the area of $\triangle ABC$ plus the area of a square whose side has length $b - a$. Translating that observation into an algebraic equation gives

$$c^2 = 4 \times \frac{1}{2}(ab) + (b - a)^2 = 2ab + b^2 - 2ab + a^2 = a^2 + b^2.$$

$\square$

Variation on second proof of Theorem 9.2.8. The last equation in the proof above may also be seen geometrically. The four triangles and one square can be rearranged to exactly cover an $a \times a$ square and a $b \times b$ square—see the right hand part of Fig. 9.10. $\square$

The next proof is of uncertain, but ancient, origin. It is difficult to know for certain whether or not the historical person Pythagoras actually discovered a rigorous proof of the theorem that is named for him, but there is at least some evidence that Pythagoras himself may have used a proof like the next one [30, Vol. 1, page 354].

Third proof of Theorem 9.2.8. We use the same notation as in the preceding proof. Construct a square whose side has length $a + b$ and place a copy of $\triangle ABC$ at each vertex as shown. The area not covered by the triangles is a $c \times c$ square. (The Euclidean angle sum theorem is required to prove that the corners of the shaded region are right angles.) The triangles can then be moved so that the area not covered consists of two square, one $a \times a$ and the other $b \times b$. $\square$

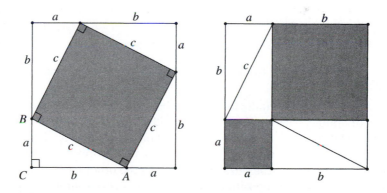

FIGURE 9.11: Third proof of the Pythagorean Theorem

Our final proof of the Pythagorean Theorem is again taken from Euclid's *Elements*, but this time from Book VI. It is in many ways the most beautiful and decisive proof of all (see [30, Vol. 1, page 350]). The proof is based on the following proposition, which is Proposition VI.31 in the *Elements*, and is actually a significant generalization of the Pythagorean Theorem. We will allow ourselves to use some algebraic notation to express the relationships in the proof more succinctly even though Euclid described the entire argument in purely geometric terms.

Proposition 9.2.9. *Let $\triangle ABC$ be a right triangle with right angle at C. If D, E, and F are points such that the triangles $\triangle ABD$, $\triangle BCE$, and $\triangle CAF$ are similar, then*

$$\alpha(\triangle ABD) = \alpha(\triangle BCE) + \alpha(\triangle CAF).$$

Proof. The proof has two steps. First we show that if the conclusion of the Proposition holds for one set of similar triangles built of the sides of $\triangle ABC$, then it holds for every such set. Second, we exhibit one set of triangles for which the conclusion holds.

Suppose D, E, and F are points such that the triangles $\triangle ABD$, $\triangle BCE$, and $\triangle CAF$ are similar. Let α_0 denote the area of $\triangle ABD$. By Theorem 9.2.6, $\alpha(\triangle BCE) = (a/c)^2\alpha_0$ and $\alpha(\triangle CAF) = (b/c)^2\alpha_0$. Hence the equation $\alpha(\triangle ABD) = \alpha(\triangle BCE) + \alpha(\triangle CAF)$ can be rewritten as $\alpha_0 = (a/c)^2\alpha_0 + (b/c)^2\alpha_0$. Since $\alpha_0 > 0$, the last equation holds if and only if $c^2 = a^2 + b^2$.

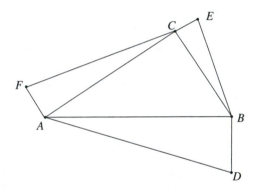

FIGURE 9.12: The three small triangles are similar

We have shown the following: The conclusion of the theorem holds for a given set of similar triangles based on the sides of $\triangle ABC$ if and only if $a^2 + b^2 = c^2$. The relationship $a^2 + b^2 = c^2$ either holds for the triangle $\triangle ABC$ or it does not. If $a^2 + b^2 = c^2$, then the conclusion of the Proposition is true for every set of similar triangles on the sides of $\triangle ABC$. On the other hand, if $a^2 + b^2 \neq c^2$, then the conclusion of the proposition is false for every set of similar triangles on the sides of $\triangle ABC$.

To complete the proof of the proposition, we must exhibit one set of similar triangles for which the relationship $\alpha(\triangle ABD) = \alpha(\triangle BCE) + \alpha(\triangle CAF)$ is valid. Drop a perpendicular from C to $\overleftrightarrow{AB}$ and call the foot of that perpendicular G. As in the proof of Theorem 7.5.1, $\triangle ABC \sim \triangle CBG \sim \triangle ACG$. Hence we can take $F = E = G$ and $D = C$ to form a set of similar triangles for which the conclusion of the proposition holds. (See Fig. 9.13.) $\square$

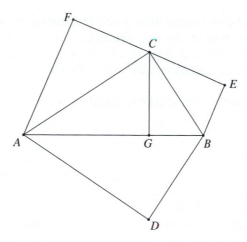

FIGURE 9.13: Flip the right triangles across the edges of $\triangle ABC$

Fourth proof of Theorem 9.2.8. Let $\triangle ABC$ be a right triangle. By Proposition 9.2.9, the area of the right-angled isosceles triangle on the hypotenuse of $\triangle ABC$ is equal to the sum of the areas of the right-angled isosceles triangles on the legs of $\triangle ABC$. The area of the right-angled isosceles triangle on a segment is half the area of the square on that segment. Hence the area of the square on the hypotenuse is equal to the sum of the areas of the squares on the legs. □

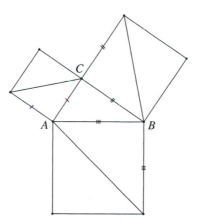

FIGURE 9.14: Fourth proof of the Pythagorean Theorem

This section contains five proofs of the Pythagorean Theorem (counting the two variations on Proof 2 as different proofs). In Chapter 7 we gave yet another proof based on the theory of similar triangles. Thus we have seen a total of six different proofs of the theorem. Even so we have barely scratched the surface;

over the centuries people have given literally hundreds of different proofs of the Pythagorean Theorem. In 1927 Elisha Loomis compiled 367 different proofs of the theorem and published them in a book [44]. Exercises 9.12 and 9.13 contain two additional proofs, including one by U.S. President James Garfield (1831–1881).

9.3   DISSECTION THEORY IN NEUTRAL GEOMETRY

After examining some of the consequences of the Euclidean Area Postulate in the last section, we now step back and look again at what we can do with just the Neutral Area Postulate. The two parts of the Neutral Area Postulate may be summarized in this way: If two polygonal regions can be decomposed into nonoverlapping triangular regions in such a way that there is a one-to-one correspondence between the triangles in one region and the triangles in the other and corresponding triangles are congruent, then the two regions have the same area.

The main objective of this section and the next two sections is to prove that the converse of the Neutral Area Postulate is a theorem in both Euclidean and hyperbolic geometries. We will prove that any two polygonal regions that have the same area can be decomposed into nonoverlapping triangular regions in such a way that there is a one-to-one correspondence between the triangles in the first region and the triangles in the second region, and corresponding triangles are congruent. The proof of the theorem is quite complicated and it will occupy most of three sections. In this section we will state the fundamental theorem and examine the part of the proof that is valid in neutral geometry. The second part of the proof requires different arguments in Euclidean and hyperbolic geometries. The hyperbolic theorem is attributed to Bolyai; its proof nicely clarifies the relationship between defect and area.

Let us begin with a careful definition of the terminology and a statement of the problem.

Definition 9.3.1. Let R and R' be two polygonal regions. We say that R and R' are *equivalent by dissection* (written $R \equiv R'$) if there exist triangulations $R = T_1 \cup \cdots \cup T_n$ and $R' = T'_1 \cup \cdots \cup T'_n$ such that the two triangulations contain the same number of triangles and, for each i, the triangle corresponding to the triangular region T_i is congruent to the triangle corresponding to T'_i. See Fig. 9.15 for an example. Various other names are used for this relationship in the literature; two of the more descriptive terms are "scissors congruent" and "equidecomposable."

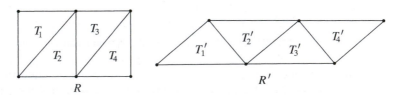

FIGURE 9.15: Equivalent by dissection

The process of cutting a region into triangles and then reassembling the the triangles to form a different region is usually called *dissection* and the theory of how and when it can be done is called *dissection theory*. We can now state the problem we will be addressing.

Dissection Problem. *Given two regions R and R′ with $\alpha(R) = \alpha(R′)$, find triangulations T and T′ which show that $R \equiv R′$.*

Dissection theory has been studied since ancient times by mathematicians of various cultures. The theory is quite rich and includes many specialized techniques for dissecting Euclidean regions. A few of them are described in the exercises at the end of the chapter. We will focus most of our attention on a proof of the following theorem.

Theorem 9.3.2 (Fundamental Theorem of Dissection Theory). *If R and R′ are two polygonal regions such that $\alpha(R) = \alpha(R′)$, then $R \equiv R′$.*

The statement of the theorem is deceptively simple. This is, in fact, a deep result that is not at all obvious. The reader is encouraged to work Exercise 9.14 [especially part (c)] at this time in order to make that point. Another indication of the depth of the theorem is the fact that the analogous result is not true in three dimensions. Max Dehn proved in 1902 that it is not possible to cut a tetrahedron into a finite number of polyhedral pieces and reassemble the pieces to form a cube of equal volume.[1]

We need the following technical facts about equivalence by dissection.

Theorem 9.3.3. *Equivalence by dissection is an equivalence relation; that is, $\equiv$ has the following properties.*

1. *(Reflexivity) $R \equiv R$ for every polygonal region R.*
2. *(Symmetry) If $R_1 \equiv R_2$, then $R_2 \equiv R_1$.*
3. *(Transitivity) If $R_1 \equiv R_2$ and $R_2 \equiv R_3$, then $R_1 \equiv R_3$.*

Sketch of proof. Reflexivity and symmetry are obvious from the definition, so we concentrate on transitivity.

Suppose R_1, R_2, and R_3 are three polygonal regions such that $R_1 \equiv R_2$ and $R_2 \equiv R_3$. We must prove that $R_1 \equiv R_3$. Fig. 9.16 illustrates the problem. The two equivalences correspond to two different triangulations of R_2. Let us say that the triangular regions in the first subdivision of R_2 are $T_1, \ldots, T_n$ while the triangular regions in the second subdivision are $T'_1, \ldots, T'_m$. In order to get one common subdivision we subdivide R_2 into regions of the form $T_i \cap T'_j$ where T_i and T'_j overlap. Each will be either a triangular region or a convex quadrilateral region. In the latter case, draw in a diagonal to subdivide the quadrilateral into two triangles.

The result of that operation is a subdivision of R_2 into small triangles such that each small triangle T''_k is contained in both a T_i and a T'_j. Since T''_k is contained in

---

[1] A nice exposition of Dehn's theorem and proof may be found in Chapter 5 of the book by Eves [25].

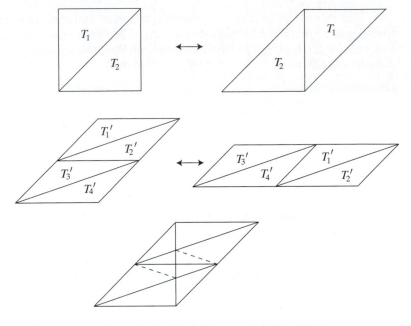

FIGURE 9.16: Proof of Theorem 9.3.3, Transitivity

T_i, there is a corresponding congruent triangle in R_1. Because T_k'' is contained in T_j', there is a corresponding congruent triangle in R_3. In this way the triangles T_k'' induce triangulations of R_1 and R_3 that show $R_1 \equiv R_3$. $\qquad\square$

Terminology. In order to simplify terminology, we will continue the trend towards not distinguishing between a triangle and the corresponding triangular region. In particular we will say that two triangular regions are congruent when what we really mean is that the triangles to which the regions correspond are congruent. We will also speak of triangles and quadrilaterals being equivalent by dissection when, strictly speaking, what we mean is that the corresponding regions are equivalent.

The following construction is basic to the proof of the Fundamental Theorem of Dissection Theory.

Construction. Let $\triangle ABC$ be a triangle. Let M be the midpoint of $\overline{AC}$ and let N be the midpoint of $\overline{BC}$. Drop perpendiculars from A and B to the line $\overleftrightarrow{MN}$ and name the feet of those two perpendiculars E and D, respectively. The next theorem asserts that the quadrilateral $\square ABDE$ is a Saccheri quadrilateral. Hence we will call $\square ABDE$ the *Saccheri quadrilateral associated with the triangle* $\triangle ABC$. Note that the Saccheri quadrilateral is determined by the order in which the vertices of the triangle are listed; if the vertices of the the triangle are listed in a different order, then a different quadrilateral results. Note too that the segment $\overline{AB}$ is the base of the triangle but the summit of the Saccheri quadrilateral. In order to keep the triangle right side up we are forced to draw the Saccheri quadrilateral upside down (base on top).

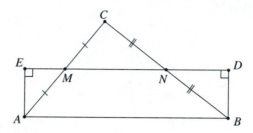

FIGURE 9.17: Construction of associated Saccheri quadrilateral

Theorem 9.3.4. *If $\triangle ABC$ is a triangle and $\square ABDE$ is the associated Saccheri quadrilateral, then $\square ABDE$ is a Saccheri quadrilateral with base $\overline{DE}$ and summit $\overline{AB}$.*

Proof. Let $\triangle ABC$ and $\square ABDE$ be as in the construction above. It is clear from the construction that angles $\angle BDE$ and $\angle AED$ are right angles. In order to show that $\square ABDE$ is a Saccheri quadrilateral we must prove that $\overline{AE} \cong \overline{BD}$. Drop a perpendicular from C to $\overleftrightarrow{MN}$ and call the foot F. Since $\overline{AM} \cong \overline{MC}$ (M is the midpoint) and $\angle AME \cong \angle CMF$ (Vertical Angles Theorem), we have that $\triangle AME \cong \triangle CMF$ (AAS). In a similar way we see that $\triangle BND \cong \triangle CNF$. It follows that $\overline{AE} \cong \overline{CF} \cong \overline{BD}$ and so $\square ABDE$ is a Saccheri quadrilateral. $\square$

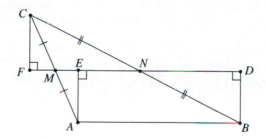

FIGURE 9.18: Another possible diagram of the triangle and its associated Saccheri quadrilateral

What we really want to prove is that the Saccheri quadrilateral we have constructed is equivalent by dissection to the triangle. The following theorem is the main result of the section.

Theorem 9.3.5. *If $\triangle ABC$ is a triangle and $\square ABDE$ is the associated Saccheri quadrilateral, then $\triangle ABC \equiv \square ABDE$.*

Proof of a special case of Theorem 9.3.5. As in the preceding proof, drop a perpendicular from C to $\overleftrightarrow{MN}$ and call the foot F. We will prove directly that

$\triangle ABC \equiv \square ABDE$ in the case in which F is between M and N as illustrated in Fig. 9.19. The general case requires a more complicated argument that uses an intermediate parallelogram, but it is helpful to see why the theorem is true in this simple case first.

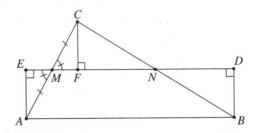

FIGURE 9.19: Proof of equivalence in case F is between M and N

Suppose F is between M and N. Let R denote the quadrilateral region corresponding to $\square ABNM$, let T_1 denote the triangular region corresponding to $\triangle CMF$, let T_2 denote the triangular region corresponding to $\triangle CNF$, let T_1' denote the triangular region corresponding to $\triangle AME$, and let T_2' denote the triangular region corresponding to $\triangle BND$. Then $\triangle ABC$ can be decomposed as $R \cup T_1 \cup T_2$ while $\square ABDE$ can be decomposed as $R \cup T_1' \cup T_2'$. Since $T_1 \cong T_1'$ and $T_2 \cong T_2'$ (by AAS as in the preceding proof), it follows that $\triangle ABC \equiv \square ABDE$. $\square$

The proof of the general case is less straightforward. We will first construct a parallelogram $\square ABGM$ associated with $\triangle ABC$. We will prove that $\triangle ABC \equiv \square ABGM$ and then that $\square ABGM \equiv \square ABDE$. By transitivity of equivalence, this completes the proof. The details are contained in the next two lemmas.

Construction. Let $\triangle ABC$ be a triangle, let M denote the midpoint of $\overline{AC}$, and let N denote the midpoint of $\overline{BC}$. Choose a point G such that $M * N * G$ and $\overline{MN} \cong \overline{NG}$. The quadrilateral $\square ABGM$ is called the *parallelogram associated with* $\triangle ABC$. The name is justified by the next lemma.

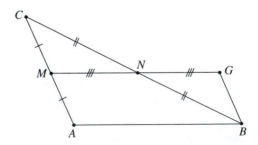

FIGURE 9.20: The parallelogram associated with $\triangle ABC$

Lemma 9.3.6. *If $\triangle ABC$ is a triangle and $\square ABGM$ is the associated parallelogram, then $\square ABGM$ is a parallelogram and $\triangle ABC \equiv \square ABGM$. Furthermore, $\overline{AM} \cong \overline{BG}$ and $\mu(\angle AMG) + \mu(\angle MGB) = 180°$.*

Proof. Exercise 9.19.   □

We want to prove that the associated Saccheri quadrilateral is equivalent by dissection to the parallelogram just constructed. This is relatively easy to do in case the parallelogram is short; but tall, thin parallelograms present a challenge.

Lemma 9.3.7. *Let $\triangle ABC$ be a triangle, let $\square ABGM$ be the associated parallelogram and let $\square ABDE$ be the associated Saccheri quadrilateral. Then $\square ABDE \equiv \square ABGM$.*

Proof. The proof of the case in which G lies on the segment $\overline{ED}$ is left as an exercise (Exercise 9.20).

FIGURE 9.21: G lies on segment $\overline{ED}$

Now consider the case in which G does not lie on $\overline{ED}$. In that case we may assume that $G * E * D$. (If this is not already true, simply interchange the labels on vertices A and B.) The strategy is not to prove directly that $\square ABGM$ is equivalent to $\square ABDE$, but instead to replace the parallelogram $\square ABGM$ with an equivalent parallelogram $\square ABHG$ that is closer to $\overline{ED}$. That operation is repeated until one vertex of the parallelogram lies in $\overline{ED}$ and the first case applies.

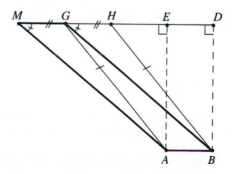

FIGURE 9.22: Moving G toward $\overline{ED}$

Choose a point H on $\overrightarrow{MG}$ such that $M * G * H$ and $MG = GH$. By Lemma 9.3.6, $\overline{AM} \cong \overline{BG}$ and $\angle AMG \cong \angle BGH$. Therefore $\triangle AMG \cong \triangle BGH$

(SAS). The quadrilateral regions $\square ABGM$ and $\square ABHG$ are equivalent by dissection because they have the triangle $\triangle ABG$ in common and $\triangle AMG \cong \triangle BGH$. Notice that $\square ABHG$ shares the important properties of $\square ABGM$, namely $\overline{AG} \cong \overline{BH}$ and $\mu(\angle AGH) + \mu(\angle GHB) = 180°$ (because $\angle MGA \cong \angle GHB$).

By the Archimedean Property of Real Numbers, there exists a positive integer n such that $n \cdot MG \geq ME$. We will assume that n is the smallest such integer. Apply the construction in the previous paragraph a total of n times. The result is a new parallelogram $\square ABG'M'$ such that $\square ABG'M'$ is equivalent to $\square ABGM$ and angles $\angle AM'G'$ and $\angle BG'M'$ are supplements. We claim that G' must lie in the segment $\overline{ED}$. If so, the final parallelogram is equivalent to $\square ABDE$ by the first case in this proof (the exercise). Transitivity of equivalence then completes the proof of the theorem.

To prove the claim, we first observe that $M' * E * D$ since M' is the "G" of the previous stage and n was chosen to be the smallest integer such that $n \cdot MG \geq ME$. Suppose $E * D * G'$. Since $\angle AEM'$ is a right angle, the Saccheri-Legendre Theorem implies that $\angle AM'E$ is an acute angle. In the same way, $\angle BG'D$ is acute. But angles $\angle AM'G'$ and $\angle BG'M'$ are supplements, so they cannot both be acute (see Figure 9.23). Therefore, $E * D * G'$ is impossible. On the other hand, $n \cdot MG \geq ME$ implies that G' must be on the ray $\overrightarrow{ED}$ and therefore G' lies on $\overline{ED}$. □

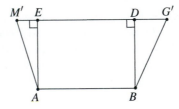

FIGURE 9.23: $E * D * G'$ is impossible because $\angle AM'G'$ and $\angle BG'M'$ are supplements

Proof of Theorem 9.3.5. By Lemma 9.3.6, $\triangle ABC \equiv \square ABGM$, and by Lemma 9.3.7, $\square ABDE \equiv \square ABGM$. Therefore, $\triangle ABC \equiv \square ABDE$ (Theorem 9.3.3). □

In the proofs later in this chapter we will need one additional technical fact about associated Saccheri quadrilaterals. It is the final theorem in the section.

Theorem 9.3.8. *Let $\triangle ABC$ be a triangle and let $\square ABDE$ be its associated Saccheri quadrilateral. If H is a point such that $\overline{AH}$ crosses $\overleftrightarrow{DE}$ at the midpoint of $\overline{AH}$, then $\square ABDE$ is also the Saccheri quadrilateral associated with $\triangle ABH$.*

Proof. Exercise 9.21. □

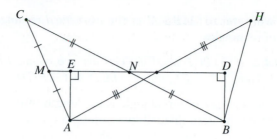

FIGURE 9.24: $\triangle ABC$ and $\triangle ABH$ share the Saccheri quadrilateral $\square ABDE$

9.4 DISSECTION THEORY IN EUCLIDEAN GEOMETRY

In this section we prove the Fundamental Theorem of Dissection Theory in Euclidean geometry. The proof we give is a characteristically Euclidean proof, but it will be presented in such a way as to make it as transparent as possible how to modify the proof to make it work in hyperbolic geometry. In order to accomplish this, we will carefully spell out the steps in the proof as separate lemmas. All theorems in this section assume the Euclidean Parallel Postulate.

There are many ways to prove the dissection theorem in Euclidean geometry. In order to illustrate this, Exercises 9.15, 9.16, and 9.17 outline an alternative proof. The alternative proof is based on a dissection theoretic proof of the Pythagorean Theorem (such as Bhaskara's). It is somewhat simpler than the proof in the section, but it does not generalize to hyperbolic geometry.

Our proof of Theorem 9.3.2 is divided into three steps. In this section we will see how to accomplish the three steps in Euclidean geometry. In the next section we will use the same three steps to prove the theorem in hyperbolic geometry.

Step 1. Use the results of the previous section to show that if $\triangle ABC$ and $\triangle DEF$ are two triangles such that $\alpha(\triangle ABC) = \alpha(\triangle DEF)$ and $\overline{AB} \cong \overline{DE}$, then $\triangle ABC \equiv \triangle DEF$.

Step 2. Use Step 1 to show that if $\triangle ABC$ and $\triangle DEF$ are any two triangles such that $\alpha(\triangle ABC) = \alpha(\triangle DEF)$, then $\triangle ABC \equiv \triangle DEF$.

Step 3. Use Step 2 and an inductive argument to show that if $\alpha(R_1) = \alpha(R_2)$, then $R_1 \equiv R_2$.

In the previous section we proved that every triangle is equivalent by dissection to its associated Saccheri quadrilateral. In Euclidean geometry a Saccheri quadrilateral is a rectangle. We begin the proof by stating in Euclidean terms what we have already proved.

Lemma 9.4.1. *If $\triangle ABC$ is a triangle, then there are points A' and B' such that $\square ABB'A'$ is a rectangle and $\triangle ABC \equiv \square ABB'A'$.*

Proof. This is just a restatement of Theorem 9.3.5 for Euclidean geometry. □

We will refer to $\square ABB'A'$ as the *associated rectangle*. Next we need an easy property of rectangles.

Lemma 9.4.2. *If $\square ABCD$ and $\square EFGH$ are two rectangles such that $\alpha(\square ABCD) = \alpha(\square EFGH)$ and $AB = EF$, then $\square ABCD \cong \square EFGH$.*

Proof. Since the area of a rectangle is the length of the base times the height, the height must be the area divided by the length of the base. Thus the length of the base and the height of the first rectangle must equal the length of the base and the height of the second rectangle, and so the rectangles are congruent. □

Those two ingredients allow us to complete Step 1 of the proof.

Theorem 9.4.3. *If $\triangle ABC$ and $\triangle DEF$ are two triangles such that $\alpha(\triangle ABC) = \alpha(\triangle DEF)$ and $\overline{AB} \cong \overline{DE}$, then $\triangle ABC \equiv \triangle DEF$.*

Proof. Let $\triangle ABC$ and $\triangle DEF$ be triangles such that $\alpha(\triangle ABC) = \alpha(\triangle DEF)$ and $\overline{AB} \cong \overline{DE}$. Suppose $\square ABB'A'$ and $\square DEE'D'$ are the associated rectangles (Lemma 9.4.1). Since the two rectangles have the same area and congruent bases, Lemma 9.4.2 implies that they are congruent. Therefore, $\triangle ABC \equiv \square ABB'A' \equiv \square DEE'D' \equiv \triangle DEF$, and the theorem follows from transitivity of equivalence (Theorem 9.3.3). □

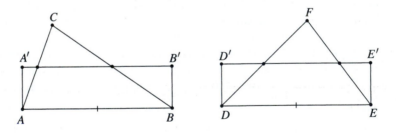

FIGURE 9.25: Step 1, Theorem 9.4.3

The only additional ingredient needed for Step 2 is Continuity of Distance, which is a theorem in neutral geometry.

Theorem 9.4.4. *If $\triangle ABC$ and $\triangle DEF$ are two triangles such that $\alpha(\triangle ABC) = \alpha(\triangle DEF)$, then $\triangle ABC \equiv \triangle DEF$.*

Proof. Let $\triangle ABC$ and $\triangle DEF$ be two triangles with $\alpha(\triangle ABC) = \alpha(\triangle DEF)$. Without loss of generality, we may assume that $DF \geqslant AC$. (If this is not already the case, simply relabel the vertices of the triangles to make it true.) Let M be the midpoint of $\overline{AC}$ and let N be the midpoint of $\overline{BC}$. (See Fig. 9.26.)

Note that $AM \leqslant (1/2)DF$ and that there are points P on $\overrightarrow{MN}$ such that $MP > (1/2)DF$. By Continuity of Distance (Theorem 6.4.8) and the Intermediate Value Theorem, there must be a point G on $\overrightarrow{MN}$ such that $AG = (1/2)DF$. Choose a point H on $\overrightarrow{AG}$ such that $\overline{GH} \cong \overline{AG}$. By Lemma 9.3.8, $\triangle ABC$ and $\triangle ABH$ share the same associated rectangle, and so $\triangle ABC \equiv \triangle ABH$ (Theorem 9.3.5). Furthermore, $\overline{AH} \cong \overline{DF}$, so $\triangle ABH \equiv \triangle DEF$ (Theorem 9.4.3). Therefore, $\triangle ABC \equiv \triangle DEF$ (Theorem 9.3.3, Part 3). □

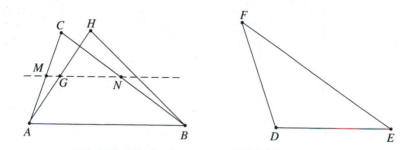

FIGURE 9.26: Step 2, Theorem 9.4.4

For Step 3 in the proof we need one final easy fact about areas of Euclidean triangles. It is a type of continuity for area.

Lemma 9.4.5. *If $\triangle ABC$ is a triangle and a is a number such that $0 < a < \alpha(\triangle ABC)$, then there exists a point D on $\overline{AB}$ such that $\alpha(\triangle ADC) = a$.*

Proof. Use the Ruler Postulate to choose a point D on $\overline{AB}$ such that

$$AD = \frac{a}{\alpha(\triangle ADC)} AB.$$

Since $\triangle ABC$ and $\triangle ADC$ share the same height, it follows from Theorem 9.2.5 that $\alpha(\triangle ADC) = a$. □

We can now complete the final step in the proof of the Fundamental Theorem of Dissection Theory in Euclidean geometry. Let us restate the theorem first.

Theorem 9.4.6. *If R and R' are two polygonal regions such that $\alpha(R) = \alpha(R')$, then $R \equiv R'$.*

Proof. Let R and R' be two polygonal regions such that $\alpha(R) = \alpha(R')$ (hypothesis). Choose triangulations $T_1, \ldots, T_n$ and $T'_1, \ldots, T'_m$ of R and R', respectively (definition of polygonal region).

Consider the two triangles T_1 and T'_1. Relabel, if necessary, so that it is T_1 that has the smaller area. If it should happen that $\alpha(T_1) = \alpha(T'_1)$, then delete T_1 from R and delete T'_1 from R'. If $\alpha(T_1) \neq \alpha(T'_1)$, then $\alpha(T_1) < \alpha(T'_1)$. By Lemma 9.4.5, T'_1 can be subdivided into two triangles T''_1 and T'''_1 such that $\alpha(T''_1) = \alpha(T_1)$. Delete T_1 from R and delete T''_1 from R'.

The result of the operation above is a new pair of polygonal regions R_1 and R'_1. Because the deleted triangles have the same area, we know that $\alpha(R_1) = \alpha(R'_1)$. If we could prove that $R_1 \equiv R_1'$, then it would follow that $R \equiv R'$ because $T_1 \equiv T_1''$ by Theorem 9.4.4 and the two equivalences could be combined to give one equivalence from R to R'. Thus we can work with R_1 and R_1' for the remainder of the proof.

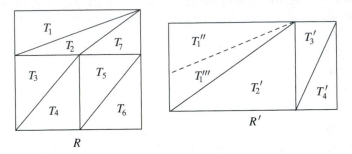

FIGURE 9.27: Remove one triangle from each region

Not only do the new regions R_1 and R_1' have smaller areas than the originals, but, more importantly, the total number of triangles in the two triangulations has been reduced. (The number is reduced by 2 in the first case and by 1 in the second case.) By applying the operation of the previous paragraph a finite number of times, we eventually reach regions that are triangular. We can apply Theorem 9.4.4 to those triangular regions. Successively adding the triangles back in and applying Theorem 9.4.4 each time eventually results in an equivalence between the original regions. □

Remark. The last proof really should be formulated as a proof by mathematical induction. But we have not studied that proof form, so we present the argument in a more informal way instead.

9.5 AREA AND DEFECT IN HYPERBOLIC GEOMETRY

In this section we prove the Fundamental Theorem of Dissection Theory in hyperbolic geometry. Along the way we also prove that area and defect are equivalent for hyperbolic triangles in the sense that the area of a triangle is proportional to its defect. The proofs of the two results are intimately entangled.

In order to understand the connection, consider the three steps in the Euclidean proof of the Fundamental Theorem. The main ingredient in Step 1 of the Euclidean proof was Lemma 9.4.2, which asserts that two rectangles with the same base and area are congruent. We do not know enough about area in hyperbolic geometry to prove a comparable theorem about Saccheri quadrilaterals in hyperbolic geometry. However, we do know a lot about defect of Saccheri quadrilaterals; in particular, we know that two Saccheri quadrilaterals with the same defect and summit are congruent (Theorem 8.2.12). So our plan will be to follow the same outline in the proof of Step 1, but to replace Lemma 9.4.2 with Theorem 8.2.12. The result will be that we prove a different theorem than expected: we will prove that two triangles that have equal defects and congruent bases are equivalent by dissection. Since the proof of Step 2 in the previous section used only neutral results, we can go on to remove the hypothesis that the two triangles share a common base.

The theorem we prove by this method is attributed to J. Bolyai. From it follow the other important results of the section. In particular, it is the key to understanding the relationship between area and defect in hyperbolic geometry.

Theorem 9.5.1 (Bolyai's Theorem). *If $\triangle ABC$ and $\triangle DEF$ are two triangles such that $\delta(\triangle ABC) = \delta(\triangle DEF)$, then $\triangle ABC \equiv \triangle DEF$.*

Note. The Hyperbolic Parallel Postulate is assumed in Bolyai's Theorem and in all the theorems in the remainder of the section.

Terminology. In this section we will be studying area and defect at the same time. Strictly speaking, defect applies to triangles and quadrilaterals while area applies to triangular regions and quadrilateral regions. In order to simplify the terminology, we will not make this distinction. For example, we will speak of the defect of a quadrilateral region.

The proof of Bolyai's Theorem requires the following simple lemma.

Lemma 9.5.2. *If $\triangle ABC$ is a triangle and $\square ABDE$ is the associated Saccheri quadrilateral, then $\delta(\square ABDE) = \delta(\triangle ABC)$.*

Proof. Exercise 9.22. □

As in the Euclidean theorem, it is convenient to break the proof of Bolyai's Theorem into two steps. The following lemma corresponds to Step 1 in the Euclidean proof.

Lemma 9.5.3. *If $\triangle ABC$ and $\triangle DEF$ are two triangles such that $\delta(\triangle ABC) = \delta(\triangle DEF)$ and $\overline{AB} \cong \overline{DE}$, then $\triangle ABC \equiv \triangle DEF$.*

Proof. Let $\triangle ABC$ and $\triangle DEF$ be two triangles such that $\delta(\triangle ABC) = \delta(\triangle DEF)$ and $\overline{AB} \cong \overline{DE}$ (hypothesis). Let $\square ABB'A'$ and $\square DEE'D'$ be the Saccheri quadrilaterals corresponding to $\triangle ABC$ and $\triangle DEF$, respectively. By Theorem 9.5.2, $\delta(\square ABB'A) = \delta(\square DEE'D')$. Since $\overline{AB}$ is the summit of $\square ABB'A$ and $\overline{DE}$ is the summit of $\square DEE'D'$, the two quadrilaterals have congruent summits (hypothesis). Hence Theorem 8.2.12 implies that $\square ABB'A' \cong \square DEE'D'$. Therefore, the conclusion follows from Theorem 9.3.5 and Transitivity of Equivalence (Theorem 9.3.3, Part 3). □

Proof of Theorem 9.5.1. The proof is essentially the same as that of Theorem 9.4.4. Let $\triangle ABC$ and $\triangle DEF$ be two triangles such that $\delta(\triangle ABC) = \delta(\triangle DEF)$ (hypothesis). Without loss of generality, we may assume that $DF \geqslant AC$. Let M be the midpoint of $\overline{AC}$ and let N be the midpoint of $\overline{BC}$. Choose a point G on $\overleftrightarrow{MN}$ such that $AG = (1/2)DF$. (Use Continuity of Distance and the Intermediate Value Theorem exactly as in the proof of Theorem 9.4.4.) Choose a point H on $\overrightarrow{AG}$ such

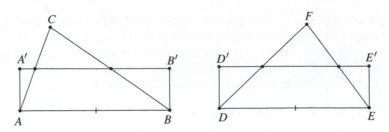

FIGURE 9.28: Two triangles in the proof of Lemma 9.5.3

that $A * G * H$ and $\overline{GH} \cong \overline{AG}$. By Theorem 9.3.8, $\triangle ABC$ and $\triangle ABH$ share the same Saccheri quadrilateral. Hence $\triangle ABC \equiv \triangle ABH$ (Theorem 9.3.5). In particular, $\delta(\triangle ABH) = \delta(\triangle ABC) = \delta(\triangle DEF)$ (Lemma 9.5.2). Furthermore, $\overline{AH} \cong \overline{DF}$, so $\triangle ABH \equiv \triangle DEF$ by Lemma 9.5.3. Therefore, $\triangle ABC \equiv \triangle DEF$ (Theorem 9.3.3, Part 3).   □

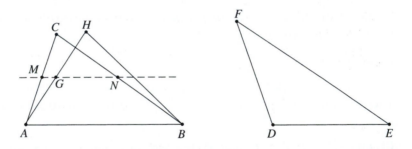

FIGURE 9.29: Proof of Bolyai's Theorem

The following theorem is one of the most important consequences of Bolyai's Theorem. It says that, in hyperbolic geometry, defect and area are essentially the same. The fact that there is a relationship between the two quantities is a surprising and beautiful result. Their definitions do not suggest any relationship at all. The main hint that they might be related is the fact that both of them satisfy the same basic axioms: additivity and invariance under congruence.

Theorem 9.5.4 (Area and defect are proportional). *There exists a constant k such that*

$$\alpha(\triangle ABC) = k\delta(\triangle ABC)$$

for every triangle $\triangle ABC$.

Here is one surprising, but immediate, consequence.

Corollary 9.5.5. *There is an upper bound on the areas of triangles.*

Proof. Since $\delta(\triangle ABC) < 180°$ for every triangle, it must be the case that $\alpha(\triangle ABC) < 180k$.   □

Theorem 9.5.4 allows us to reformulate Bolyai's Theorem. This second version is Step 2 in the proof of the Fundamental Theorem of Dissection Theory in hyperbolic geometry.

Corollary 9.5.6. *Let $\triangle ABC$ and $\triangle DEF$ be two triangles. If $\alpha(\triangle ABC) = \alpha(\triangle DEF)$, then $\triangle ABC \equiv \triangle DEF$.*

Proof. If $\alpha(\triangle ABC) = \alpha(\triangle DEF)$, then $\delta(\triangle ABC) = \delta(\triangle DEF)$ (Theorem 9.5.4). Thus $\triangle ABC \equiv \triangle DEF$ by Bolyai's Theorem (Theorem 9.5.1).   □

We need one more lemma before we are ready to tackle the proof of Theorem 9.5.4.

Lemma 9.5.7. *For any two triangles $\triangle ABC$ and $\triangle DEF$,*

$$\frac{\alpha(\triangle ABC)}{\delta(\triangle ABC)} = \frac{\alpha(\triangle DEF)}{\delta(\triangle DEF)}.$$

Proof. First consider the case in which $\delta(\triangle ABC) = \delta(\triangle DEF)$. By Theorem 9.5.1, $\triangle ABC \equiv \triangle DEF$ and so $\alpha(\triangle ABC) = \alpha(\triangle DEF)$. Therefore, the lemma holds in this special case.

Second, consider the case in which $\delta(\triangle ABC)/\delta(\triangle DEF)$ is a rational number. Without loss of generality we may assume that

$$\frac{\delta(\triangle ABC)}{\delta(\triangle DEF)} = \frac{a}{b},$$

where both a and b are positive integers and $a < b$. Use Theorem 8.6.5 to locate points $P_0, P_1, \ldots, P_b$ on $\overline{DE}$ such that $D = P_0$, $P_{i-1} * P_i * P_{i+1}$ for every i, $P_b = E$, and $\delta(\triangle FP_iP_{i+1}) = (1/b)\delta(\triangle DEF)$ for each i. Note that $\delta(\triangle ABC) = \delta(\triangle FP_0P_a)$ (by Additivity of Defect). By the first case in the proof, all the small triangles $\triangle FP_iP_{i+1}$ have the same area. Hence we can use Additivity of Defect and Additivity of Area to reach our conclusion in this case.

In the general case $\delta(\triangle ABC)/\delta(\triangle DEF)$ is a real number, possibly irrational. The general case follows from the rational case and the Comparison Theorem (Theorem 4.2.4). The application of the Comparison Theorem is exactly like that in the proof of the Parallel Projection Theorem, so we will not repeat it.   □

Proof of Theorem 9.5.4. Fix one triangle $\triangle DEF$ and let

$$k = \frac{\alpha(\triangle DEF)}{\delta(\triangle DEF)}.$$

By the lemma,

$$\frac{\alpha(\triangle ABC)}{\delta(\triangle ABC)} = k$$

for every $\triangle ABC$ and so $\alpha(\triangle ABC) = k\delta(\triangle ABC)$ (same k) for every $\triangle ABC$.   □

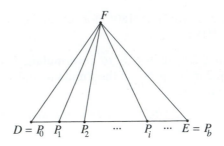

FIGURE 9.30: Proof of Lemma 9.5.7

Remark. We have not actually defined the area of a triangle in hyperbolic geometry. Given the results of this section, the natural way in which to do so is simply to define $\alpha(\triangle ABC) = \delta(\triangle ABC)$. In other words, the simplest thing to do is to take the constant k in Theorem 9.5.4 to be 1. This is yet another indication that there is a universal unit of measurement built into hyperbolic geometry.

We are now ready to complete the proof of the Fundamental Theorem of Dissection Theory in hyperbolic geometry. The third step in the proof is essentially the same as in Euclidean geometry. The only difference is that we use the continuity of defect in place of Lemma 9.4.5. Let us state the theorem one more time.

Theorem 9.5.8. *Let R_1 and R_2 be two polygonal regions. If $\alpha(R_1) = \alpha(R_2)$, then $R_1 \equiv R_2$.*

Proof. Start with triangulations $T_1, \ldots, T_n$ and $T'_1, \ldots, T'_m$ of R_1 and R_2, respectively.

Consider the two triangles T_1 and T'_1. Relabel, if necessary, so that it is T_1 that has the smaller area. If it should happen that $\alpha(T_1) = \alpha(T'_1)$, then delete T_1 from R_1 and delete T'_1 from R_2. If $\alpha(T_1) \neq \alpha(T'_1)$, then $\alpha(T_1) < \alpha(T'_1)$. By Theorems 8.6.5 and 9.5.4, T'_1 can be subdivided into two triangles T''_1 and T'''_1 such that $\alpha(T''_1) = \alpha(T_1)$. Delete T_1 from R_1 and delete T''_1 from R_2. The result of this operation is a new pair of polygonal regions R'_1 and R'_2. Because the deleted triangles have the same area, we know that $\alpha(R'_1) = \alpha(R'_2)$. Note also that the total number of triangles in the two regions combined has been reduced by at least one. The proof is completed using exactly the same inductive argument as in the proof of Theorem 9.4.6. □

The reader has probably been wondering why we don't define area in hyperbolic geometry in a way that is more analogous to the way in which area is defined in Euclidean geometry. Of course we can't begin by defining the area of a rectangle since there are no rectangles in hyperbolic geometry, but why not define the area of a triangle to be one-half the length of the base times the height? Our final theorem shows that this obvious approach does not work in hyperbolic geometry.

Theorem 9.5.9. *There exist triangles that have the same base and height but different areas.*

Proof. Let ℓ be a line and let P be a point such that $d(P, \ell) = 1$. Let Q_0 be the foot of the perpendicular from P to ℓ. Choose points $Q_1, Q_2, \ldots$ on ℓ such that $Q_{i-1} * Q_i * Q_{i+1}$ and $Q_i Q_{i+1} = 1$ for each i. Let T_i denote the triangular region corresponding to $\triangle PQ_{i-1}Q_i$. Then for each i, the length of the base and the height of T_i are both 1. We will show that it is impossible for all the T_i to have the same area.

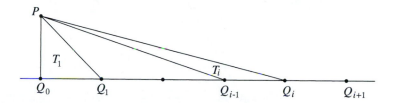

FIGURE 9.31: Proof of Theorem 9.5.9

Suppose all the T_i have the same area (RAA hypothesis). Then all the T_i have the same defect (Theorem 9.5.4). Let $\delta_i = \delta(T_i)$. Then $\delta_1 + \delta_2 + \cdots + \delta_n = \delta(\triangle PQ_0Q_n) < 180°$. But the defects are all equal to δ_1, so we see that $n\delta_1 < 180°$ for every n. But this contradicts the Archimedean Property of the real numbers. Hence the defects cannot all be equal and therefore the areas cannot all be equal either (Theorem 9.5.4). □

EXERCISES

9.1. Use the definitions of interior of an angle and interior of a triangle to prove the following fact in neutral geometry: *A point P is in the interior of $\triangle ABC$ if and only if P is on the same side of $\overleftrightarrow{AB}$ as C, P is on the same side of $\overleftrightarrow{BC}$ as A, and P is on the same side of $\overleftrightarrow{AC}$ as B.* Conclude that Int($\triangle ABC$) is the intersection of the three half planes determined by the three points A, B, and C.

9.2. Prove Theorem 9.1.7.

9.3. Let $\square ABCD$ be a convex quadrilateral in neutral geometry. Prove that $\blacktriangle ABC \cup \blacktriangle CDA = \blacktriangle DAB \cup \blacktriangle BCD$ and that each pair of triangles is nonoverlapping.

9.4. Prove the Euclidean formula for the area of a right triangle (Theorem 9.2.3).

9.5. Prove the Euclidean formula for the area of a triangle (Theorem 9.2.5).

9.6. Let $\triangle ABC$ be a Euclidean triangle. Give a direct proof, using theorems of Euclidean geometry but not the Euclidean Area Postulate, that the quantity

$$(1/2) \text{ base} \times \text{height}$$

is the same regardless of which side of $\triangle ABC$ is used as the base.

9.7. Prove Theorem 9.2.6.

9.8. Let $\square ABCD$ be a Euclidean parallelogram. Choose one side as base and define the corresponding height for the parallelogram. Prove that the area of the parallelogram is the length of the base times the height.

9.9. Let $\square ABCD$ be a Euclidean parallelogram. Prove that the diagonal $\overline{BD}$ divides the parallelogram into two triangles of equal area; that is $\alpha(\triangle ABD) = \alpha(\triangle BCD)$. (This is Euclid's Proposition I.34.)

9.10. Let $\square ABCD$ be a Euclidean parallelogram. Choose a point I on the diagonal $\overline{BD}$. Let E and G be the points on segments $\overline{AB}$ and $\overline{CD}$, respectively, such that I lies on $\overline{EG}$ and $\overleftrightarrow{EG} \parallel \overleftrightarrow{BC}$. Let F and H be the points on segments $\overline{BC}$ and $\overline{AD}$ respectively such that I lies on $\overline{FH}$ and $\overleftrightarrow{FH} \parallel \overleftrightarrow{AB}$. (See Fig. 9.32.) Prove that $\alpha(\square AEIH) = \alpha(\square IFCG)$. (This is Euclid's Proposition I.43.)

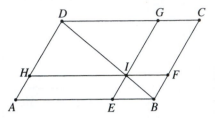

FIGURE 9.32: A parallelogram subdivided as in Exercise 9.10

9.11. Recall that *trapezoid* was defined on page 133. The sides that are contained in parallel lines are called the *bases* of the trapezoid; in Euclidean geometry the distance between the two parallel lines is called the *height* of the trapezoid. Prove that the area of a Euclidean trapezoid is the height times the average of the lengths of the bases.

9.12. President James Garfield gave a proof of the Pythagorean Theorem that is based on the formula in the previous exercise. (It is proof number 231 in [44].) Reproduce that proof by equating the area of the trapezoid in Fig. 9.33 to the sum of the areas of the three triangles shown in the diagram.

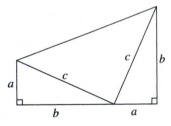

FIGURE 9.33: President Garfield's proof of the Pythagorean Theorem

9.13. Give another proof of the Pythagorean Theorem, this one based on Fig. 9.34. (Be sure to check that pieces that appear to be congruent really are.) Your proof should show explicitly that the square on the hypotenuse is equivalent by dissection to the union of the two squares on the legs.

9.14. This is an exercise in Euclidean geometry. For each of the following pairs of rectangles, find *explicit triangulations* of each such that corresponding triangles are congruent.
   (a) The 1×1 square and the $3 \times (1/3)$ rectangle.
   (b) The 1×1 square and the $\sqrt{2} \times (1/\sqrt{2})$ rectangle.
   (c) The 1×1 square and the $\pi \times (1/\pi)$ rectangle.

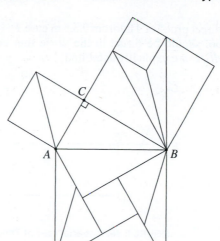

FIGURE 9.34: A dissection proof of the Pythagorean Theorem

9.15. This is another exercise in Euclidean geometry. The two parts together give a constructive proof that every rectangle is equivalent by dissection to a square.

   (a) Prove that every rectangle is equivalent by subdivision to a rectangle $\square ABCD$ for which $BC \leqslant AB \leqslant 4BC$.

   (b) Prove that every rectangle with $\square ABCD$ for which $BC \leqslant AB \leqslant 4BC$ is equivalent to a square. (See Fig. 9.35.)

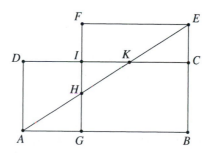

FIGURE 9.35: A rectangle is equivalent to a square

9.16. Suppose R is a polygonal region that is the union of two nonoverlapping squares. Prove that R is equivalent by dissection to a single square. Illustrate with a diagram.

9.17. Combine the following three results to give a second proof of the Fundamental Theorem of Dissection Theory in Euclidean geometry.

   (a) Every triangle is equivalent to a rectangle (Lemma 9.4.1).

   (b) Every rectangle is equivalent to a square (Exercise 9.15).

   (c) Two squares are equivalent to one square (Exercise 9.16).

9.18. Give a direct proof of Theorem 9.3.5 in case $F * M * E * N * D$ (see Fig. 9.36). The proof should be direct in the sense that equivalent subdivisions of $\triangle ABC$ and $\square ABDE$ are explicitly exhibited.

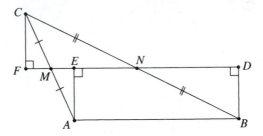

FIGURE 9.36: A special case of Theorem 9.3.5

9.19. Prove Lemma 9.3.6.

9.20. Prove Lemma 9.3.7 in case G lies on segment $\overline{ED}$.

9.21. Prove Theorem 9.3.8.

9.22. Prove Lemma 9.5.2.

C H A P T E R   10

Circles

Circles have always played a prominent role in geometry. Euclid worked with both circles and lines right from the beginning of his *Elements* and gave them roughly equal treatment. In particular, one of his postulates addressed the existence of circles and he used that postulate in his constructions in Book I. Despite the fact that circles are part of the *Elements* from the start, Euclid postponed most of his theorems about circles until Book III or later. In much the same way we have delayed consideration of circles until now. For us, lines are more fundamental than circles because we have taken "line" to be the most important undefined term in our development of geometry. None of the axioms we are using says anything directly about circles. As a result, we have not yet mentioned circles. In this chapter we attempt to give circles their due.

We begin with the basic definitions. After that we follow Euclid in investigating the ways in which circles interact with lines and triangles. Most of the theorems regarding intersections of circles and lines come from Book III of the *Elements* while the theorems about inscribed and circumscribed triangles come from Book IV. One surprise is the fact that many of Euclid's theorems about circles turn out to be neutral despite the fact that they come relatively late in his development of geometry. There is also one key theorem that is equivalent to the Euclidean Parallel Postulate.

After studying circles in neutral geometry, we will prove some specialized theorems regarding circles in Euclidean geometry. We look at angles inscribed in Euclidean circles and also investigate the familiar Euclidean formulas for the circumference and area of a circle. In addition, the chapter includes a section in which more specialized results regarding circles in Euclidean geometry are explored using dynamic software. The theorems in that section are "modern" in the sense that they were discovered long after Euclid wrote the *Elements*.

We will also finally fill the gap in the proof of Euclid's Proposition 1 by proving two fundamental theorems about circular continuity. The three sections that include the word *Euclidean* in their titles are part of Euclidean geometry; everything else in the chapter is neutral.

10.1 BASIC DEFINITIONS

Let us begin with the definition of a circle. Unlike *line*, *circle* can be defined quite simply.

Definition 10.1.1. Given a point O and a positive real number r, the *circle with center O and radius r* is defined to be the set of all points P such that the distance from O to P is r. In symbols,

$$C(O, r) = \{P \mid OP = r\}.$$

The number r is called *the radius* of the circle, but we often refer to one of the segments $\overline{OP}$ as *a radius* as well.

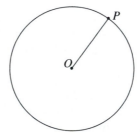

FIGURE 10.1: A circle and a radius

Notation. We will consistently use letters from the beginning of the Greek alphabet to denote circles. Thus we will usually use α, β, or γ in place of the more cumbersome $C(O, r)$.

Definition 10.1.2. Let $C(O, r)$ be a circle. A *chord* of $C(O, r)$ is a segment joining two points P and Q on $C(O, r)$. Two points P and Q on $C(O, r)$ are said to be *antipodal* points if $P * O * Q$. In case P and Q are antipodal points, the chord $\overline{PQ}$ is called *a diameter* of the circle.

Again the word *diameter* can refer to either the segment $\overline{PQ}$ or the length $d = PQ$. It is clear that the diameter is twice the radius; that is, if P and Q are endpoints of a diameter and O is the center of the circle, then $PQ = 2OP$, or $d = 2r$.

Definition 10.1.3. Points A such that $OA < r$ are said to be *inside* the circle $C(O, r)$ while points B such that $OB > r$ are said to be *outside* the circle.

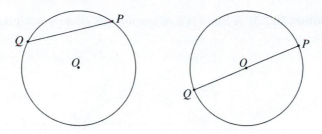

FIGURE 10.2: A chord and a diameter

10.2 CIRCLES AND LINES

In this section we investigate the ways in which a circle and a line can intersect. All the theorems in the section are neutral. We begin by showing that the intersection of a line and a circle can consist of at most two points.

Theorem 10.2.1. *If γ is a circle and ℓ is a line, then the number of points in $\gamma \cap \ell$ is 0, 1, or 2.*

Proof. Let $\gamma = C(O, r)$ be a circle and let ℓ be a line. Suppose there exist three distinct points A, B, and C that lie on both $C(O, r)$ and ℓ (RAA hypothesis).

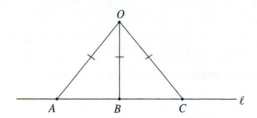

FIGURE 10.3: Three points lie on both C and ℓ

Since A, B, and C are three distinct collinear points, they can be ordered; let us assume that $A * B * C$. Because all three points lie on γ, $OA = OB = OC = r$. Applying the Isosceles Triangle Theorem three times gives $\mu(\angle ABO) = \mu(\angle BAO) = \mu(\angle BCO) = \mu(\angle CBO)$. Since $\mu(\angle BAO) + \mu(\angle ABO) < 180°$ (Lemma 6.6.3), we can conclude that $\mu(\angle ABO) < 90°$. But it is impossible for both $\angle ABO$ and $\angle CBO$ to have measure less than $90°$ because they are supplementary angles. Thus it is impossible for three distinct points to lie on both ℓ and γ. $\square$

The theorem allows us to classify lines that intersect a circle. If a line intersects a circle, it does so at either one or two points. Each kind of line has a name.

Definition 10.2.2. A line ℓ is *tangent* to circle γ if ℓ intersects γ in exactly one point. If P is the point at which ℓ intersects γ, then we say that ℓ *is tangent to γ at P.* A segment $\overline{AB}$ is said to be tangent to a circle γ if $\overline{AB}$ is contained in a line that is tangent to γ and the point of tangency is an interior point of $\overline{AB}$.

Definition 10.2.3. A line ℓ is a *secant line* for circle γ if ℓ intersects γ in two distinct points.

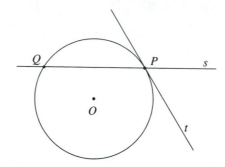

FIGURE 10.4: t is a tangent line, s is a secant line

For each type of intersecting line there is a theorem. The theorem we call the Tangent Line Theorem combines Euclid's Propositions III.18 and III.19 while the theorem we call the Secant Line Theorem is Euclid's Proposition III.3.

Theorem 10.2.4 (Tangent Line Theorem). *Let t be a line, $\gamma = C(O, r)$ a circle, and P a point of $t \cap \gamma$. The line t is tangent to the circle γ at the point P if and only if $\overleftrightarrow{OP} \perp t$.*

Proof. Let t be a line and let $\gamma = C(O, r)$ be a circle such that t and γ intersect at P. Suppose, first, that t is tangent to γ at the point P (hypothesis). We must prove that $\overleftrightarrow{OP} \perp t$. Drop a perpendicular from O to t and call the foot Q. It suffices to prove that $Q = P$.

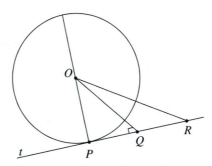

FIGURE 10.5: Proof of the Tangent Line Theorem

Suppose $P \neq Q$ (RAA hypothesis). Then we can choose a point R on t such that $P * Q * R$ and $\overline{PQ} \cong \overline{QR}$. By SAS, $\triangle OPQ \cong \triangle ORQ$ and therefore $OP = OR$. Thus R lies on γ (definition of circle). This contradicts the fact that P is the only point that lies on both t and γ and so we must reject the RAA hypothesis and conclude that $P = Q$. Therefore $t \perp \overleftrightarrow{OP}$.

The proof of the converse is left as an exercise (Exercise 10.1). □

Theorem 10.2.5. *If γ is a circle and t is a tangent line, then every point of t except for P is outside γ.*

Proof. Exercise 10.2. □

Theorem 10.2.6 (Secant Line Theorem). *If $\gamma = C(O, r)$ is a circle and ℓ is a secant line that intersects γ at distinct points P and Q, then O lies on the perpendicular bisector of the chord $\overline{PQ}$.*

Proof. Exercise 10.3. □

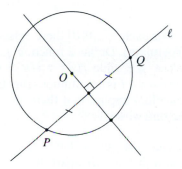

FIGURE 10.6: The center of the circle lies on the perpendicular bisector of a chord

The following theorem is Euclid's Proposition III.2.

Theorem 10.2.7. *If γ is a circle and ℓ is a secant line such that ℓ intersects γ at points P and Q, then every point on the interior of $\overline{PQ}$ is inside γ and every point of $\ell \smallsetminus \overline{PQ}$ is outside γ.*

Proof. Exercise 10.4. □

According to the last theorem, a secant line contains points that are inside the circle as well as points that are outside. The next theorem is a type of converse; it asserts that any line that contains a point from the inside and point from the outside is a secant line. The theorem is also a form of continuity for circles in that it asserts that a line cannot get from the inside to the outside of a circle without crossing the circle; there can be no gaps in a circle.

Theorem 10.2.8 (Elementary Circular Continuity). *If γ is a circle and ℓ is a line such that ℓ contains a point A that is inside γ and a point B that is outside γ, then ℓ is a secant line for γ.*

Proof. Let $\gamma = C(O, r)$ be a circle and ℓ be a line such that ℓ contains a point A that is inside γ and a point B that is outside γ (hypothesis). Since a tangent line does not contain points that are inside γ (Theorem 10.2.5), it is not possible for ℓ to be a tangent line. It is therefore enough to prove that $\ell \cap \gamma \neq \emptyset$ (Theorem 10.2.1). Specifically, we will prove that there exists a point between A and B that lies on γ.

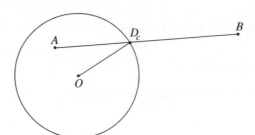

FIGURE 10.7: A point D_c whose distance from O is exactly equal to r

Let $d = AB$. For each $x \in [0, d]$ there exists a unique point $D_x \in \overline{AB}$ such that $AD_x = x$ (Ruler Postulate). Define a function $f : [0, d] \to [0, \infty)$ by $f(x) = OD_x$. Note that $f(0) = OA < r$ while $f(d) = OB > r$. (See Fig. 10.7.) By Continuity of Distance (Theorem 6.4.8), f is a continuous function. Hence the Intermediate Value Theorem (from calculus) implies that there exists a number c such that $f(c) = r$. The point D_c is the one we seek. □

Since a line contains points that are arbitrarily far apart and points inside $\mathcal{C}(O, r)$ are at most a distance $2r$ apart, no line can be completely inside a circle. Therefore, we have the following corollary.

Corollary 10.2.9. *If γ is a circle and ℓ is a line such that ℓ contains a point A that is inside γ, then ℓ is a secant line for γ.*

The last theorem is an "elementary" form of circular continuity in that it tells us that a line must cross a circle. The more advanced form of circular continuity, which is what Euclid needs in his proof of Proposition 1, asserts that two circles must intersect. That form of continuity will be proved later in the chapter. It is interesting to observe, however, that the rational plane fails even the elementary form of continuity just proved for neutral geometry.

EXAMPLE 10.2.10 Circular continuity fails in the rational plane

Consider the rational plane, Example 5.4.15. Let γ be the circle with center at the origin $(0,0)$ and radius 3; thus $\gamma = \{(x, y) \mid x^2 + y^2 = 9\}$. Let ℓ be the line with equation $x = 2$. Simple calculation shows that γ and ℓ do not intersect in the rational plane. Of course the circle and line described by the same equations in $\mathbb{R}^2$ do intersect at $(2, \pm\sqrt{5})$. ■

We conclude the section with a theorem about tangent circles. It combines Euclid's Propositions III.11 and III.12.

Definition 10.2.11. Two circles $\gamma_1 = \mathcal{C}(O_1, r_1)$ and $\gamma_2 = \mathcal{C}(O_2, r_2)$ are *tangent* if $\gamma_1 \cap \gamma_2$ consists of precisely one point.

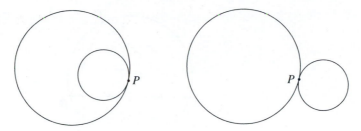

FIGURE 10.8: Two pairs of tangent circles

Theorem 10.2.12 (Tangent Circles Theorem). *If the circles* $\gamma_1 = \mathcal{C}(O_1, r_1)$ *and* $\gamma_2 = \mathcal{C}(O_2, r_2)$ *are tangent at P, then the centers O_1 and O_2 are distinct and the three points O_1, O_2, and P are collinear. Furthermore, the circles share a common tangent line at P.*

Proof. Exercise 10.5.   □

10.3 CIRCLES AND TRIANGLES

In the preceding section we proved that three distinct, collinear points cannot all lie on one circle. Now we ask whether or not three noncollinear points lie on a circle. The question is this: Given three noncollinear points, is there a circle that contains all three of them? The answer, in neutral geometry at least, is "not necessarily." This may surprise you since you probably remember from high school geometry that three noncollinear points always determine a circle. We will prove that the assertion that three noncollinear points always lie on a circle is equivalent to the Euclidean Parallel Postulate.

Of course three noncollinear points do determine a triangle. So the question we are really asking is whether or not the three vertices of a given triangle must lie on a circle. A circle that contains all the vertices of a triangle has a special name.

Definition 10.3.1. A circle that contains all three vertices of the triangle $\triangle ABC$ is said to *circumscribe* the triangle. The circle is called the *circumcircle* and its center is called the *circumcenter* of the triangle. If a triangle has a circumcircle, we say that *the triangle can be circumscribed.*

The basic theorem tells us that a triangle can be circumscribed if and only if the perpendicular bisectors of the sides of the triangle have a point in common. The theorem also says that if there is a circumcircle, it is unique. The uniqueness justifies the use of the definite article in the definition above. Note that the theorem is a theorem in neutral geometry.

Theorem 10.3.2 (Circumscribed Circle Theorem). *A triangle can be circumscribed if and only if the perpendicular bisectors of the sides of the triangle are concurrent. If a triangle can be circumscribed, then the circumcenter and the circumcircle are unique.*

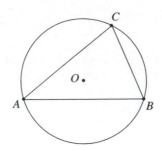

FIGURE 10.9: A circumcircle for $\triangle ABC$

Proof. Let $\triangle ABC$ be a triangle and let ℓ, m, and n be the perpendicular bisectors of sides $\overline{AB}, \overline{AC}$ and $\overline{BC}$, respectively. Suppose, first, that ℓ, m, and n have a point O of concurrency. By Theorem 6.4.7, $AO = BO = CO$. Let $r = AO$; then $\gamma = \mathcal{C}(O, r)$ circumscribes $\triangle ABC$.

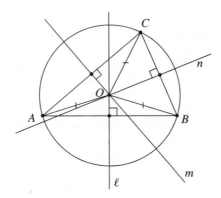

FIGURE 10.10: Proof of Theorem 10.3.2

Now suppose $\triangle ABC$ is circumscribed by $\gamma = \mathcal{C}(O, r)$ and let ℓ, m, and n be the perpendicular bisectors of the sides as above. The fact that $AO = BO$ implies that O lies on ℓ (Theorem 6.4.7). It follows in a similar way that O must lie on m and n. Therefore, ℓ, m, and n are concurrent and O is the point of concurrency.

If $\triangle ABC$ can be circumscribed, then the previous paragraph shows that the center of the circumcircle must be the point O of concurrency of the three perpendicular bisectors. Hence the circumcenter is unique. The radius of the circumcircle must be OA, so the circumcircle itself is unique. □

Theorem 10.3.2 does not directly answer the question of whether or not every triangle can be circumscribed. It does relate the existence of a circumcircle to parallelism of lines, which sets us up for the following theorem. The statement that every triangle has a circumcircle can be added to our list of statements equivalent to the Euclidean Parallel Postulate.

Theorem 10.3.3. *The Euclidean Parallel Postulate is equivalent to the assertion that every triangle can be circumscribed.*

Proof. We must prove two things: First, if the Euclidean Parallel Postulate holds, then every triangle can be circumscribed. Second, if the Euclidean Parallel Postulate does not hold (and therefore the Hyperbolic Parallel Postulate does hold), then there exists a triangle that cannot be circumscribed. Each of these statements is interesting and worthy of separate notice, so we state them as separate theorems below.   □

The Euclidean half of the theorem is Euclid's Proposition IV.5.

Theorem 10.3.4. *If the Euclidean Parallel Postulate holds, then every triangle can be circumscribed.*

Proof. Assume that the Euclidean Parallel Postulate holds and that $\triangle ABC$ is a triangle. Let $m = \overleftrightarrow{AB}$ and let k be the perpendicular bisector of $\overline{AB}$. In the same way let $n = \overleftrightarrow{AC}$ and let ℓ be the perpendicular bisector of $\overline{AC}$. We want to show that k and ℓ must intersect. If $k \parallel \ell$, then by Theorem 6.8.3, Part 3, either $m = n$ or $m \parallel n$. Since the lines through two sides of a triangle are neither parallel nor equal, we can conclude that $k \nparallel \ell$. Hence there exists a point O that lies on both k and ℓ. Let $r = OA$. Note that $OA = OB = OC$ (Theorem 6.4.7), so all three vertices lie on the circle $\mathcal{C}(O, r)$.   □

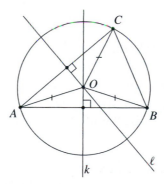

FIGURE 10.11: The two perpendicular bisectors k and ℓ meet at the circumcenter

In the proof of the theorem we saw that the perpendicular bisectors of sides $\overline{AB}$ and $\overline{AC}$ meet at the circumcenter. We could just as well have used the perpendicular bisectors of a different pair of sides. Since the circumcircle is unique, those two bisectors must intersect at the very same point.

Corollary 10.3.5. *In Euclidean geometry the three perpendicular bisectors of the sides of any triangle are concurrent and meet at the circumcenter of the triangle.*

We now turn to hyperbolic geometry where the situation is quite different.

Theorem 10.3.6. *If the Euclidean Parallel Postulate fails, then there exists a triangle that cannot be circumscribed.*

Proof. Assume that the Euclidean Parallel Postulate fails. In that case the Hyperbolic Parallel Postulate holds (Theorem 6.10.1), so we may use any of the theorems of hyperbolic geometry.

Let $\square ABCD$ be any Saccheri quadrilateral with base $\overline{AB}$ and summit $\overline{CD}$. Let M be the midpoint of $\overline{AB}$ and N be the midpoint of $\overline{CD}$. By Theorem 8.2.9, $MN < AD$. Choose a point E on $\overrightarrow{MN}$ such that $ME = AD = BC$. We will show that $\triangle CDE$ cannot be circumscribed. In particular, we will show that the perpendicular bisectors of $\overline{DE}$ and $\overline{CE}$ are parallel. By Theorem 10.3.2, we conclude that there cannot be a circumscribed circle for this triangle.

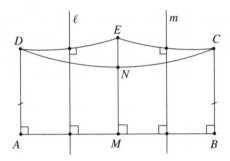

FIGURE 10.12: The triangle $\triangle CDE$ cannot be circumscribed because ℓ and m are parallel

Notice that $\square AMED$ and $\square BMEC$ are Saccheri quadrilaterals. Let ℓ be the line through the midpoints of $\overline{AM}$ and $\overline{DE}$ and let m be the line through the midpoints of $\overline{MB}$ and $\overline{CE}$. By Theorem 6.9.10, $\ell \perp \overleftrightarrow{AB}$ and $m \perp \overleftrightarrow{AB}$, so $\ell \parallel m$ (Alternate Interior Angles Theorem). Also by Theorem 6.9.10, $\ell \perp \overleftrightarrow{DE}$ and $m \perp \overleftrightarrow{CE}$. Hence ℓ is the perpendicular bisector of $\overline{DE}$ and m is the perpendicular bisector of $\overline{EC}$, so the proof is complete. ☐

The theorem asserts that there exist some hyperbolic triangles that cannot be circumscribed. Of course there are lots of triangles, even in hyperbolic geometry, that can be circumscribed. To construct one, simply start with a circle and choose the three vertices of the triangle to lie on the circle. Such a triangle is said to be *inscribed* in the circle. Later in this section we will observe that it is possible to inscribe other figures in a circle.

Let $\triangle ABC$ be a triangle. The lines determined by the edges of $\triangle ABC$ are all secant lines for the circumscribed circle. One might also ask if there exits a circle that is tangent to all the edges of a triangle. Such a circle is called an inscribed circle. In view of Theorem 10.3.3, it is surprising that we can prove (in neutral geometry) that every triangle has an inscribed circle. The key observations are that, by Theorems 10.2.4 and 6.4.6, the center of the inscribed circle must lie on all the

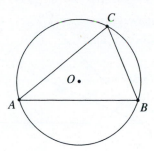

FIGURE 10.13: The triangle $\triangle ABC$ is inscribed in the circle

angle bisectors, and that the Crossbar Theorem implies that the angle bisectors must intersect.

Definition 10.3.7. Let $\triangle ABC$ be a triangle. A circle $\mathcal{C}(O, r)$ is called the *inscribed circle* for $\triangle ABC$ if each of the segments $\overline{AB}$, $\overline{AC}$, and $\overline{BC}$ is tangent to $\mathcal{C}(O, r)$. The center of the inscribed circle is called the *incenter* of the triangle.

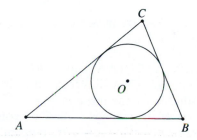

FIGURE 10.14: An inscribed circle for $\triangle ABC$

The next theorem is Euclid's Proposition IV.4.

Theorem 10.3.8 (Inscribed Circle Theorem). *Every triangle has a unique inscribed circle. The bisectors of the interior angles in any triangle are concurrent and the point of concurrency is the incenter of the triangle.*

Proof. Let $\triangle ABC$ be a triangle. By the Crossbar Theorem, the bisector of $\angle CAB$ intersects $\overline{BC}$ in a point D. Applying the Crossbar Theorem again, this time to $\triangle ABD$, gives a point E at which the bisector of $\angle ABD$ intersects $\overline{AD}$. Observe that E is in the interior of $\angle ACB$. We will show that E is the incenter of $\triangle ABC$.

Let F, G, and H be the feet of the perpendiculars from E to the lines $\overleftrightarrow{AB}$, $\overleftrightarrow{BC}$, and $\overleftrightarrow{AC}$, respectively. Angles $\angle EAB$ and $\angle EBA$ are both acute since each has measure that is half the measure of an interior angle of $\triangle ABC$. Thus F is between A and B (Lemma 6.9.6). An application of the Exterior Angle Theorem, just like that in the proof of Lemma 6.9.6, shows that G is on the ray $\overrightarrow{BC}$ and H is on the ray $\overrightarrow{AC}$.

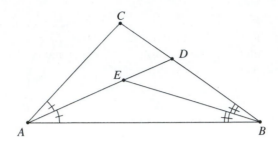

FIGURE 10.15: The point E lies on two angle bisectors

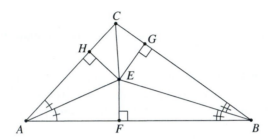

FIGURE 10.16: The point E is the incenter

Now $\triangle AEH \cong \triangle AEF$ (AAS) and $\triangle BEF \cong \triangle BEG$ (AAS again), so $\overline{EH} \cong \overline{EF} \cong \overline{EG}$. Let $r = EH$. Then all three of the points F, G, and H lie on the circle $\gamma = \mathcal{C}(E, r)$. Furthermore, the three radii $\overline{EF}$, $\overline{EG}$, and $\overline{EH}$ are perpendicular to the respective sides of the triangle $\triangle ABC$ by construction. Thus the three lines determined by the sides of the triangle are tangent to γ (Theorem 10.2.4).

Since E is equidistant from lines $\overleftrightarrow{AC}$ and $\overleftrightarrow{BC}$ and lies in the interior of $\angle ACB$, E is also on the bisector of $\angle ACB$ (Theorem 6.4.6). In particular, the three bisectors of the interior angles of $\triangle ABC$ are concurrent at E. Furthermore, we could have defined E to be the point at which any two of the bisectors intersect, so the argument above which showed that F is between A and B can equally well be used to show that G is between B and C and H is between A and C. Therefore, γ is an inscribed circle for $\triangle ABC$.

The proof of uniqueness is left as an exercise (Exercise 10.10). $\qquad\square$

We end this section with a theorem that asserts the existence of polygons inscribed in a circle. The theorem is neutral, but we will refer to it later in this chapter when we investigate circumference and area for circles in Euclidean geometry and in the following chapter when we study compass and straightedge constructions. We must first define polygon.

Definition 10.3.9. Fix an integer $n \geqslant 3$. Suppose $P_1, P_2, \ldots, P_n$ are n distinct points such that no three of the points are collinear. Suppose further that any two of the segments $\overline{P_1P_2}$, $\overline{P_2P_3}$, ..., $\overline{P_{n-1}P_n}$, and $\overline{P_nP_1}$ are either disjoint or share an endpoint. If those conditions are satisfied, the points $P_1, P_2, ..., P_n$ determine a *polygon*. The polygon is defined to be the union of the segments $\overline{P_1P_2}$, $\overline{P_2P_3}$, ..., $\overline{P_{n-1}P_n}$ and is denoted by $P_1P_2\cdots P_n$. In symbols,

$$P_1P_2\cdots P_n = \overline{P_1P_2} \cup \overline{P_2P_3} \cup \cdots \cup \overline{P_{n-1}P_n} \cup \overline{P_nP_1}.$$

The points $P_1, P_2, ..., P_n$ are called the *vertices* of the polygon and the segments $\overline{P_1P_2}$, $\overline{P_2P_3}$, ..., $\overline{P_{n-1}P_n}$, and $\overline{P_nP_1}$ are called the *sides* of the polygon. A polygon with n sides is called an n-sided polygon or simply an n-gon.

FIGURE 10.17: Polygons

Obviously polygon is a generalization of triangle and quadrilateral: A triangle is a three-sided polygon and a quadrilateral is a four-sided polygon. Other familiar names that are used for polygons with a specified number of sides are pentagon (five sides), hexagon (six sides), heptagon (seven sides), octagon (eight sides), nonagon (nine sides), decagon (ten sides), dodecagon (twelve sides), and so on.

Definition 10.3.10. A polygon $P_1P_2\cdots P_n$ is a *regular polygon* if all the segments $\overline{P_1P_2}$, $\overline{P_2P_3}$, ..., $\overline{P_{n-1}P_n}$, and $\overline{P_nP_1}$ are congruent to each other and all the angles $\angle P_1P_2P_3, \angle P_2P_3P_4, ..., \angle P_{n-2}P_{n-1}P_n, \angle P_{n-1}P_nP_1$, and $\angle P_nP_1P_2$ are congruent to each other.

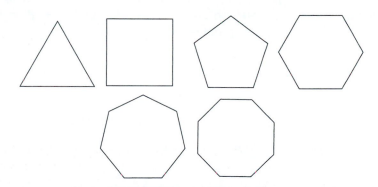

FIGURE 10.18: Regular polygons

Definition 10.3.11. A polygon $P_1P_2\cdots P_n$ is said to be *inscribed in the circle* γ if all the vertices of $P_1P_2\cdots P_n$ lie on γ.

Theorem 10.3.12. *Let γ be a circle and P_1 a point on γ. For each $n \geqslant 3$ there is a regular polygon $P_1 P_2 \cdots P_n$ inscribed in γ.*

Proof. Exercise 10.11.   □

Note that the theorem merely asserts that the inscribed polygon exists. When we prove a theorem like this we usually say we have "constructed" the polygon. That is true in the sense that we construct it using a protractor. In the next chapter we will study constructions using only a compass and straightedge. That kind of construction is much more delicate and only certain regular special polygons can be constructed in that sense.

10.4 CIRCLES IN EUCLIDEAN GEOMETRY

In this section we investigate some of the special properties of angles inscribed in Euclidean circles. *The Euclidean Parallel Postulate is assumed for the remainder of the section.* The first two theorems show that a given angle in a triangle is right if and only if the side opposite the angle is a diameter of the circumscribed circle.

Theorem 10.4.1. *Let $\triangle ABC$ be a triangle and let M be the midpoint of $\overline{AB}$. If $AM = MC$, then $\angle ACB$ is a right angle.*

Corollary 10.4.2. *If the vertices of triangle $\triangle ABC$ lie on a circle and $\overline{AB}$ is a diameter of that circle, then $\angle ACB$ is a right angle.*

The corollary is usually paraphrased this way: An angle inscribed in a semicircle is a right angle. It is Euclid's Proposition III.31.

Proof of Theorem 10.4.1. Let $\alpha = \mu(\angle BAC)$ and let $\beta = \mu(\angle ABC)$. By the Isosceles Triangle Theorem we have $\alpha = \mu(\angle ACM)$ and $\beta = \mu(\angle MCB)$ as well. Since M is between A and B, $\mu(\angle ACB) = \mu(\angle ACM) + \mu(\angle MCB)$. Now $\sigma(\triangle ABC) = 180°$, so $2\alpha + 2\beta = 180°$ and $\alpha + \beta = 90°$. Therefore, $\mu(\angle ACB) = 90°$ and the proof is complete.   □

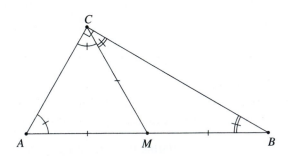

FIGURE 10.19: Theorem 10.4.1

The converse of Theorem 10.4.1 is also a theorem.

Theorem 10.4.3. *Let* $\triangle ABC$ *be a triangle and let M be the midpoint of* $\overline{AB}$. *If* $\angle ACB$ *is a right angle, then* $AM = MC$.

Proof. Exercise 10.12.   □

Corollary 10.4.4. *If* $\angle ACB$ *is a right angle, then* $\overline{AB}$ *is a diameter of the circle that circumscribes* $\triangle ABC$.

Theorem 10.4.5 (The 30-60-90 Theorem). *If the interior angles in triangle* $\triangle ABC$ *measure* $30°, 60°,$ *and* $90°$, *then the length of the side opposite the* $30°$ *angle is one half the length of the hypotenuse.*

Proof. Exercise 10.13.   □

Theorem 10.4.6 (Converse to the 30-60-90 Theorem). *If* $\triangle ABC$ *is a right triangle such that the length of one leg is one-half the length of the hypotenuse, then the interior angles of the triangle measure* $30°, 60°,$ *and* $90°$.

Proof. Exercise 10.14.   □

Definition 10.4.7. Let $\gamma = \mathcal{C}(O, r)$ be a circle. An *inscribed angle* for γ is an angle of the form $\angle PQR$, where P, Q, and R all lie on γ. The *arc intercepted* by the inscribed angle $\angle PQR$ is the set of points on γ that lie in the interior of $\angle PQR$. A *central angle* for γ is an angle of the form $\angle POR$, where P and R lie on γ.

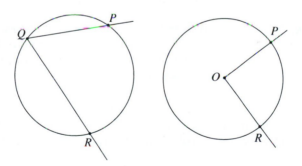

FIGURE 10.20: An inscribed angle and a central angle

There is a natural correspondence between inscribed angles and central angles.

Definition 10.4.8. Suppose $\angle PQR$ is an inscribed angle for $\mathcal{C}(O, r)$ such that either Q and R lie on opposite sides of $\overleftrightarrow{OP}$ or P and Q lie on opposite sides of $\overleftrightarrow{OR}$. In that case, $\angle POR$ is called the *corresponding* central angle.

The following theorem is the main result of the section. It is Euclid's Proposition III.20; the corollary is Euclid's Proposition III.21.

Theorem 10.4.9 (Central Angle Theorem). *The measure of an inscribed angle for a circle is one half the measure of the corresponding central angle.*

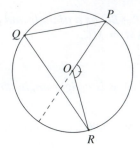

FIGURE 10.21: Corresponding central angle

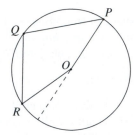

FIGURE 10.22: The central angle does not correspond

Remark. It is fairly clear from the definition of corresponding central angle that an inscribed angle has a corresponding central angle if and only if the inscribed angle has measure less than $90°$. This fits with the theorem since the corresponding central angle must have measure less than $180°$. The present situation is one in which we would be better off if we allowed angles of measure $180°$ and greater. If we allowed that, then we would easily be able to identify a central angle corresponding to every inscribed angle and the Central Angle Theorem would hold in all cases. We would also be able to see that Corollary 10.4.2 is really just a special case of Theorem 10.4.9.

Before getting to the proof of the Central Angle Theorem, let us look at an important corollary.

Corollary 10.4.10 (Inscribed Angle Theorem). *If two inscribed angles intercept the same arc, then the angles are congruent.*

Proof of Corollary 10.4.10. In case there is a corresponding central angle, the corollary follows immediately from the Central Angle Theorem since the intercepted arc determines the central angle uniquely. (See Fig. 10.23.) In general, let $\angle PQR$ be an inscribed angle. Choose S to be the antipodal point for Q. The inscribed angles $\angle PQS$ and $\angle SQR$ both have corresponding central angles. (See Fig. 10.24.) The first case of the proof can be applied to them. □

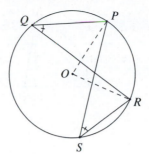

FIGURE 10.23: The inscribed angles $\angle PQR$ and $\angle PSR$ are congruent because they share the same central angle

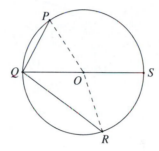

FIGURE 10.24: The inscribed angles $\angle PQS$ and $\angle SQR$ both have corresponding central angles

Proof of the Central Angle Theorem. Let $\angle PQR$ be inscribed in circle $\gamma = \mathcal{C}(O, r)$ (hypothesis). There are three cases to consider: O lies on one of the sides of $\angle PQR$, O lies in the interior of $\angle PQR$, or O is neither on the angle nor in its interior.

Assume, first, that $Q * O * R$ (see Fig. 10.25). The proof of this case is similar to the proof of Theorem 10.4.1. Let $\alpha = \mu(\angle OQP)$ and let $\beta = \mu(\angle ORP)$. By the Isosceles Triangle Theorem, $\alpha = \mu(\angle QPO)$ and $\beta = \mu(\angle OPR)$. Therefore, $\sigma(\triangle QPR) = 2\alpha + 2\beta$ and so $2\alpha + 2\beta = 180°$ (Angle Sum Theorem). Also by the Angle Sum Theorem, $\mu(\angle POR) = 180° - 2\beta$. Replacing $180°$ by $2\alpha + 2\beta$ in the last equation gives $\mu(\angle POR) = 2\alpha$, which is the desired result.

Next assume that O lies in the interior of $\angle PQR$ (see Fig. 10.26). Let S be the point on γ such that $Q * O * S$. (The points Q and S are antipodal points.) Then $\mu(\angle PQR) = \mu(\angle PQS) + \mu(\angle RQS)$ and $\mu(\angle POR) = \mu(\angle POS) + \mu(\angle ROS)$ (Protractor Postulate). Hence this case follows from the previous case by addition.

Finally, assume that O is not on the angle $\angle PQR$ nor in the interior of $\angle PQR$ (see Fig. 10.27). Again let S be the point on the circle such that $Q * O * S$. Either P is in the interior of $\angle RQS$ or R is in the interior of $\angle PQS$. Change the notation, if necessary, so that R is in the interior of $\angle PQS$. Then $\mu(\angle PQR) = \mu(\angle PQS) - \mu(\angle RQS)$ (Protractor Postulate). Therefore, the conclusion follows from the first case by subtraction. □

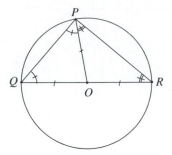

FIGURE 10.25: Central Angle Theorem, Case 1

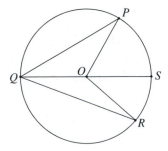

FIGURE 10.26: Central Angle Theorem, Case 2

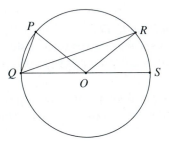

FIGURE 10.27: Central Angle Theorem, Case 3

The final topic in this section is one that will prove to be very useful in a later chapter when we study inversions in circles. In order to be consistent with the application, we will change notation just a little.

Definition 10.4.11. Let β be a circle and let O be a point that does not lie on β. The *power* of O with respect to β is defined as follows: Choose any line ℓ through O that intersects β. If ℓ is a secant line that intersects β at points Q and R, define the power of O to be the product $(OQ)(OR)$. In case ℓ is tangent to β at P, define the power of O to be $(OP)^2$.

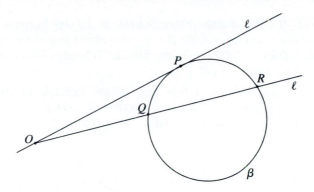

FIGURE 10.28: The power of a point with respect to a circle

The power of a point appears to depend on which line ℓ is used in the definition. However, the next theorem asserts that the number will always be the same, regardless of which line ℓ is used. The theorem is Euclid's famous Proposition III.36.

Theorem 10.4.12. *The power of a point is well defined; that is, the same value is obtained regardless of which line ℓ is used in the definition as long as the line has at least one point of intersection with the circle.*

Proof. Let β be a circle, let r be the radius of β, and let O be a point that does not lie on β. If O is the center of β, then every line ℓ passing through O intersects β in two points Q and R. Furthermore, $(OQ)(OR) = r^2$ regardless of which line ℓ is used, so the power of O with respect to β is well defined in this case.

Now assume that $O \neq B$ where B is the center of β. Let S and T be the two points of intersection of $\overleftrightarrow{OB}$ and β. The proof of the theorem will be completed by proving two claims. First, if ℓ is any secant line passing through O such that ℓ intersects β in the two points Q and R, then $(OQ)(OR) = (OS)(OT)$. Second, if ℓ is a line that passes through O and is tangent to β at P, then $(OP)^2 = (OS)(OT)$.

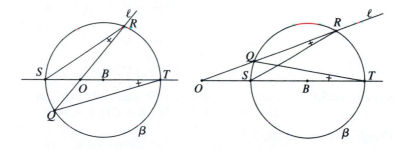

FIGURE 10.29: ℓ is a secant line for β; either O is inside β or O is outside β

If ℓ is a secant line that passes through two distinct points Q and R of β, then $\angle ORS \cong \angle OTQ$ because both angles intercept the same arc on β (Inscribed Angle

Theorem 10.4.10). Furthermore, $\angle ROS \cong \angle TOQ$ because the angles are either identical (in case O is outside β) or are vertical angles (in case O is inside β). As a result, $\triangle ORS \sim \triangle OTQ$. By the Similar Triangles Theorem, $OR/OS = OT/OQ$ and so $(OR)(OQ) = (OS)(OT)$.

Now suppose ℓ is a line that passes through O and is tangent to β at P. By the Pythagorean Theorem, $(OP)^2 = (OB)^2 - (BP)^2$. Thus $(OP)^2 = (OB - BP)(OB + BP) = (OB - BS)(OB + BT) = (OS)(OT)$. □

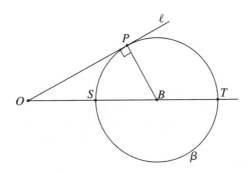

FIGURE 10.30: ℓ is tangent to β

10.5   CIRCULAR CONTINUITY

In this section we return to neutral geometry to prove another version of circular continuity. The theorems in the section will be needed in a minor way in the next section when we investigate the circumference and area of a circle. They will be needed in a major way in the next chapter when we explore some of Euclid's compass-and-straightedge constructions. In such constructions, points are specified by the intersection of two circles. In Chapter 1, for example, we looked at Euclid's construction of an equilateral triangle—the very first proof in Book I of the *Elements*. We found that Euclid made the (unjustified) assumption that the two circles he constructed must intersect. This section supplies the justification for that assumption: We will prove that if γ and γ' are two circles and there is one point of γ that is inside γ' and another point of γ that is outside γ', then γ and γ' must intersect in exactly two points. *The theorems in this section are all part of neutral geometry.*

We will need the following simple lemma about triangles.

Lemma 10.5.1. *If $\triangle ABC$ is a right triangle with right angle at vertex C and D is a point such that $A * D * B$ and $AD = AC$, then $CD < CB$.*

Proof. Exercise 10.20. □

Construction. Let $\gamma = \mathcal{C}(O, r)$ be a circle and let C be a point on γ. Let t be the line that is tangent to γ at C and choose a point B on t. For each $x, 0 < x < \mu(\angle BOC)$, define P_x to be the point on γ such that P_x lies in the interior of $\angle BOC$ and $\mu(\angle P_x OC) = x°$. (See Fig. 10.31.)

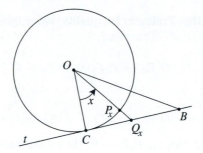

FIGURE 10.31: The construction of P_x

Lemma 10.5.2. $\lim_{x \to 0^+} CP_x = 0$.

Proof. Assume the notation of the construction, above. For each x, $0 < x < \mu(\angle BOC)$, the ray $\overrightarrow{OP_x}$ must intersect $\overline{CB}$ in a point Q_x (Crossbar Theorem). By Lemma 10.5.1, $0 < CP_x < CQ_x$ and by the Continuity Axiom (Theorem 5.7.27), $\lim_{x \to 0^+} CQ_x = 0$. Therefore, $\lim_{x \to 0^+} CP_x = 0$ by the Sandwich Theorem for Limits (calculus). □

Next we prove that distances vary continuously as a point moves around a circle. In order to make that assertion precise, we define a function whose domain is angle measure and whose range is distance.

Construction. Let $\gamma = \mathcal{C}(O, r)$ be a circle, let C be a point on γ, and let O' be a point not on γ. Choose one side of $\overleftrightarrow{OC}$. For each number $x \in [0, 180)$, define P_x to be the point on the specified side of $\overleftrightarrow{OC}$ such that $OP_x = r$ and $\mu(\angle COP_x) = x°$. Define a function $f : (0, 180) \to [0, \infty)$ by $f(x) = O'P_x$. (See Fig. 10.32.)

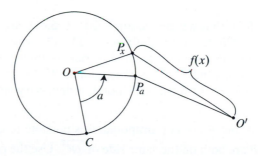

FIGURE 10.32: The construction of $f(x)$

Theorem 10.5.3. *The function f (defined above) is continuous.*

Proof. Fix $a \in (0, 180)$. In order to prove that f is continuous at a we must show that $\lim_{x \to a} f(x) = f(a)$.

Applying the Triangle Inequality twice gives $O'P_x < O'P_a + P_aP_x$ and $O'P_a < O'P_x + P_aP_x$. Therefore,

$$O'P_a - P_aP_x < O'P_x < O'P_a + P_aP_x$$

or

$$f(a) - P_aP_x < f(x) < f(a) + P_aP_x.$$

By Lemma 10.5.2, $\lim_{x \to a} P_aP_x = 0$. Hence the Sandwich Theorem for Limits (from calculus) gives $\lim_{x \to a} f(x) = f(a)$. ☐

We can now prove the main result of the section.

Theorem 10.5.4 (Circular Continuity Principle). *Let $\gamma = C(O, r)$ and $\gamma' = C(O', r')$ be two circles. If there exists a point of γ that is inside γ' and there exists another point of γ that is outside γ', then $\gamma \cap \gamma'$ consists of exactly two points.*

Proof. Let γ and γ' be as in the hypotheses of the theorem. There exists a point $A \in \gamma$ such that A is outside γ' and there exists a point $B \in \gamma$ such that B is inside γ'.

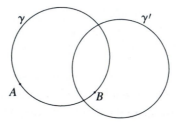

FIGURE 10.33: A is inside γ' and B is outside γ'

Note, first, that we may assume that A and B are not antipodal points on γ. Suppose A and B are antipodal points. The distance AO' is strictly greater than r'. The fact that the distance from O' varies continuously as a point moves around the circle γ means that $A'O' > r'$ for points A' on γ that are close to A. We can therefore replace A with a nearby point, if necessary, and ensure that A and B are not antipodal.

Since A and B are not antipodal, it is possible to choose a point C on γ such that A and B are both on the same side of $\overleftrightarrow{OC}$. Use the point C and that side of $\overleftrightarrow{OC}$ to define a function f as in Theorem 10.5.3. There exist numbers a and b between 0 and 180 such that $P_a = A$ and $P_b = B$. Then f is continuous, $f(a) > r'$, and $f(b) < r'$, so we can apply the Intermediate Value Theorem to conclude that there must be a number c such that $f(c) = r'$. The point $Q = P_c$ is a point of $\gamma \cap \gamma'$.

The proof that there must be a second point in $\gamma \cap \gamma'$ is left as an exercise (Exercise 10.21). ☐

10.6 CIRCUMFERENCE AND AREA OF EUCLIDEAN CIRCLES

In this section we investigate the familiar formulas $C = 2\pi r$ and $A = \pi r^2$ for the circumference and area of a circle of radius r. There are several things about these formulas that should be understood at the outset.

First, these are Euclidean formulas. In Euclidean geometry the circumference of a circle is proportional to the radius, but that is not true in hyperbolic geometry. In hyperbolic geometry the circumference of a circle grows exponentially with the radius. The hyperbolic formulas for circumference and area give a lot of insight into the structure of hyperbolic space, but developing such formulas would require going too far beyond the techniques expounded here so we restrict our study to Euclidean geometry. The fact that the familiar formulas are Euclidean will be apparent from our use of both the Similar Triangles Theorem and the Euclidean formula for the area of a triangle in the proofs in this section.

Second, before we can understand and appreciate these formulas we must define what we mean by the circumference and area of a circle. The Ruler Postulate allows us to measure the length of a straight line segment and it is easy to generalize to length of a polygonal path (one made up of a finite number of straight line segments laid end to end). But generalizing length to curved paths, such as circles, requires a limiting process. You have already encountered the relevant limiting process in your calculus course when you used integrals to compute lengths of curves. It is interesting to observe, however, that the idea of using a limiting process to calculate the length of a curve is due to the ancient Greeks and predates the invention of calculus by two millennia.

Finally, it should be noted that the formulas for the circumference and area of a circle cannot be separated from the definition of the real number π. In fact the main reason for the inclusion of this discussion of circumference and area is the need to clarify that relationship. After we have defined what we mean by the circumference of a circle we will prove that the ratio of the circumference to the diameter is the same for all circles. That will allow us to define π to be that common ratio. In other words, the formula $C = 2\pi r$ is just a definition and the theorem is the assertion that C/d is the same for all circles.

The origins of the ideas in this section are ancient. The so-called "method of exhaustion" is usually attributed to Eudoxus of Cnidus (408–355 B.C.). Using it, the circumference of a circle can be defined to be the least upper bound of the perimeters of polygons inscribed in the circle. The Greeks knew that if two circles of diameters d_1 and d_2 had circumferences C_1 and C_2, respectively, then the ratio of the circumferences would be the same as the ratio of the diameters; that is,

$$\frac{C_1}{C_2} = \frac{d_1}{d_2}.$$

From this it follows that

$$\frac{C_1}{d_1} = \frac{C_2}{d_2}$$

for any two circles, so the ratio of circumference to diameter is the same for all circles. That ratio is the number we will define to be π.

Euclid does not state any propositions about the circumference of a circle, but he does state a proposition about the areas of circles.[1] Proposition XII.2 states that if the circles have areas A_1 and A_2, then the ratio of the areas is equal to the ratio of the squares of the diameters; that is,

$$\frac{A_1}{A_2} = \frac{d_1^2}{d_2^2}.$$

From this it follows that there is a constant k such that for every circle $A = kd^2$.

There is no indication that Euclid connected this constant k with the constant in the formula relating circumference to diameter. The fact that there is a relationship between the two constants is quite remarkable since one constant relates lengths while the other relates areas. The connection between the constants in the two formulas was first discovered by Archimedes of Syracuse (287–212 B.C.). He wrote a short book entitled *Measurement of a Circle* whose first proposition states that the area of a circle is equal to that of a right triangle whose height is the radius of the circle and whose base is the circumference of the circle. He proved this using the method of exhaustion. Archimedes also found accurate approximations to the numerical value of π by carefully calculating the perimeters of polygons inscribed in the circle. Using inscribed and circumscribed polygons with as many as 96 sides he was able to prove that $3\frac{10}{71} < \pi < 3\frac{1}{7}$.

Even more accurate approximations to the numerical value of π were obtained by Tsu Ch'ung Chi (430–501) and by al-Khwarizmi (780–850). In 1761 Johann Lambert (1728–1777) proved that π is irrational, so these numerical values for π can never be more than approximations. Not only is π an irrational number, but it is *transcendental*, which means that it is not the solution to any polynomial equation with integer coefficients.[2] The fact that π is transcendental was first established by Ferdinand von Lindenmann in 1882.

This section contains only an outline of the definition of π and the proof of Archimedes' formula. In particular, we will define the circumference to be the limit of the perimeters of a particular sequence of inscribed polygons. We prefer not to get into the technicalities of showing that other sequences of inscribed polygons produce the same limit or that inscribed and circumscribed polygons converge to a common limit; instead we will leave such technicalities regarding limits to courses in calculus and real analysis. For historical reasons it seems appropriate to follow Archimedes and use regular polygons with $3 \cdot 2^n$ sides.

The Euclidean Parallel Postulate is assumed for the remainder of this section.

Definition 10.6.1. The *perimeter* L of polygon $P_1 P_2 \cdots P_n$ is defined by

$$L = P_1 P_2 + P_2 P_3 + \cdots + P_{n-1} P_n + P_n P_1.$$

---

[1] It is interesting to note that Euclid begins Book XII with two propositions about the area of circles, but all the remaining propositions in Book XII concern volumes of solid figures.

[2] By contrast, $x = \sqrt{2}$ satisfies the equation $x^2 - 2 = 0$.

Construction. Let $\gamma = \mathcal{C}(O, r)$ be a circle and let $\overline{P_1 P_4}$ be a diameter of γ. Choose one side of $\overleftrightarrow{P_1 P_4}$ and let P_2 and P_3 be two points on that side of $\overleftrightarrow{P_1 P_4}$ such that $\mu(\angle P_1 O P_2) = \mu(\angle P_2 O P_3) = 60°$ and $OP_2 = OP_3 = r$. Finally, choose points P_5 and P_6 on the opposite side of $\overleftrightarrow{P_1 P_4}$ such that $\mu(\angle P_4 O P_5) = \mu(\angle P_5 O P_6) = 60°$ and $OP_5 = OP_6 = r$. It is easy to check that $H_1 = P_1 P_2 \cdots P_6$ is a regular hexagon inscribed in γ. Let L_1 be the perimeter of H_1.

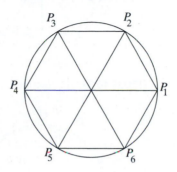

FIGURE 10.34: The inscribed hexagon H_1

Observe that H_1 is not uniquely determined by this construction. The initial choice of diameter $\overline{P_1 P_4}$ is arbitrary, as is the choice of a side of $\overleftrightarrow{P_1 P_4}$. If some other diameter $\overline{P_1' P_4'}$ is chosen, or if a different side of $\overleftrightarrow{P_1 P_4}$ is chosen, then all of the triangles $\triangle OP_i P_{i+1}$ and $\triangle OP_i' P_{i+1}'$ are congruent. Thus L_1, the perimeter of H_1, is completely determined by the construction.

Starting with H_1, we now construct a sequence $H_1, H_2, H_3, \ldots$ in which H_n is a regular $3 \cdot 2^n$-sided polygon inscribed in γ in such a way that all the vertices of H_{n-1} are also vertices of H_n. Here is the construction of H_2: For $i = 1, \ldots, 5$, take Q_i to be the point on the bisector of angle $\angle P_i O P_{i+1}$ such that $OQ_i = r$ and take Q_6 to be the point on the bisector of angle $\angle P_6 O P_1$ such that $OQ_6 = r$. Define H_2 to be the twelve-sided polygon $P_1 Q_1 P_2 Q_2 \cdots P_6 Q_6$ and define L_2 to be the perimeter of H_2. This process is continued in the obvious way to define $H_n, n \geq 3$. Define L_n to be the perimeter of H_n. Notice that $H_n, n \geq 2$, is uniquely determined by H_1 so L_n is a well-defined number associated with γ.

The following theorem is an easy consequence of the Triangle Inequality. (See Fig. 10.36.)

Theorem 10.6.2. *The sequence $L_1, L_2, L_3, \ldots$ is a strictly increasing sequence.*

Proof. Exercise 10.23.   □

We want to define the circumference of γ to be the limit of the sequence $\{L_1, L_2, L_3, \ldots\}$. We already know that this is an increasing sequence; in order to be certain that it converges we also need to know that it is bounded above. To prove

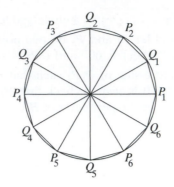

FIGURE 10.35: The inscribed dodecagon H_2

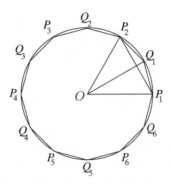

FIGURE 10.36: $L_1 < L_2$

that, we make use of a circumscribed square whose perimeter is greater than that of every H_n.

Construction. Let $\gamma = \mathcal{C}(O, r)$ be a circle and assume the notation of the previous construction. Choose points R_1, R_2, R_3, and R_4 so that $\square O P_1 R_1 Q_2$, $\square O Q_2 R_2 P_4$, $\square O P_4 R_3 Q_5$, and $\square O Q_5 R_4 P_1$ form squares. It is easy to see that $\square R_1 R_2 R_3 R_4$ is also a square, that each side of $\square R_1 R_2 R_3 R_4$ is tangent to γ, and that the perimeter of $\square R_1 R_2 R_3 R_4$ is $8r$. The square $S = \square R_1 R_2 R_3 R_4$ is called the *circumscribed square*.

Theorem 10.6.3. *For each n, L_n is less than the perimeter of the circumscribed square.*

Proof. Exercise 10.24.   ☐

The last two theorems tell us that the sequence $\{L_n\}$ is increasing and bounded. By a standard theorem from calculus, this implies that the sequence converges. The circumference is the limit of that sequence. This should seem intuitively right since the inscribed polygons are getting closer and closer to the circle.

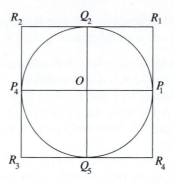

FIGURE 10.37: The circumscribed square S

Definition 10.6.4. Let $\gamma = C(O, r)$ be a circle. The *circumference* of $\gamma = C(O, r)$ is defined by

$$C = \lim_{n \to \infty} L_n.$$

Theorem 10.6.5. *The circumference of $C(O, r)$ is independent of the center point O and depends only on the radius r.*

Proof. Exercise 10.25. □

Theorem 10.6.6. *If r and r' are two positive numbers and C and C' are the circumferences of the circles $\gamma = C(O, r)$ and $\gamma' = C(O, r')$, respectively, then $C/r = C'/r'$.*

Proof. The triangles $\triangle OP_1 P_2$ and $\triangle OP_1' P_2'$ are similar triangles. (See Figure 10.38.) By the Similar Triangles Theorem, $P_1 P_2 / OP_1 = P_1' P_2' / OP_1'$ and so $P_1 P_2 / r = P_1' P_2' / r'$. Since this is true of each of the sides of the inscribed hexagons, it follows by addition that $L_1/r = L_1'/r'$. The Similar Triangles Theorem can be used in exactly the same way to show that $L_n/r = L_n'/r'$ for every n. Taking the limit as $n \to \infty$ gives the desired result. □

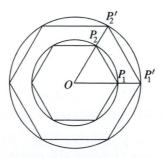

FIGURE 10.38: The triangles are similar, with common ratio r/r'

Definition 10.6.7. The number π is one-half the common ratio; that is, if $C(O, r)$ is a circle of radius r and circumference C, then $\pi = C/2r$.

Since this ratio is the same for all circles, regardless of the radius (Theorem 10.6.6), π is a well-defined real number.

Be sure to notice that the formula $C = 2\pi r$ is a definition, not a theorem. The fact that C/r is the same for all circles was proved using the Similar Triangles Theorem. Therefore, we have only proved that the formula $C = 2\pi r$ is valid in Euclidean geometry.

Since each of the triangles $\triangle OP_i P_{i+1}$ is equilateral, we see that $L_1 = 6r$ and therefore $\pi > 3$. On the other hand, the perimeter of the circumscribed square is $8r$, so $\pi < 4$. These numerical estimates are very crude.

Before we can define the area of a circle we need some preliminary definitions.

Definition 10.6.8. Let $\gamma = C(O, r)$ be a circle. The *interior* of γ consists of all points inside γ. In other words,

$$\text{Int}(\gamma) = \{X \mid OX < r\}.$$

The *circular region* determined by γ is the union of the circle γ and its interior $\text{Int}(\gamma)$.

FIGURE 10.39: A circular region

It is easy to identify a polygonal region $\overline{H}_n$ associated with the inscribed polygon H_n. For each side of H_n, there is a triangle whose vertices are O and the two endpoints of the edge. Define $\overline{H}_n$ to be the union of those n triangles.

Theorem 10.6.9. *For each n, $\alpha(\overline{H}_n) < \alpha(\overline{H}_{n+1})$ and $\alpha(\overline{H}_n) < \alpha(S)$.*

Proof. Exercise 10.26. ☐

The last theorem allows us to define the area of the circular region to be the limit of the areas of the inscribed polygonal regions.

Definition 10.6.10. Let $\gamma = C(O, r)$ be a circle and let $R(\gamma)$ be the associated circular region. The *area of the circular region* associated with γ is defined by

$$\alpha(R(\gamma)) = \lim_{n \to \infty} \alpha(\overline{H}_n).$$

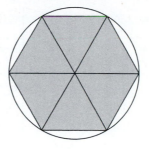

FIGURE 10.40: The inscribed hexagonal region $\overline{H}_1$

We come now to Archimedes' Theorem.

Theorem 10.6.11 (Archimedes' Theorem). *If γ is a circle of radius r, C is the circumference of γ, and A is the area of the associated circular region, then*

$$A = (1/2)rC.$$

Note that this formula is also Euclidean; it is just a manifestation of the fact that the area of a Euclidean triangle is 1/2 base times height. The following corollary is immediate.

Corollary 10.6.12. *If γ is a circle of radius r, C is the circumference of γ, and A is the area of the associated circular region, then $A = \pi r^2$.*

Archimedes' Theorem can be understood intuitively. Start with a rectangle of size $r \times (1/2)C$. Subdivide it into triangles as indicated in Figure 10.41. In the diagram there are eleven congruent isosceles triangles. The two remaining triangles together form a twelfth isosceles triangle which is congruent to the others. These twelve triangles can be cut apart and reassembled to approximately cover a circle of radius r and circumference C as in Fig. 10.42. If the rectangle were subdivided into more triangles, they could be cut apart and reassembled to more exactly cover the circle.

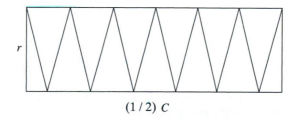

$(1/2)\ C$

FIGURE 10.41: The rectangle is dissected into triangles

Before proving Archimedes' Theorem we state a definition and make some simple observations about it.

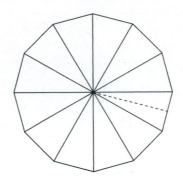

FIGURE 10.42: The triangles are reassembled to approximately cover the circle

Definition 10.6.13. Let $P_1 P_2 \cdots P_n$ be a regular polygon inscribed in the circle $C(O, r)$. The distance from O to one of the sides of the polygon is called the *apothem* of the polygon.

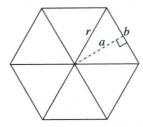

FIGURE 10.43: a is the apothem of the hexagon

Since the polygon is regular, O is the same distance from all the sides. It follows from Theorem 6.4.4 that the apothem is measured along the perpendicular from O to one of the sides.

Let $P_1 P_2 \cdots P_n$ be a regular polygon inscribed in the circle $C(O, r)$ and let a be the apothem. We can make the following observations.

1. If b is the length of one of the sides of the polygon, then

$$r - (1/2)b < a < r.$$

2. If A is the area of the polygonal region determined by $P_1 P_2 \cdots P_n$ and L is the perimeter of the polygon, then

$$A = (1/2)aL.$$

The first observation follows from the Triangle Inequality and the Scalene Inequality (Theorems 6.4.2 and 6.4.1). The second observation follows immediately from the Euclidean formula for the area of a triangle (Theorem 9.2.5) and the definitions of apothem and perimeter.

We now have all the ingredients we need to prove the theorem.

Proof of Archimedes' Theorem. Let γ be a circle of radius r, let C be the circumference of γ, and let A be the area of the associated circular region. As before, H_n, $\overline{H}_n$, and L_n denote the inscribed polygon, the associated polygonal region, and the perimeter, respectively. By definition, $A = \lim_{n\to\infty} \alpha(\overline{H}_n)$. It follows from the second observation, above, that $\alpha(\overline{H}_n) = (1/2)a_n L_n$, where a_n is the apothem of H_n. By the first observation and Lemma 10.5.2, $\lim_{n\to\infty} a_n = r$. Therefore

$$
\begin{aligned}
A &= \lim_{n\to\infty} \alpha(\overline{H}_n) \\
&= \lim_{n\to\infty} (1/2)a_n L_n \\
&= (1/2)\left(\lim_{n\to\infty} a_n\right)\left(\lim_{n\to\infty} L_n\right) \\
&= (1/2)rC.
\end{aligned}
$$

$\square$

10.7 EXPLORING EUCLIDEAN CIRCLES

In this section we explore some additional circles that are important in advanced Euclidean geometry. Most of the circles we will study are associated with triangles. The section does not contain any proofs, but proofs of several of the theorems are outlined in the exercises.

This section is another opportunity to explore advanced Euclidean geometry using dynamic software. It builds on the results of Section 7.7, and that section is a necessary prerequisite. In addition to the operations listed at the beginning of Section 7.7, you should also know how to use your software to construct circles.

If you have not already done so, you should use your software to construct the circumscribed and inscribed circles associated with a triangle $\triangle ABC$ and you should experiment with what happens to those circles when the vertices of the triangle are moved. Locate the center of the circumscribed circle as the point at which two of the perpendicular bisectors of the sides intersect. The circle itself should be defined using that center point and one of the vertices of the triangle.

The inscribed circle is the only circle that is tangent to all three sides of a triangle, but it is not the only circle that is tangent to the three lines determined by the sides. In order to simplify terminology, let us refer to the line determined by a side of a triangle as a *sideline* of the triangle. Any circle that is tangent to all three sidelines of a triangle is called an *equicircle*. There are a total of four equicircles. One of them, the incircle, is inside the triangle; the other three are called *excircles* (or sometimes *escribed circles*) and they are outside the triangle. (See Fig. 10.44.) Try to construct the three excircles using your dynamic software.

There is one excircle for each side of the triangle. Let us denote the excircle opposite vertex A by γ_A. Then γ_A must be tangent to $\overline{BC}$ and also tangent to the

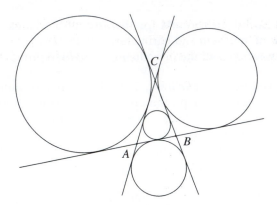

FIGURE 10.44: The four equicircles for $\triangle ABC$

rays $\overrightarrow{AB}$ and $\overrightarrow{AC}$. Therefore, the center of γ_A is equidistant from $\overline{BC}$, $\overrightarrow{AB}$, and $\overrightarrow{AC}$. It follows (Theorem 6.4.6) that the center of γ_A must lie on the bisectors of the interior angle at A and the two exterior angles at B and C. Verify that these three angle bisectors are concurrent and use this observation to construct the three excircles.

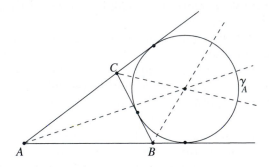

FIGURE 10.45: The excircle γ_A

Perhaps the most famous of all the circles associated with a triangle is the *nine-point circle*. As the name suggests, this is a single circle that passes through nine separate points that are important for the triangle. The nine points are the midpoints of the three sides, the feet of the three altitudes, and the midpoints of the three segments joining the orthocenter to the vertices. This should be enough of a hint to allow you to construct the nine-point circle. In fact you can use the tools you constructed in Chapter 7: Start with $\triangle ABC$, construct the medial triangle for $\triangle ABC$, and then find the circumscribed circle for the medial triangle. You should perform this construction and then verify that the other six points really do lie on this same circle. Here is the theorem.

Theorem 10.7.1 (Nine-Point Circle Theorem). *If △ABC is any triangle, then the midpoints of the sides of △ABC, the feet of the altitudes of △ABC, and the midpoints of the segments joining the orthocenter of △ABC to the three vertices all lie on a single circle.*

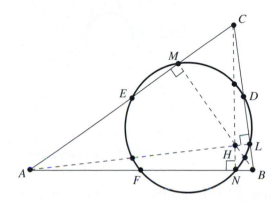

FIGURE 10.46: The nine-point circle for △ABC

In 1765, Leonard Euler proved that the first six points listed in the theorem all lie on a circle. It was not until 1820 that Charles-Julien Brianchon (1783–1864) and Jean-Victor Poncelet (1788–1867) proved that the remaining points (the midpoints of the segments joining the orthocenter to the vertices) also lie on the same circle.

Hide the nine points and keep just the nine-point circle itself visible. Now add in the four equicircles that you constructed earlier in this section. The following amazing fact about the nine-point circle was discovered in 1822 by the German mathematician Karl Wilhem Feuerbach (1800–1834).

Theorem 10.7.2 (Feuerbach's Theorem). *The nine-point circle is tangent to each of the four equicircles.*

Let us look at one more theorem regarding circles and triangles. This theorem is not as well known as Feuerbach's Theorem but is also quite surprising. Start with a triangle △ABC. Choose arbitrary points D, E, and F in the interiors of the sides $\overline{BC}$, $\overline{AC}$, and $\overline{AB}$, respectively. Now construct the three circumscribed circles for the triangles △AEF, △BDF, and △CDE. These circles are called *Miquel circles* for △ABC. You should observe that the three Miquel circles have a point in common.

Theorem 10.7.3 (Miquel's Theorem). *If △ABC is any triangle and points D, E, and F are chosen in the interiors of the sides $\overline{BC}$, $\overline{AC}$, and $\overline{AB}$, respectively, then the circumcircles for △AEF, △BDF, and △CDE intersect in a point M.*

This theorem was discovered by the French mathematician Augueste Miquel in 1832. The point M in the theorem is called a *Miquel point* for the triangle.

Polygons inscribed in Euclidean circles have some surprising properties. One of the most interesting was discovered by Blaise Pascal (1623–1662). Pascal discovered

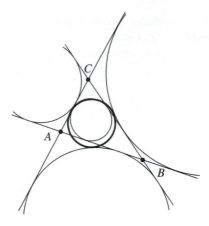

FIGURE 10.47: Feuerbach's Theorem

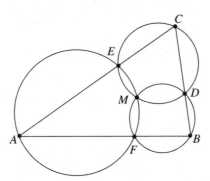

FIGURE 10.48: Miquel's Theorem

the next theorem when he was only sixteen years old and he gave it the colorful Latin title *mysterium hexagrammicum*. For that reason the theorem is still referred to as *Pascal's mystic hexagram*. Construct a circle and six points A, B, C, D, E, and F, cyclically ordered around the circle. Adjust the points so that no pair of opposite sides of the hexagon $ABCDEF$ are parallel. Draw the lines through the sides and mark the points at which the opposite sidelines intersect. Pascal's Theorem asserts that these three points are collinear.

Theorem 10.7.4 (Pascal's Mystic Hexagram). *If a hexagon with no pair of opposite sides parallel is inscribed in a circle, then the three points at which opposite sides intersect are collinear.*

SUGGESTED READING

Chapter 4 of *Journey Through Genius*, [23].

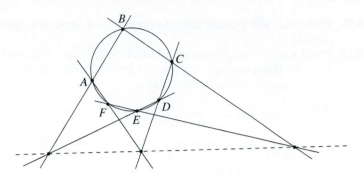

FIGURE 10.49: Pascal's Mystic Hexagram

EXERCISES

10.1. Complete the proof of the Tangent Line Theorem (Theorem 10.2.4).

10.2. Prove that points on a tangent line lie outside the circle (Theorem 10.2.5).

10.3. Prove the Secant Line Theorem (Theorem 10.2.6).

10.4. Prove that some points on a secant line lie outside the circle and some lie inside (Theorem 10.2.7).

10.5. Prove the Tangent Circles Theorem (Theorem 10.2.12).

10.6. Let $\gamma = \mathcal{C}(O, r)$ be a circle, let ℓ and m be two nonparallel lines that are tangent to γ at the points P and Q, and let A be the point of intersection of ℓ and m. Prove the following (in neutral geometry.)
   (a) O lies on the bisector of $\angle PAQ$.
   (b) $PA = QA$.
   (c) $\overleftrightarrow{PQ} \perp \overleftrightarrow{OA}$.

10.7. This is an exercise in neutral geometry.
   (a) Let a and c be two numbers such that $0 < a < c$. Prove that there exists a triangle $\triangle ABC$ such that $\angle BCA$ is a right angle, $BC = a$, and $AB = c$.
   (b) Let γ be a circle and let P be a point that is outside γ. Prove that there exist two lines through P that are tangent to γ.

10.8. Use the Pythagorean Theorem to prove Elementary Circular Continuity (Theorem 10.2.8) in Euclidean geometry.

10.9. Recall that an altitude of a triangle is a line through one vertex that is perpendicular to the line determined by the other two vertices. Prove that the Euclidean Parallel Postulate is equivalent (in neutral geometry) to the assertion that the three altitudes of any triangle are concurrent.

10.10. Complete the proof of the Inscribed Circle Theorem (Theorem 10.3.8) by showing that the inscribed circle is unique.

10.11. Prove that a regular n-sided polygon can be inscribed in a given circle (Theorem 10.3.12).

10.12. Prove that the hypotenuse of a Euclidean right triangle is a diameter of the circumscribed circle (Theorem 10.4.3).

10.13. Prove the 30-60-90 Theorem (Theorem 10.4.5).

10.14. Prove the converse to the 30-60-90 Theorem (Theorem 10.4.6).

10.15. Use Theorem 10.4.12 to prove the Pythagorean Theorem.

10.16. Prove that every quadrilateral inscribed in a circle is convex. (This is a neutral theorem.)

10.17. Prove the following theorem in Euclidean geometry (Euclid's Proposition III.22). *If □ABCD is a quadrilateral inscribed in the circle γ, then the sum of the measures of the opposite angles is 180°; that is,*

$$\mu(\angle ABC) + \mu(\angle CDA) = 180° = \mu(\angle BCD) + \mu(\angle DAB).$$

10.18. Prove the following theorem in Euclidean geometry (Euclid's Proposition III.32). *Let △ABC be a triangle inscribed in the circle γ and let t be the line that is tangent to γ at A. If D and E are points on t such that D * A * E and D is on the same side of $\overleftrightarrow{AC}$ as B, then ∠DAB ≅ ∠ACB and ∠EAC ≅ ∠ABC.*

10.19. Define □ABCD to be a *circumscribed quadrilateral* for circle γ if each side of □ABCD is tangent to γ. Prove the following theorem in neutral geometry. *If γ is a circle and □ABCD is a circumscribed quadrilateral for γ, then AB + CD = BC + DA.*

10.20. Prove Lemma 10.5.1.

10.21. Complete the proof of the Circular Continuity Principle (Theorem 10.5.4) by showing that there must be a second point $R \in \gamma \cap \gamma'$, $R \neq Q$.

10.22. Let a, b, and c be three positive numbers such that a is the largest and $a < b + c$. Prove that there exists a triangle △ABC such that $BC = a$, $AC = b$, and $AB = c$. (This is Euclid's Proposition I.22.)

10.23. Prove Theorem 10.6.2.

10.24. Prove Theorem 10.6.3.

10.25. Prove Theorem 10.6.5.

10.26. Prove Theorem 10.6.9.

10.27. Prove that the interior of a circle is a convex set.

10.28. Circles in spherical geometry. The distance between two points on the sphere is defined to be the length of the great circular segment joining them. Consider a circle of radius r on the sphere of radius 1. Use analytic geometry and calculus to verify that the circumference and area satisfy $C = 2\pi \sin r$ and $A = 2\pi(1 - \cos r)$. What happens to the circumference of the circle as r increases? Use the Maclaurin series expansions of $\sin r$ and $\cos r$ to compare these formulas with the Euclidean formulas.

10.29. Find the circumference of a circle of radius r in the taxicab metric and the square metric. (Refer to Exercises 5.26 and 5.27 in Chapter 5.)

10.30. Let △ABC be a Euclidean triangle and let △A'B'C' be its orthic triangle. For simplicity, assume that the points are labeled as in Fig. 7.17 and that the orthic triangle is inside the original as in that figure.
  (a) Prove that the circle with $\overline{AB}$ as diameter passes through A' and B'.
  (b) Use part (a) and Exercise 10.17, above, to prove that angles ∠ABC and ∠A'B'A are supplements.
  (c) Use a similar argument to prove that angles ∠ABC and ∠C'B'C are supplements.
  (d) Prove that ∠A'B'A ≅ ∠C'B'C.
  (e) Prove that $\overrightarrow{B'B}$ is the bisector of ∠A'B'C'.
  (f) Prove that the orthocenter of △ABC is the same as the incenter of △A'B'C'.

10.31. Let $\triangle ABC$ be a Euclidean triangle and let L be a point on $\overleftrightarrow{BC}$. Use the Law of Sines and a trigonometric identity to prove that

$$\frac{BL}{LC} = \frac{AB\sin(\angle BAL)}{AC\sin(\angle LAC)}.$$

10.32. Use the previous exercise to prove the following trigonometric form of Ceva's Theorem: *Let $\triangle ABC$ be a Euclidean triangle. The proper Cevian lines $\overleftrightarrow{AL}$, $\overleftrightarrow{BM}$, and $\overleftrightarrow{CN}$ are concurrent if and only if*

$$\frac{\sin(\angle BAL)}{\sin(\angle LAC)} \cdot \frac{\sin(\angle CBM)}{\sin(\angle MBA)} \cdot \frac{\sin(\angle ACN)}{\sin(\angle NCB)} = 1$$

and either 0 or 2 of the points L, M, and N lie outside triangle.

10.33. Use the trigonometric form of Ceva's Theorem to prove that the bisectors of the three interior angles of any Euclidean triangle concur.

10.34. Use the trigonometric form of Ceva's Theorem and a trigonometric identity to prove that the bisector of an interior angle and the bisectors of the two remote exterior angles of any Euclidean triangle concur. The point of concurrence is the center of one of the excircles of the triangle—see Fig. 10.45.

10.35. Let P, Q, and R be the points at which the incircle is tangent to the sides of the Euclidean triangle $\triangle ABC$. Use Ceva's Theorem (original form) to prove that the lines $\overleftrightarrow{AP}$, $\overleftrightarrow{BQ}$, and $\overleftrightarrow{CR}$ are concurrent. The point of concurrency is called the *Gergonne point* of the circle. It is named after the French mathematician Joseph Diaz Gergonne (1771–1859). Note that the Gergonne point is usually not the same as the incenter.

10.36. The purpose of this exercise is to prove the Nine-Point Circle Theorem. Let $\triangle ABC$ be a Euclidean triangle and let points D, E, F, L, M, N, and H be as in Fig. 10.46. Let γ be the circumscribed circle for $\triangle DEF$.

(a) Prove that $\square EDBF$ is a parallelogram. Prove that $DB = DN$. Use a symmetry argument to show that N lies on γ. Prove, in a similar way, that L and M lie on γ.

(b) Let K be the midpoint of $\overline{HC}$. Prove that $\overleftrightarrow{EK} \parallel \overleftrightarrow{AL}$ and $\overleftrightarrow{EF} \parallel \overleftrightarrow{BC}$. Conclude that $\overleftrightarrow{EF} \perp \overleftrightarrow{EK}$. Prove, in a similar way, that $\overleftrightarrow{FD} \perp \overleftrightarrow{DK}$. Prove that D and E lie on the circle with diameter $\overline{FK}$. Conclude that γ equals the circle with diameter $\overline{FK}$. In particular, K lies on γ. Similar proofs show that the midpoints of the segments joining H to the other two vertices also lie on γ.

TECHNOLOGY EXERCISES

T10.1. Make a tool (macro) that finds the circumcircle of a triangle. Move the vertices of your triangle around and observe how the circumcircle changes. What must you do to the vertices of the triangle to make the diameter of the circumcircle approach infinity? Does the triangle have to be large in order for the circumcircle to be large?

T10.2. Make a tool (macro) that finds the incircle of a triangle.

T10.3. Make a tool that draws all three excircles for a given triangle.

T10.4. Make a tool that constructs the nine-point circle for a given triangle $\triangle ABC$.

    (a) Verify that all nine points really do lie on the circle. Vary the size and shape of $\triangle ABC$ and observe that this continues to be the case.

    (b) Use the tools you created earlier to construct the orthic and medial triangles for $\triangle ABC$.

    (c) Verify that the nine-point circle for $\triangle ABC$ is the circumcircle for both the medial triangle and the orthic triangle.

    (d) For each side of $\triangle ABC$, construct the segment from the midpoint of that side of $\triangle ABC$ to the midpoint of the segment from the orthocenter to the opposite vertex of $\triangle ABC$. What do you observe about these three segments?

T10.5. Pascal's Theorem.

    (a) Construct a circle and choose six points $A, B, C, D, E,$ and F in cyclic order on the circle. Verify that Pascal's Theorem holds for these six points. Specifically, verify that the three points determined by $\overleftrightarrow{AB} \cap \overleftrightarrow{DE}$, $\overleftrightarrow{BC} \cap \overleftrightarrow{EF}$, and $\overleftrightarrow{CD} \cap \overleftrightarrow{AF}$ are collinear (provided no two of these lines are parallel).

    (b) Investigate what happens when the six points are arranged in some other order on the circle. Specifically, choose random points $A, B, C, D, E,$ and F in any order on the circle. Are the points of $\overleftrightarrow{AB} \cap \overleftrightarrow{DE}$, $\overleftrightarrow{BC} \cap \overleftrightarrow{EF}$, and $\overleftrightarrow{CD} \cap \overleftrightarrow{AF}$ collinear? (Again, adjust the points so that no two of the opposite sidelines are parallel.)

    (c) Investigate what happens when some other conic section (such as an ellipse or a parabola) is used instead of a circle.

T10.6. A corollary of Pascal's Theorem. Construct a triangle $\triangle ABC$ and its circumcircle γ. For each vertex of the triangle, construct a line through that vertex that is tangent to γ at the vertex. Adjust the triangle so that none of these tangent lines is parallel to the opposite sideline of $\triangle ABC$. Finally, find the points of intersection between each tangent line and the opposite sideline of the triangle. Verify that these three points are collinear. Explain how this result can be viewed as a limiting case of Pascal's theorem. Illustrate your explanation by making a Sketchpad drawing of Pascal's Mystic Hexagram and then taking a limit as certain pairs of vertices approach each other.

T10.7. Brianchon's Theorem. Brianchon proved a theorem that is dual to Pascal's. Start with a circle and construct a hexagon that is circumscribed about the circle. Now construct the segments connecting opposite vertices of the hexagon. What is true of the three segments? Write a statement of the theorem you have discovered.

T10.8. In Exercise T7.8 you made a tool that constructs the pedal triangle. Let $\triangle ABC$ be a triangle and let γ be the circumcircle for $\triangle ABC$. Choose a point P on γ and construct the pedal triangle for $\triangle ABC$ and P. What do you observe? Verify the following theorem of Simson:[3] *The feet of the perpendiculars dropped from a point P on the circumcircle of $\triangle ABC$ to the sidelines of $\triangle ABC$ are collinear.* So the pedal triangle degenerates to a line segment when the point P lies on the circumcircle. The line determined by the feet of the perpendiculars is called a *Simson line* for the triangle. There is not just one Simson line associated with

---

[3]This theorem is usually attributed to the Scottish mathematician Robert Simson (1687–1768), but there is no record that he ever published it. The theorem was published in 1799 by William Wallace (1768–1843).

$\triangle ABC$, but a different Simson line for each point on the circumcircle. Move the point P around the circumcircle and observe how the Simson line varies. Are the feet of the perpendiculars ever collinear for a point P that does not lie on the circumcircle?

CHAPTER 11

Constructions

11.1 COMPASS AND STRAIGHTEDGE CONSTRUCTIONS
11.2 NEUTRAL CONSTRUCTIONS
11.3 EUCLIDEAN CONSTRUCTIONS
11.4 CONSTRUCTION OF REGULAR POLYGONS
11.5 AREA CONSTRUCTIONS
11.6 THREE IMPOSSIBLE CONSTRUCTIONS

This chapter completes our coverage of the propositions in Books I through VI of Euclid's *Elements*. The one important topic from the *Elements* that we have not yet studied is compass and straightedge construction; we intend to take a look at such constructions in this chapter. Consideration of constructions was delayed until now because we need the circular continuity principles of the last chapter to build the constructions on a rigorous foundation.

We will do as many of the constructions as possible in the context of neutral geometry, but there are some that require the Euclidean Parallel Postulate and are therefore part of Euclidean geometry. The entire chapter relies on the neutral geometry of Chapter 6 as well as some results regarding circles from §10.2. The Euclidean constructions require Chapter 7 and some theorems about Euclidean circles from §10.4. The Circular Continuity Principle (Theorem 10.5.4) is needed for all the constructions, but it can be treated as an axiom and simply assumed for use in this chapter. The area constructions in §11.5 require just the first two sections of Chapter 9.

The constructions can only be truly understood and appreciated by those who work them out for themselves, so most of the actual constructions in this chapter are left as exercises. The constructions should be done using a real compass and straightedge and should not be worked out by drawing rough freehand sketches. An excellent alternative is to use the dynamic computer software that was discussed in the Technology section of the Preface. This chapter provides a great opportunity to put that technology to good use. Even though this is an appropriate place in which to make use of modern technology, the reader should take the time to do at least some of the early constructions using the classic tools of Euclid. Only in that way can one experience what these constructions were like for those who first devised them.

11.1   COMPASS AND STRAIGHTEDGE CONSTRUCTIONS IN GEOMETRY

Compass and straightedge constructions are the constructions that can be carried out using Euclid's Postulates 1, 2, and 3. Specifically, Postulates 1 and 2 allow the straightedge to be used either to construct a segment joining two given points or to extend a given line segment. Postulate 3 allows the use of a compass to draw a circle that has a given point as center and passes through a second given point.

As explained in the first two chapters of this book, the geometry of Euclid and the ancient Greeks is based on a love of pure, simple forms. This view is reflected in the tools that the Greeks chose to use in their geometry. The straightedge is simply that: a straight edge. It does not have any marks on it whatsoever and is used only for the two purposes explicitly permitted by Postulates 1 and 2. Thus a straightedge can be used to draw or extend segments but it cannot remember any distances when it is moved from one position to another. In a similar way Euclid's compass is a "collapsing compass" that can be used to draw the circle that has a given point as center and passes through a second point, but cannot be used to move a given radius from one location to another. When the compass is lifted out of the plane, it collapses and the distance between the two points is lost to it.

By contrast, the tools we have used in our development of geometry, the protractor and the ruler, have numerical scales on them and can be used to measure distances and angles. This difference is not simply due to the fact that Euclid lived a long time ago and had only primitive tools available to him. The Greeks considered measurement to be one of the practical arts, which were left to slaves, and beneath the dignity of an educated person. Thus considerations of beauty and elegance played a role in their decision to use only a simple straightedge and compass.

Euclid's propositions are of two kinds. Most of them are theorems in the sense in which we use the term today. They have the standard logical "if ..., then ..." form that is characteristic of a theorem. Others of Euclid's propositions are not theorems in that sense but instead assert that it is possible to start with certain given data and to construct a particular figure. What is not stated but assumed is that the construction is to be carried out using only the compass and straightedge. More than one-fourth of the propositions in Book I and all the propositions in Book IV, as well as many propositions in other Books of the *Elements*, are constructions.

One clue to the fact that one of Euclid's propositions is actually a construction is the use of the infinitive in the statement. For example, Proposition I.1 contains the phrase "...to construct ..." while Proposition I.2 begins "To place at ..." and Proposition I.3 says "...to cut off ...." All three of those propositions are constructions. Another clue, at least in the translations available to us, is the fact that the proof of a construction usually ends with QEF (*quod erat faciendum*—that which was to have been done) rather than QED.

Euclid essentially identified constructibility with existence. While we are comfortable making use of a point if its existence is implied by one of our axioms (perhaps by the one-to-one correspondence in the Ruler Postulate), Euclid would only use a point if he could construct it using compass and straightedge. More precisely, Euclid used his postulates to construct certain line segments and circles;

he made the unstated assumption that these lines and circles intersected as indicated in his diagrams and that they defined certain points of intersection.

Compass and straightedge constructions have been an important part of geometry since long before Euclid. Over the years, construction problems have inspired many new mathematical ideas and have led to the discovery of a great deal of interesting and important mathematics. As we will see later in this chapter, several of the construction problems studied by the ancient Greeks were not completely solved until the eighteenth and nineteenth centuries. For all these reasons, constructions remain a standard part of the geometry curriculum.

In order to understand just what is meant by a construction, let us look at an example. The example is the very first Proposition in Book I. It is a proposition whose proof we have examined before, but in the meantime we have proved a theorem that fills in the gap in Euclid's argument.

Construction Problem 11.1.1 (I.1). *Given a segment $\overline{AB}$, construct an equilateral triangle $\triangle ABC$ with base $\overline{AB}$.*

Solution. Let α be the circle with center A that passes through B and let β be the circle with center B that passes through A. The two circles α and β meet at two points; let C be one of the two points. Construct the triangle $\triangle ABC$ by drawing the segments $\overline{AC}$ and $\overline{BC}$.

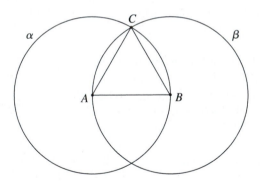

FIGURE 11.1: Constructing an equilateral triangle

We must prove that $\triangle ABC$ is an equilateral triangle. First observe that $AB = AC$ because both B and C lie on α (definition of circle) and then observe that $BA = BC$ because both A and C lie on β. Therefore, $AB = AC = BC$ and $\triangle ABC$ is equilateral (definition of equilateral triangle). □

Notice that the solution to the construction problem has two parts. The first part consists of a sequence of instructions indicating how to use the compass and straightedge to construct various segments, circles, and points. The second part consists of a proof that the objects constructed have the properties claimed for them. The solutions you give to the construction problems posed in this chapter should follow the same two-part pattern.

Another feature of the solution that should be noted is the fact that circular continuity is used to prove the existence of points of intersection. In every construction, either the Circular Continuity Principle (Theorem 10.5.4) or the Elementary Circular Continuity Theorem (Theorem 10.2.8) will be needed to prove the existence of points of intersection. Since it is an expected part of every construction, the use of circular continuity will not be mentioned explicitly, but it will be implicit in all the constructions. Strictly speaking, we should also verify the fact that the hypotheses of the Circular Continuity Principle are satisfied in the proof above, but again that will be left unstated.

All but one of the constructions we will consider in this chapter are taken from Euclid. In order to help locate them in the *Elements* we will include the number of the corresponding Proposition from the *Elements* in parentheses after the number of the construction. The construction above is labeled (I.1) because it is Proposition 1 in Book I of the *Elements*. This chapter includes all the constructions from Book I as well as selected constructions from Books II, III, IV, and VI.

Like many of Euclid's constructions, the example construction uses only neutral geometry. The triangle constructed has three congruent sides and three congruent angles. If we wanted to go further and assert that each of the three angles measures $60°$, we would need the Euclidean Parallel Postulate.

In this and other constructions it is a good idea to count the number of steps involved. This helps to maintain awareness of the complexity of the construction and to understand it better. The usual way to count is to count the number of times a straightedge is used to draw a new line or segment and the number of times the compass is used to construct a new circle. The use of the straightedge to extend a segment is not ordinarily counted as a separate step nor is the use of circular continuity to identify a point of intersection. Thus the number of steps needed to construct the point C in the example is two because it was necessary to draw two circles in order to identify C as a point of intersection. Constructing the triangle $\triangle ABC$ requires two more steps: Use the straightedge twice to draw the segments $\overline{AC}$ and $\overline{BC}$. When you work the exercises in this chapter you should try to do the constructions as efficiently as possible.

While it is not customary to count the identification of a point of intersection as a separate step when analyzing the complexity of a construction, such identification is an important part of the process. One of the nice aspects of the dynamic geometry software packages is the fact that they have the user "mark" a point of intersection before it is available to be used in subsequent constructions. This raises the user's awareness of the way in which the various points are defined.

11.2 NEUTRAL CONSTRUCTIONS

We begin with some neutral constructions, mostly from Book I. Euclid's first order of business is to prove that a segment can be copied from one location to another. He breaks the proof into three steps. The first step, which we have already studied, is to construct an equilateral triangle.[1] In the second construction the first construction is put to work to move one endpoint of the segment to the new location. After that it is quite easy to move the other endpoint so that the segment lies along a specified ray. Note that once a construction has been completed, it may be used as part of a subsequent construction.

Construction Problem 11.2.1 (I.2). *Given a point A and a segment $\overline{BC}$, construct a point D such that $\overline{AD} \cong \overline{BC}$.*

Solution. If $A = B$, take $D = C$. Assume $A \neq B$. Construct a point E such that $\triangle ABE$ is an equilateral triangle (previous construction). Draw the circle α with center B and radius $\overline{BC}$. Let F be the point at which the ray $\overrightarrow{EB}$ meets α. Draw the circle β with center E and radius $\overline{EF}$. Let D be the point at which the ray $\overrightarrow{EA}$ crosses β. (See Fig. 11.2.)

The proof that $\overline{AD} \cong \overline{BC}$ is left as an exercise (Exercise 11.1).   □

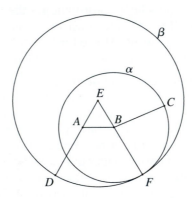

FIGURE 11.2: Moving one endpoint of a segment

The complexity of the last construction is 6. If you count more steps than that, remember that once the step of constructing a segment has been counted, the segment can be extended without increasing the complexity. When one construction is used as part of another, all the steps in the earlier construction must be counted.

Construction Problem 11.2.2 (I.3). *Given segments $\overline{AB}$ and $\overline{CD}$ with $CD < AB$, construct a point E on $\overline{AB}$ such that $\overline{AE} \cong \overline{CD}$.*

Solution. Exercise 11.2.   □

The last construction shows that it is possible to move a segment from one location to another. Once this construction is completed, the difference between a collapsible and a noncollapsible compass becomes unimportant. Thus Euclid could

---

[1]Now we finally see why Euclid began with the construction of an equilateral triangle; he needed that construction as a step in the proof that he could copy a segment.

have assumed a noncollapsible compass from the start, but he rightly chose not to do so because the assumption is unnecessary.

Next we construct the bisector of an angle and the midpoint of a segment.

Construction Problem 11.2.3 (I.9). *Given an angle* $\angle BAC$, *construct a point D such that* $\overrightarrow{AD}$ *is the bisector of* $\angle BAC$.

Solution. Exercise 11.3.   □

Construction Problem 11.2.4 (I.10). *Given a segment* $\overline{AB}$, *construct a point M on* $\overline{AB}$ *such that* $AM = MB$

Solution. Exercise 11.4.   □

We are now ready to construct perpendiculars. There is one construction for the perpendicular through an internal point and a different construction for the perpendicular through an external point.

In the next construction we are given a line ℓ. When a line is one of the givens in a construction problem, that really means that two points on the line are given. In other words, if line ℓ is given in the statement of a construction, then the solution can contain a statement like "let P and Q be two points on ℓ."

Construction Problem 11.2.5 (I.11). *Given a line* ℓ *and a point A that lies on* ℓ, *construct a point B such that* $\overleftrightarrow{AB}$ *is perpendicular to* ℓ.

Solution. Exercise 11.5.   □

Construction Problem 11.2.6 (I.12). *Given a line* ℓ *and an external point P, construct the line through P that is perpendicular to* ℓ.

Solution. Exercise 11.6.   □

The next construction allows us to construct a triangle where we need it. That construction, in turn, will allow us to copy an angle at a new location.

Construction Problem 11.2.7 (I.22). *Given three positive numbers* $a, b,$ *and* c *such that c is the largest and* $c < a + b$ *and points A and B such that* $AB = c$, *construct a triangle* $\triangle ABC$ *such that* $a = BC$ *and* $b = AC$.

Solution. Exercise 11.7.   □

To say that c is the largest of the three numbers means that $c \geqslant a$ and $c \geqslant b$. Note that the conditions $a + b > c, a + c > b$ and $b + c > a$ are all necessary in order for the triangle to exist (Triangle Inequality). If we assume that c is the largest and $c < a + b$, then the other two inequalities are automatically satisfied.

Construction Problem 11.2.8 (I.23). *Given an angle* $\angle BAC$ *and a ray* $\overrightarrow{DE}$, *construct a point F such that* $\angle FDE \cong \angle BAC$.

Solution. Exercise 11.8.   □

We come now to the familiar construction of a parallel line.

Construction Problem 11.2.9 (I.31). *Given a line ℓ and an external point P, construct a line m such that P lies on m and m ∥ ℓ.*

Solution. Exercise 11.9. □

When a circle is given in the statement of a construction problem, it is the points on the circle itself that are given. Thus the location of the center of the circle is not directly part of the given information; instead the center point can be constructed. It takes three distinct points to determine a circle, so the fact that a circle is given should be interpreted to mean that three (noncollinear) points are known. In other words, if the circle γ is one of the givens in the statement of a construction problem, the solution may include the statement "let P, Q, and R be three distinct points on γ." Euclid's next construction uses those three points to locate the center of the circle.

Construction Problem 11.2.10 (III.1). *Given a circle, construct the center of the circle.*

Solution. Exercise 11.10. □

Our final neutral construction is the construction of the circle inscribed in a triangle.

Construction Problem 11.2.11 (IV.4). *Given a triangle, construct the inscribed circle.*

Solution. Exercise 11.11. □

11.3 EUCLIDEAN CONSTRUCTIONS

In this section we include a few selected constructions that require the Euclidean Parallel Postulate. Euclid placed the next construction immediately before his statement of the Pythagorean Theorem. True to form, he does not state a theorem regarding the area of squares on the sides of a triangle until after he has first proved that the squares can be constructed.

Construction Problem 11.3.1 (I.46). *Given a segment, construct a square on it.*

Solution. Exercise 11.13. □

Here is a construction that is not explicitly stated in Euclid. It is interesting because the construction is relatively simple in Euclidean geometry, but the circle described need not exist in hyperbolic geometry.

Construction Problem 11.3.2. *Given a line ℓ, an external point P, and a point Q that lies on ℓ, construct a circle that passes through P and is tangent to ℓ at Q.*

Solution. Exercise 11.14. □

In Exercise 10.7 you proved that for every point P that lies outside a circle γ there are two lines through P that are tangent to γ. The procedure suggested in the hint for that exercise could be turned into a (relatively complex) neutral construction of tangent lines, but we prefer to include a Euclidean construction because of its simplicity and because of its relevance to the theorems about inversions in circles in the next chapter.

Construction Problem 11.3.3 (III.17). *Given a circle γ and point P that lies outside γ, construct two lines through P that are tangent to γ.*

Solution. Let O be the center of γ (Construction Problem 11.2.10). Draw the circle α that passes through P and whose center is the midpoint of $\overline{OP}$. The remainder of the construction is suggested by Fig. 11.3. The details are left as an exercise (Exercise 11.15). □

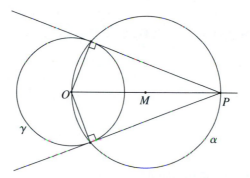

FIGURE 11.3: Construction of two tangent lines

Warning. You may be tempted to take the straightedge, place it at P, and then rotate it until it first intersects γ. That construction is not allowable because the way the straightedge is used in it is not one of the two explicitly permitted uses of a straightedge.

Construction Problem 11.3.4 (IV.5). *Given a triangle, construct the circumscribed circle.*

Solution. Exercise 11.16. □

One of the neutral constructions allows us to construct the midpoint of a segment, so every segment can be bisected in neutral geometry. In Euclidean geometry it is quite easy to generalize that construction and divide a segment into any number of equal pieces. In particular, every segment can be trisected. Later in the chapter we will discuss the problem of trisecting an angle. As you probably already know, there are some angles that cannot be trisected with compass and straightedge.

Construction Problem 11.3.5 (VI.9). *Given a segment and a positive integer n, construct points that divide the segment into n segments of equal length.*

Solution. Let $\overline{AB}$ be a segment. Construct a point C that does not lie on $\overleftrightarrow{AB}$. Define $C_0 = A$. Lay off segments $\overline{C_0C_1}, \overline{C_1C_2}, ..., \overline{C_{n-1}C_n}$ such that all the segments $\overline{C_iC_{i+1}}$ have the same length. Let ℓ be the line $\overleftrightarrow{C_nB}$. Draw lines through the points $C_1, C_2, ..., C_{n-1}$ that are parallel to ℓ. By the Parallel Projection Theorem, those lines subdivide $\overline{AB}$ into n equal pieces.   □

Fig. 11.4 shows how to trisect $\overline{AB}$.

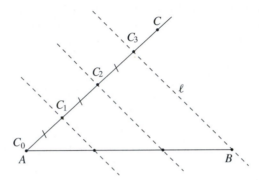

FIGURE 11.4: Trisecting a segment

11.4 CONSTRUCTION OF REGULAR POLYGONS

In Chapter 10 we proved that for every circle γ and for every positive integer n there exists a regular n-sided polygon inscribed in γ. We now consider the question of whether or not such inscribed regular polygons can be constructed using compass and straightedge. That problem has historically been one of the most studied construction problems. Euclid constructed regular n-sided polygons for $n = 3, 4, 5, 6, 8$, and 15. The construction of the regular 15-sided polygon is the culmination of Book IV of the *Elements*.

All the constructions in this section are Euclidean.

Construction Problem 11.4.1 (IV.6). *Given a circle γ, construct a square inscribed in γ.*

Solution. Exercise 11.17.   □

Euclid's construction of an inscribed regular pentagon is fairly complicated, so we will omit it. There is a much simpler construction that is due to H. W. Richmond (1863–1948). While Richmond's construction of the pentagon itself is relatively simple, some substantial trigonometric analysis is required to prove that the constructed pentagon is regular.

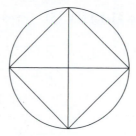

FIGURE 11.5: Constructing an inscribed square

Construction Problem 11.4.2 (IV.11). *Given a circle γ, construct a regular pentagon inscribed in γ.*

Richmond's Solution. Let *A* be the center and let $\overline{AB}$ be a radius of the given circle γ. Construct the line that is perpendicular to $\overleftrightarrow{AB}$ at *A*. Let *C* be one of the two points at which the perpendicular intersects the circle. Construct the midpoint *D* of $\overline{AC}$. Construct the bisector of angle ∠*ADB* and mark the point *E* at which the angle bisector intersects the segment $\overline{AB}$. Construct a perpendicular to $\overleftrightarrow{AB}$ at *E* and let *F* be one of the two points at which the perpendicular intersects the circle.

Segment $\overline{BF}$ is one side of the inscribed regular pentagon. The next side could be constructed by repeating the same construction on $\overline{AF}$.   □

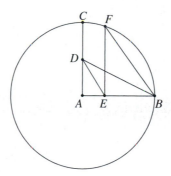

FIGURE 11.6: Richmond's construction of an inscribed regular pentagon

We will not prove that $\overline{BF}$ in the construction above is one side of the regular pentagon, but it is worthwhile to carry out the construction using dynamic software and to verify that empirically by measuring the angle ∠*BAF*. If you do this experiment, you will find that the angle measures precisely 72° (Exercise 11.18).

The regular hexagon can be constructed by building equilateral triangles on the radii of the circle (see Fig. 11.7). Thus there is an easy solution to the following construction problem.

Construction Problem 11.4.3 (IV.15). *Given a circle γ, construct a regular hexagon inscribed in γ.*

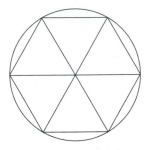

FIGURE 11.7: Constructing an inscribed regular hexagon by constructing six equilateral triangles on a radius

Once a regular n-gon has been constructed, it is often possible to construct other regular polygons from it. For example, if a regular n-gon has been constructed and n can be factored as $n = mk$, then both the regular m-gon and the regular k-gon can also be constructed by simply using some of the vertices of the regular n-gon. We illustrate this principle by using the construction of the regular inscribed hexagon to construct an inscribed equilateral triangle; see Fig. 11.8.

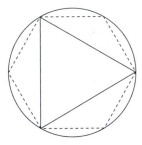

FIGURE 11.8: Constructing an inscribed equilateral triangle from an inscribed regular hexagon

We have, therefore, solved the following construction problem.

Construction Problem 11.4.4. *Given a circle γ, construct an equilateral triangle inscribed in γ.*

Observe that an inscribed regular n-gon can be constructed if and only if an angle of measure $(360/n)^\circ$ can be constructed. Since every angle can be bisected, this means that if an inscribed regular n-gon can be constructed, then a regular $2n$-gon can be constructed as well. For example, an inscribed square was constructed above. Bisecting each of the angles formed by the center of the circle and the vertices of the square results in an inscribed regular octagon.

Hence we have also solved the following construction problem.

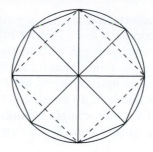

FIGURE 11.9: Constructing an inscribed regular octagon from an inscribed square

Construction Problem 11.4.5 (IV.15). *Given a circle γ, construct a regular octagon inscribed in γ.*

It should now be more clear why Archimedes chose to use inscribed regular polygons with $3 \cdot 2^k$ sides to compute the circumference and area of a circle. It is relatively easy to construct an inscribed regular hexagon and a regular $(3 \cdot 2^k)$-gon can then be constructed by repeatedly bisecting the angles.

One final observation is that once two angles have been constructed, the angle whose measure is the difference between the measures of the two constructed angles can also be constructed. This is exactly how Euclid constructed the regular 15-gon. The central angles for an inscribed regular pentagon have measure $(360/5)° = 72°$ while the central angles for an inscribed regular hexagon measure $(360/6)° = 60°$. It is therefore possible to construct an angle of measure 12° and thus to inscribe a regular 30-gon. Using every other vertex gives a regular 15-gon. Alternatively, the construction of an inscribed equilateral triangle and an inscribed regular pentagon allow the construction of an angle of measure $120° - 72° = 48°$ and 15 such angles make two complete revolutions of the circle.

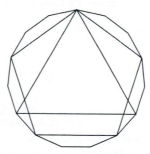

FIGURE 11.10: Constructing an inscribed regular 15-gon from an inscribed regular pentagon and an inscribed equilateral triangle

Construction Problem 11.4.6 (IV.16). *Given a circle γ, construct a regular 15-sided polygon inscribed in γ.*

Solution. Exercise 11.20.   □

While Euclid constructed regular n-gons for many different n, he never did construct a regular heptagon ($n = 7$). This raises the obvious question: For which n can the regular n-sided polygon be constructed? Very little progress was made on that question from the time of Euclid until the late eighteenth century. At that time the question was completely answered by Carl Friedrich Gauss (1777–1855). As a teenager, Gauss discovered a construction of a regular 17-sided polygon. Soon thereafter he proved an astonishingly beautiful theorem that completely describes all the constructible regular n-gons. Before we can state Gauss's theorem we need a definition.

Definition 11.4.7. A *Fermat prime* is a prime number p that can be written in the form $p = 2^{(2^k)} + 1$ for some nonnegative integer k.

Fermat wondered whether all numbers of the form $2^{(2^k)} + 1$ might be prime. Since he had to do his calculations by hand, he could only check this for small values of k and he found that each of those numbers is prime for $k = 0, 1, 2, 3$, and 4. Using a computer, we can easily check that $2^{(2^5)} + 1$ is not prime. In fact, no one has ever found a value of k greater than 4 for which $2^{(2^k)} + 1$ is prime. Thus the only known Fermat primes are 3, 5, 17, 257, and 65537. Even today computers are kept busy round the clock searching for Fermat primes. New examples of Fermat numbers (i.e., numbers of the form $2^{(2^k)} + 1$) that are composite continue to be found, but it seems unlikely that another Fermat prime will ever be discovered.

Gauss proved the following amazingly comprehensive theorem.

Theorem 11.4.8 (Gauss). *A regular n-sided polygon can be constructed if and only if n is a power of 2 times a product of distinct Fermat primes.*

Since there are only five known Fermat primes, there are only a relatively small number of regular polygons with an odd number of sides that are known to be constructible. Once those have been constructed, the other polygons in Gauss's Theorem can be be constructed by repeatedly bisecting angles.

11.5 AREA CONSTRUCTIONS

This section contains a sampling of area constructions. There are a large number of such constructions in the *Elements* and there is no way a book like this could cover all of them. Instead we will select just a few for study. The ones chosen were selected primarily because they relate directly to topics covered earlier in the book and because they explain the background of one of the unsolved construction problems to be discussed at the end of the chapter. The constructions in this section are all part of Euclidean geometry.

Construction Problem 11.5.1 (I.42). *Given a triangle $\triangle ABC$ and an angle $\angle DEF$, construct a parallelogram that contains angle $\angle DEF$ and has the same area as triangle $\triangle ABC$.*

Solution. Euclid's construction is indicated in Fig. 11.11. In that figure, $\alpha(\square MCHG) = \alpha(\triangle ABC)$. The details are left as an exercise (Exercise 11.22). $\qquad\square$

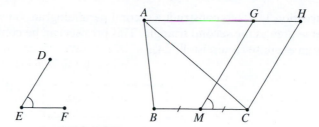

FIGURE 11.11: Constructing a parallelogram that contains a specified angle and has a specified area

In the last construction, a parallelogram is constructed that has the same area as the given triangle and contains a given angle. The next construction goes one step further and constructs the parallelogram to contain both a specified angle and a specified side. This will allow several parallelograms to be stacked to form one parallelogram.

Construction Problem 11.5.2 (I.44). *Given a triangle* $\triangle ABC$ *and an angle* $\angle DEF$, *construct a parallelogram that contains both angle* $\angle DEF$ *and side* $\overline{EF}$ *and such that the parallelogram has the same area as triangle* $\triangle ABC$.

Outline of solution. Use the preceding construction to construct a parallelogram $\square FGHI$ that contains an angle congruent to $\angle DEF$ and has the same area as triangle $\triangle ABC$. This can be done in such a way that G lies on $\overrightarrow{EF}$ and $E * F * G$. The remaining parts of Fig. 11.12 can then be constructed. By Exercise 9.10, $\alpha(\square FGHI) = \alpha(\square JKFE)$. The details are left as an exercise (Exercise 11.23). $\square$

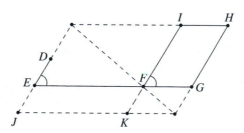

FIGURE 11.12: Constructing a parallelogram that contains a specified angle and base and has a specified area

Construction Problem 11.5.3 (I.45). *Given a polygonal region* R *and an angle* $\angle DEF$, *construct a parallelogram that contains angle* $\angle DEF$ *and has the same area as region* R.

Solution. The polygonal region R is the union of a finite number of triangles. Construct a parallelogram that has the same area as the first triangle. Then use

Construction 11.5.2 to construct a second parallelogram on top of the first that has the same area as the second triangle. This process can be continued. The details are left as an exercise (Exercise 11.24). □

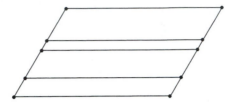

FIGURE 11.13: Construct one parallelogram on top of another so that the union forms a parallelogram

The given angle in the last several constructions could of course be chosen to be a right angle. Hence for every polygonal region it is possible to construct a rectangle whose area is equal to the area of the given region. To complete this line of construction, Euclid would like to show that it is possible to construct a square whose area is that of the given region. But for that he needs the Pythagorean Theorem. Thus he devotes the remainder of Book I to the Pythagorean Theorem and its converse. In Book II he returns to the problem and the following construction is the culmination of that Book.

Construction Problem 11.5.4 (II.14). *Given a polygonal region, construct a square whose area is the same as the area of the given polygonal region.*

Solution. First use Construction 11.5.3 to construct a rectangle $\square ABCD$ whose area is equal to that of the given region. Label the vertices of the rectangle in such a way that $AB \geqslant BC$. Then construct a point E such that $A * B * E$ and $BE = BC$. Construct the midpoint F of $\overline{AE}$ and construct the circle γ with center F and radius $\overline{FE}$. The line $\overleftrightarrow{CB}$ crosses γ at two points; let G be the one that lies on the opposite side of $\overleftrightarrow{AB}$ from C. (See Fig. 11.14.) Notice that $\triangle FBG$ is a right triangle. The Pythagorean Theorem applied to the triangle $\triangle FBG$ together with a little algebra can be used to show that the area of the square on $\overline{BG}$ is equal to the area of the rectangle $\square ABCD$. The details are left as an exercise (Exercise 11.25). □

Euclid did not use this construction anywhere else in the *Elements*. Instead he considered the solution to be an end in itself. He thought of the problem as being so important that he returned to it in Book VI when he was proving theorems regarding similar triangles. There he takes a slightly different point of view on the problem. If a given rectangle has sides of length a and b, then the square of the same area has sides of length $\sqrt{ab}$. In other words, the length of the side of the square is the geometric mean of the lengths of the sides of the rectangle.

Construction Problem 11.5.5 (VI.13). *Given two segments, construct a segment whose length is the geometric mean of the lengths of the given segments.*

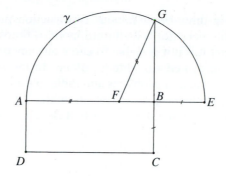

FIGURE 11.14: The area of the square on $\overline{BG}$ is equal to the area of the rectangle $\square ABCD$

Solution. Let a and b be the lengths of the two given segments. Construct points $A, B,$ and C such that $A * B * C$, $AB = a$, and $BC = b$. Construct the midpoint M of $\overline{AC}$ and construct the circle γ with center M and radius $\overline{MC}$. Construct a perpendicular to $\overleftrightarrow{AB}$ at B and let D be one of the two points at which the perpendicular crosses γ. (See Fig. 11.15.) By Corollary 10.4.2, $\triangle ACD$ is a right triangle. By Theorem 7.5.3, BD is the geometric mean of a and b. $\square$

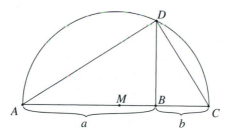

FIGURE 11.15: The length of $\overline{BD}$ is the geometric mean of a and b

11.6 THREE IMPOSSIBLE CONSTRUCTIONS

In this chapter we have studied several general classes of construction problems. None of them was solved in complete generality by the ancient Greeks. While the mathematicians of antiquity invented many ingenious methods of construction, it was beyond the power of the mathematics they developed to answer the question of what can and what cannot be constructed.

In §11.4 we considered the problem of which regular polygons can be constructed with straightedge and compass. As was explained there, that problem was completely solved by Gauss in the late eighteenth century. He proved that certain special polygons can be constructed and that the others cannot be.

There are three other famous construction problems that were formulated in antiquity but were not solved until modern times. In the nineteenth century it was proved that it is not possible to carry out any of the three constructions using compass and unmarked straightedge alone. In this section we will briefly describe those three construction problems and their solutions.

The first problem is to trisect an angle. One of the most elementary constructions allows a general angle to be bisected. We also know how to divide a segment into any number of pieces of equal length (Construction 11.2.2). So it seems reasonable to try to divide an angle into three angles of equal measure. The following Angle Trisection Problem is one of the most (in)famous problems in all of mathematics.

Trisecting an Angle. *Given an angle $\angle ABC$, construct an angle $\angle DEF$ such that* $\mu(\angle DEF) = (1/3)\mu(\angle ABC)$.

In 1837 Pierre Wantzel (1814–1848) used techniques of abstract algebra to prove that there are angles that cannot be trisected. In particular, there is no compass and straightedge construction that will trisect an angle of measure $60°$.

Many of Euclid's constructions involve a geometric form of algebra. Given a unit segment and segments of length a and b, it is possible to construct segments of length $a + b$, $|a - b|$, ab, a/b, and $\sqrt{ab}$. A basic observation (due to Descartes) is that each new length determined in a compass and straightedge construction is the root of a quadratic equation whose coefficients involve previously constructed lengths. The reason $60°$ cannot be trisected is that $\cos 20°$ satisfies a certain cubic equation with integer coefficients. Using the theory of field extensions it can be shown that solutions to this particular cubic equation cannot be roots of the required type of quadratic equations. A complete proof of Wantzel's theorem may be found in most undergraduate abstract algebra textbooks.

Of course, while some angles cannot be trisected, many angles can be. For example, it is easy to trisect a right angle: Given a right angle $\angle CAB$, construct a point D in the interior of $\angle CAB$ such that $\triangle CAD$ is an equilateral triangle. Then $\angle DAB$ has measure $30°$. (See Fig. 11.16.)

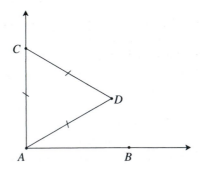

FIGURE 11.16: Trisection of a right angle

Despite the fact that Wantzel proved angle trisection is in general impossible, there are still many people who continue to attempt it. Such amateur geometers are often interesting people who come up with amazing constructions. Of course there is always a flaw in the construction (often the construction only produces an approximate trisection or it violates one of the ancient rules for the use of compass and straightedge), but the fact that so many people continue to think about the trisection problem is a testament to the fascination of Euclidean geometry and the hold it can have on those who study it. The two books [21] and [22] by Underwood Dudley make fascinating reading. They contain a collection of delightful stories about "angle trisectors" and their unconventional ideas.

The second construction problem that was left unsolved by the ancient Greeks is that of "squaring the circle." As we saw in the preceding section, Euclid was able to construct a square whose area is equal to the area of any given polygonal region. For Euclid, and the ancient Greek geometers generally, the problem of constructing a square whose area is equal to that of a given region was one of the major problems of geometry. This kind of construction is often called "squaring the region" or "quadrature." The simplest nonpolygonal region is a circle and it was natural that the Greeks would try to square the circle.

Squaring a circle. *Given a circular region, construct a square whose area is equal to that of the circular region.*

In Section 10.6 we looked at Archimedes proof that the area of a circle of radius r is πr^2. Since a circle of radius 1 has area π, squaring the circle would amount to constructing $\sqrt{\pi}$. In 1882 Ferdinand von Lindenmann proved that π is transcendental, which means that π does not satisfy any polynomial equation with integer coefficients. It follows that $\sqrt{\pi}$ cannot be constructed and therefore the circle cannot be squared.

The final unsolved construction problem from antiquity is a three-dimensional construction problem. The problem is that of doubling (or duplicating) the cube.

Doubling a cube. *Given a cube, construct a cube whose volume is twice that of the given cube.*

Doubling a cube of edge length 1 is equivalent to constructing the cube root of 2. Since $\sqrt[3]{2}$ is a root of the cubic equation $x^3 - 2 = 0$, the problem of doubling the cube can also be transformed into an algebraic problem. The same kind of reasoning as is used in the solution of the trisection problem applies to show that it is impossible to double a cube of edge length 1.

SUGGESTED READING

Chapter 1 of *Journey through Genius*, [23].

NOTE   Each solution to a construction problem in the exercises below should consist of two parts: a sequence of instructions indicating how the construction is to be carried out and a proof that the objects constructed have the required properties.

EXERCISES

11.1. Prove that $\overline{AD} \cong \overline{BC}$ in Construction Problem 11.2.1.

11.2. Solve Construction Problem 11.2.2. Count the number of steps required in the construction.

11.3. Solve Construction Problem 11.2.3.

11.4. Solve Construction Problem 11.2.4.

11.5. Solve Construction Problem 11.2.5.

11.6. Solve Construction Problem 11.2.6.

11.7. Solve Construction Problem 11.2.7.

11.8. Solve Construction Problem 11.2.8.

11.9. Solve Construction Problem 11.2.9.

11.10. Solve Construction Problem 11.2.10.

11.11. Solve Construction Problem 11.2.11.

11.12. A *rusty compass* is a compass that is stuck open so that it can only draw circles of one fixed radius. Prove that it is possible to construct angle bisectors, midpoints, and perpendicular bisectors using a straightedge and a rusty compass. Be careful when constructing the midpoint; the opening of the compass may be much smaller or much larger than the length of the given segment. Does your proof of the existence of an angle bisector work for angles whose measure is close to $180°$?

11.13. Solve Construction Problem 11.3.1.

11.14. Constructing tangent circles.
  (a) Solve Construction Problem 11.3.2.
  (b) Find an example in hyperbolic geometry of a line ℓ, a point Q that lies on ℓ, and a point P that does not lie on ℓ such that there is no circle that is tangent to ℓ at Q and passes through P.

11.15. Complete the solution to Construction Problem 11.3.3.

11.16. Solve Construction Problem 11.3.4.

11.17. Solve Construction Problem 11.4.1.

11.18. Perform Richmond's construction of a regular pentagon. (See solution to Construction Problem 11.4.2.) If you have dynamic software available, measure the angle $\angle BAF$ and confirm that it measures precisely $72°$.

11.19. Euclid did not state Construction Problem 11.4.4 as a separate proposition. Instead he solved the following more general construction problem: *Given a circle γ and a triangle $\triangle ABC$, construct a triangle inscribed in γ that is similar to $\triangle ABC$* (Proposition IV.2). Solve Euclid's construction problem. Note that this must be a Euclidean construction since it involves similar triangles.

11.20. Complete the solution to Construction Problem 11.4.6. You may assume the constructibility of a regular pentagon.

11.21. Recall that a regular polygon circumscribes a circle if each edge of the polygon is tangent to the circle. Explain why it is possible to construct a regular polygon of n sides that is inscribed in a circle γ if and only if it is possible to construct a regular polygon of n sides that circumscribes γ.

11.22. Complete the solution to Construction Problem 11.5.1. Euclid's figure (Fig. 11.11) should serve as a hint.

11.23. Complete the solution to Construction Problem 11.5.2. Again Euclid's figure (Fig. 11.12) should serve as a hint.

11.24. Complete the solution to Construction Problem 11.5.3. Once again the figure should serve as a hint.

11.25. Complete the solution to Construction Problem 11.5.4.

11.26. Given three segments whose lengths are 1, a, and b, construct segments of length $a + b$, $|a - b|$, ab, a/b and $\sqrt{ab}$. These are Euclidean constructions.

11.27. Every angle can be trisected using a compass and *marked* straightedge. The following construction is due to Archimedes. Assume the compass has two marks on it, a distance r apart. Let $\angle BAC$ be an angle. Draw a circle γ of radius r and center A. The circle will intersect both sides of the angle; in order to simplify the notation let us assume that B and C lie on the circle. Place the straightedge so that it passes through C and so that one mark is at a point D on γ and the other is at a point E on $\overleftrightarrow{AB}$. (See Fig. 11.17.) Use the Isosceles Triangle Theorem and the Euclidean Angle Sum Theorem to prove that $\mu(\angle CEB) = (1/3)\mu(\angle CAB)$.

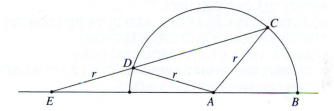

FIGURE 11.17: Archimedes' trisection using a marked straightedge

C H A P T E R   12

Transformations

12.1 THE TRANSFORMATIONAL PERSPECTIVE
12.2 PROPERTIES OF ISOMETRIES
12.3 ROTATIONS, TRANSLATIONS, AND GLIDE REFLECTIONS
12.4 CLASSIFICATION OF EUCLIDEAN MOTIONS
12.5 CLASSIFICATION OF HYPERBOLIC MOTIONS
12.6 A TRANSFORMATIONAL APPROACH TO THE FOUNDATIONS
12.7 EUCLIDEAN INVERSIONS IN CIRCLES

In this chapter we study geometry from a new point of view. Rather than looking at geometric objects directly, we study them indirectly by examining the functions, which in this context are called transformations, that preserve geometric structures. This is a thoroughly modern point of view in which the concept of function is fundamental.

The chapter begins with a brief explanation of the transformational perspective. After that, the important properties of distance-preserving transformations are studied and the basic examples are constructed. A surprising and beautiful consequence of this study is a theorem that completely describes all rigid motions of the plane. We work out this classification for Euclidean geometry first and then for hyperbolic geometry. Next the foundations of plane geometry are reformulated to reflect the transformational perspective. Finally, a class of transformations of the Euclidean plane, called inversions in circles, is studied.

The results of the last two sections of this chapter will be applied in the next chapter when we construct models for hyperbolic geometry. Inversions in circles will be the main technical tool used in those constructions. In addition, we will use the transformational reformulation of the foundations to verify that the basic model we construct satisfies all the axioms of neutral geometry.

12.1   THE TRANSFORMATIONAL PERSPECTIVE

The transformational perspective is the view that the essence of any geometry is captured by the transformations that preserve the structures of that geometry. This way of thinking originated with Felix Klein (1849–1925) in the nineteenth century. Klein's basic insight is that geometry can be understood as the study of those properties of space that are preserved by certain groups of transformations. Thus Euclidean geometry is associated with one group of transformations while hyperbolic geometry is associated with a different group of transformations. Understanding the algebraic structure of the group of transformations illuminates the geometric structure of a space. This way of unifying the different geometries became known as the *Erlanger Programm* because Klein introduced it in connection with his appointment to a professorship at Erlangen in 1872. The *Erlanger Programm* allows a synthesis of many different types of geometry, including such diverse fields as Euclidean geometry and topology. Since the time of Klein, the transformational perspective has come to permeate most branches of mathematics.

While the transformational perspective is completely different from Euclid's perspective, there is a limited sense in which transformational thinking gets us closer to Euclid. We have followed Hilbert in taking Side-Angle-Side as an axiom, but this is not at all what Euclid did. Euclid thought in terms of moving one triangle to another and in effect he made the ability to move triangles an axiom. Common Notion 4 in Book I of the *Elements* states, "Things which coincide with one another equal one another." It is clear from Euclid's proof of the Side-Angle-Side Theorem (see the proof of Proposition I.4 in Chapter 1) that he understood this to mean that triangles can be moved from one location to another in the plane and made to coincide that way. He speaks of one point being "placed on" another and commentators use the term "superposition" for the method Euclid applies in this proof. Euclid doesn't explain exactly what he means by it and he uses the method of superposition in only two other proofs (I.8 and III.24), but it is clear that he was thinking of something like what we would now call a transformation of the triangle.

The transformational approach helps to clarify the meaning of congruence. We have given a number of different definitions of "congruence"; in particular, we gave separate definitions of congruence for segments, angles, and triangles. While each separate definition is clear and intuitive, it should not be necessary to give a new definition of congruence every time we encounter a new geometric object (as we did, for example, when we defined quadrilateral for the first time). It is better to simply say that two figures are congruent if one of them can be transformed into the other by a transformation that preserves the size and shape of the figure.

The transformational point of view has found its way into most contemporary high school geometry textbooks. This perspective permeates the entire course and is even reflected in the axioms on which the course is founded. In a textbook such as [71], for example, an axiom about the existence of certain transformations takes the place of the Side-Angle-Side Postulate. That alternative approach to the foundations is explained later in the chapter.

12.2   PROPERTIES OF ISOMETRIES

In this section we look at three basic examples of transformations and study the general properties of isometries. In the next section we will greatly increase the supply of examples. Both this section and the next are part of neutral geometry.

Definition 12.2.1. A *transformation* is a function $T : \mathbb{P} \to \mathbb{P}$ that is both one-to-one and onto. A transformation is called an *isometry* if it preserves distances. In other words, an isometry is a transformation $T : \mathbb{P} \to \mathbb{P}$ such that for every pair of points A and B, $T(A)T(B) = AB$.

Notation. We will consistently use P' to denote $T(P)$, the image of P.

It is not difficult to prove that any function that preserves distances is automatically one-to-one and onto (Exercise 12.1). We now describe some of the basic examples of transformations from neutral geometry.

EXAMPLE 12.2.2   The identity transformation

The *identity* function is an isometry. The identity $\iota : \mathbb{P} \to \mathbb{P}$ is defined by $\iota(P) = P$ for every point P. ■

It is obvious that the identity transformation is an isometry, but it should not be overlooked as it is an important example of an isometry of the plane. The next example is more interesting. In fact we will see later that the isometries defined in the next example are the building blocks from which all isometries are constructed.

EXAMPLE 12.2.3   The reflection in line ℓ

Let ℓ be a line. Define $\rho_\ell : \mathbb{P} \to \mathbb{P}$, the *reflection in* ℓ, as follows. Define $P' = P$ for every P that lies on ℓ. For each external point P, drop a perpendicular from P to ℓ and call the foot Q. Let $P' = \rho_\ell(P)$ be the point on $\overrightarrow{PQ}$ such that Q is the midpoint of $\overline{PP'}$.

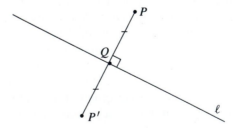

FIGURE 12.1: P' is the reflection of P in ℓ

If P is in one half-plane determined by ℓ, then P' is in the opposite half-plane. The reflection in ℓ is an isometry (Exercise 12.2). The line ℓ is often called the *mirror* for the reflection. ■

Definition 12.2.4. A point P is called a *fixed point* for transformation T if $T(P) = P$.

Each point on ℓ is a fixed point for the reflection ρ_ℓ. For each external point P, P and $\rho_\ell(P)$ lie on opposite sides of ℓ, so $P \neq \rho_\ell(P)$ and P is not a fixed point for ρ_ℓ. Every point is a fixed point for the identity transformation.

EXAMPLE 12.2.5   The dilation with center O and constant k

Let O be a point and let k be a positive number. The *dilation $D_{O,k}$ with center O and proportionality constant k* is defined as follows. Define $D_{O,k}(O) = O$. For each point $P \neq O$, define $D_{O,k}(P)$ to be the point P' on $\overrightarrow{OP}$ such that $OP' = k \cdot OP$. ■

The dilation keeps O fixed and each point $P \neq O$ is either moved toward O or away from O, depending on whether k is less than 1 or greater than 1. In the special case $k = 1$, the dilation is the identity. Every dilation is a transformation of the plane. A dilation is not an isometry, except in case $k = 1$.

Before we can increase our supply of examples, we must prove some theorems about isometries. A transformation is a function, so we can perform one transformation and then another to produce a composite transformation. The next theorem shows that the composition of two isometries is another isometry and that the inverse of an isometry is an isometry. This demonstrates that the set of isometries forms a group under the operation of composition.[1] (Composition of functions is always an associative operation, so it is not necessary to prove separately that composition of isometries is associative.)

Theorem 12.2.6. *The composition of two isometries is an isometry. The inverse of an isometry is an isometry.*

Proof. Let T_1 and T_2 be two isometries. It is clear from the definitions of one-to-one and onto that $T_2 \circ T_1$ is a transformation. Let P and Q be points. The fact that T_1 is an isometry implies that $T_1(P)T_1(Q) = PQ$ and the fact that T_2 is an isometry implies that $T_2(T_1(P))T_2(T_1(Q)) = T_1(P)T_1(Q)$. Therefore, $T_2 \circ T_1(P)T_2 \circ T_1(Q) = PQ$.

Let T be an isometry and let P and Q be two points. Since T is an isometry, $T(T^{-1}(P))T(T^{-1}((Q)) = T^{-1}(P)T^{-1}(Q)$. Therefore, $T^{-1}(P)T^{-1}(Q) = PQ$ and T is an isometry. □

Recall that an isometry is defined to be a transformation that preserves distances between points. We will now prove that an isometry preserves other geometric relationships as well. It may seem surprising that just requiring that a transformation preserve distances is enough to guarantee that it also preserves angle measures and many other geometric properties, but it should be remembered that we have assumed the Side-Angle-Side Postulate. The Side-Angle-Side Postulate asserts that distance and angle measure are not independent, but are closely linked.

---

[1] Abstract algebra is not a prerequisite for this course and it is not necessary that you understand the technical meaning of the word *group* as it is used here.

While working out the proof of the next theorem you should think about the ways in which SAS is being used, both directly and indirectly. Later in the chapter we will reverse the relationship and prove that if there exist certain transformations that preserve these properties, then SAS must be satisfied.

Theorem 12.2.7 (Properties of Isometries). *Let* $T : \mathbb{P} \to \mathbb{P}$ *be an isometry. Then* T *preserves the following geometric relationships.*

1. *T preserves collinearity; that is, if P, Q, and R are three collinear points, then $T(P)$, $T(Q)$, and $T(R)$ are collinear.*

2. *T preserves betweenness of points; that is, if P, Q, and R are three points such that $P * Q * R$, then $T(P) * T(Q) * T(R)$.*

3. *T preserves segments; that is, if A and B are points and A' and B' are their images under T, then $T(\overline{AB}) = \overline{A'B'}$ and $\overline{A'B'} \cong \overline{AB}$.*

4. *T preserves lines; that is, if ℓ is a line, then $T(\ell)$ is a line.*

5. *T preserves betweenness of rays; that is, if $\overrightarrow{OP}$, $\overrightarrow{OQ}$, and $\overrightarrow{OR}$ are three rays such that $\overrightarrow{OQ}$ is between $\overrightarrow{OP}$ and $\overrightarrow{OR}$, then $\overrightarrow{O'Q'}$ is between $\overrightarrow{O'P'}$ and $\overrightarrow{O'R'}$.*

6. *T preserves angles; that is, if $\angle BAC$ is an angle, then $T(\angle BAC)$ is an angle and $T(\angle BAC) \cong \angle BAC$.*

7. *T preserves triangles; that is, if $\triangle BAC$ is a triangle, then $T(\triangle BAC)$ is a triangle and $T(\triangle BAC) \cong \triangle BAC$.*

8. *T preserves circles; that is, if γ is a circle with center O and radius r, then $T(\gamma)$ is a circle with center $T(O)$ and radius r.*

9. *T preserves areas; that is, if R is a polygonal region, then $T(R)$ is a polygonal region and $\alpha(T(R)) = \alpha(R)$.*

Terminology. Because isometries move things around while preserving all geometric relationships, an isometry is also called a *rigid motion*. The terms *isometry of the plane* and *rigid motion* will be used synonymously.

Proof of theorem 12.2.7. We will prove Part 1. The proofs of the remaining parts of the theorem are left as exercises (Exercises 12.6–12.13).

Let P, Q, and R be three distinct collinear points and let P', Q', and R' denote their images under T. Then one of the points must be between the other two (Corollary 5.7.3). Let us say that $P * Q * R$. Since $PR = PQ + QR$ (definition of betweenness) it follows that $P'R' = P'Q' + Q'R'$ (definition of isometry). Suppose P', Q', and R' are not collinear (RAA hypothesis). Then the triangle inequality (Theorem 6.4.2) implies that $P'R' < P'Q' + Q'R'$. But this contradicts the earlier statement, so we must reject the RAA hypothesis and conclude that P', Q', and R' are collinear. □

We come now to the main theorem of the section. The following theorem will prove to be extremely useful in constructing more isometries. The proof of the theorem also offers a great deal of insight into the structure of the group of isometries of the plane. The converse will become the definition of congruence in the transformational approach to the foundations.

Theorem 12.2.8. *If $\triangle ABC$ and $\triangle DEF$ are two triangles with $\triangle ABC \cong \triangle DEF$, then there exists a unique isometry T such that $T(A) = D$, $T(B) = E$, and $T(C) = F$.*

Recall that part of the definition of triangle is the assumption that A, B, and C are noncollinear. Without that assumption on A, B, and C, the uniqueness part of the conclusion would not hold. The following corollary is a restatement of the uniqueness part of the theorem.

Corollary 12.2.9. *If f and g are two isometries and A, B, and C are three noncollinear points such that $f(A) = g(A)$, $f(B) = g(B)$, and $f(C) = g(C)$, then $f(P) = g(P)$ for every point P.*

Proof of existence in Theorem 12.2.8. Let $\triangle ABC$ and $\triangle DEF$ be two triangles such that $\triangle ABC \cong \triangle DEF$ (hypothesis). We must prove that there is an isometry T such that $T(\triangle ABC) = \triangle DEF$.

Let ℓ be the perpendicular bisector of $\overline{AD}$ and let ρ_ℓ be the associated reflection. (If it happens that $A = D$, take ℓ to be any line through A.) By the definition of reflection we have that $\rho_\ell(A) = D$. Define $B' = \rho_\ell(B)$ and $C' = \rho_\ell(C)$.

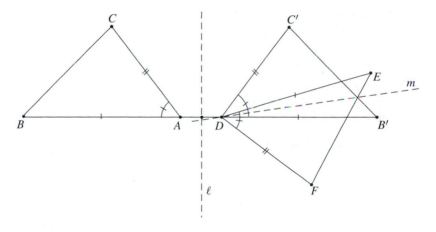

FIGURE 12.2: Proof of Theorem 12.2.8

Let m be the perpendicular bisector of $\overline{B'E}$ and let ρ_m be the associated reflection. (If it happens that $B' = E$, take m to be the line $\overleftrightarrow{DE}$.) By the definition of reflection we have that $\rho_m(B') = E$. Since $DB' = DE$, D lies on m (Theorem 6.4.7) and so $\rho_m(D) = D$. Define C'' to be $\rho_m(C')$.

Note that either $C'' = F$ or C'' is the reflection of F across $\overleftrightarrow{DE}$ (because ρ_ℓ and ρ_m preserve angles and distances—see Theorem 6.2.4). Take $n = \overleftrightarrow{DE}$. Let $f = \iota$ in case $C'' = F$ and let $f = \rho_n$ in the other case. Then $T = f \circ \rho_m \circ \rho_\ell$ is an isometry such that $T(A) = D$, $T(B) = E$, and $T(C) = F$.   □

The next lemma is the main ingredient in the proof of uniqueness.

Lemma 12.2.10. *An isometry that fixes three noncollinear points is the identity; that is, if A, B, and C are three noncollinear points and f is an isometry such that $f(A) = A$, $f(B) = B$, and $f(C) = C$, then $f = \iota$.*

Proof. Let A, B, and C be three noncollinear points and let f be an isometry such that $f(A) = A$, $f(B) = B$, and $f(C) = C$ (hypothesis). Let P be an arbitrary point. We must prove that $f(P) = P$. The proof will be divided into two cases: either P lies on $\overleftrightarrow{AB}$ or P does not lie on $\overleftrightarrow{AB}$.

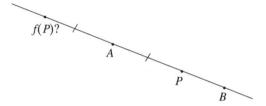

FIGURE 12.3: Two possible locations for $f(P)$ in Case 1

Suppose, first, that P lies on $\overleftrightarrow{AB}$. Since $f(A) = A$, $f(B) = B$, and f preserves collinearity (Theorem 12.2.7, Part 1), it follows that $f(P)$ lies on $\overleftrightarrow{AB}$. The distance from $f(P)$ to A must equal PA, so either $f(P) = P$ or $f(P)$ is the point on the opposite side of A that is equidistant from A. The fact that betweenness and order are preserved by f (Theorem 12.2.7, Part 2) means that $f(P) = P$.

Now suppose P is a point that does not lie on $\overleftrightarrow{AB}$. Since f preserves angles (Theorem 12.2.7, Part 5), $f(P)$ either lies on $\overrightarrow{AP}$ or on $\overrightarrow{AP'}$, where P' is the reflection of P across $\overleftrightarrow{AB}$. Since the distance from A to $f(P)$ equals the distance from A to P, we see that either $f(P) = P$ or $f(P) = P'$.

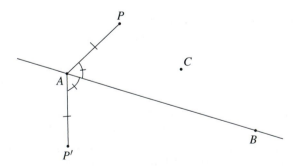

FIGURE 12.4: Two possible locations for $f(P)$ in Case 2

If P and C are on the same side of $\overleftrightarrow{AB}$, then $\overline{PC}$ does not intersect $\overleftrightarrow{AB}$. As a result, the image of $\overline{PC}$ under f does not intersect $\overleftrightarrow{AB}$ either. (The points of $\overleftrightarrow{AB}$

are fixed and f is one-to-one, so no point of $\overleftrightarrow{AB}$ can be in the image of $\overline{PC}$.) In that case $f(P)$ cannot equal P' and so we conclude that $f(P) = P$. If P and C are on opposite sides of $\overleftrightarrow{AB}$, then $\overline{PC}$ does intersect $\overleftrightarrow{AB}$. Therefore the image of $\overline{PC}$ under f must intersect $\overleftrightarrow{AB}$ as well. Again this means that $f(P) \neq P'$ and so $f(P) = P$.   ☐

Proof of uniqueness in Theorem 12.2.8. Suppose A, B, and C are three noncollinear points and f and g are two isometries such that $f(A) = g(A)$, $f(B) = g(B)$, and $f(C) = g(C)$ (hypothesis). Define $h = f^{-1} \circ g$. Then h is an isometry (Theorem 12.2.6) and, by the Lemma above, $h = \iota$. Therefore, $f \circ h = f \circ \iota$, and so $g = f$.   ☐

Several aspects of the proof of Theorem 12.2.8 will be important for later work and should be noted. First, the proof of existence does not use the full force of the assumption $\triangle ABC \cong \triangle DEF$, but only that $\overline{AB} \cong \overline{DE}$, $\angle BAC \cong \angle EDF$, and $\overline{AC} \cong \overline{DF}$. These are the hypotheses of SAS. We will use this observation later in the chapter to give a transformational proof of SAS.

A second aspect of the proof that should be explicitly mentioned is the fact that the isometry constructed in the existence proof is constructed as the composition of reflections. The number of reflections needed is at most three, one for each vertex of the triangle. (In fact the proof shows that we could take the number of reflections to be exactly two or three.) Given an isometry f, we can choose three noncollinear points A, B, and C and define $D = f(A)$, $E = f(B)$, and $F = f(C)$. The existence part of the proof gives a composition of reflections that takes $\triangle ABC$ to $\triangle DEF$ and the uniqueness part of the proof shows that this composition of reflections must be equal to f. Thus we can conclude that any transformation is a composition of reflections. We will record this second observation as a corollary. It is a corollary of the proof rather than a corollary of the theorem as stated. The corollary is the key to the classification of rigid motions of the plane in the next section.

Corollary 12.2.11. *Every isometry of the plane can be expressed as a composition of reflections. The number of reflections required is at most three.*

The decomposition into reflections is not unique and even the number of reflections can change. However, the number of reflections cannot change from even to odd or odd to even. It is far from obvious, but in fact no isometry can be written as both a composition of an even number of refections and an odd number of reflections. This is proved for Euclidean isometries in Exercise 12.32, but it is true in neutral geometry as well. Thus isometries are divided into two classes depending upon whether they can be written as a composition of an even number of reflections or an odd number of reflections. Those that can be written as a composition of an even number of reflections are called *orientation preserving* (or *direct*) and those that can be written as a odd number of reflections are *orientation reversing* (or *opposite*). Informally, an orientation preserving isometry is one that can be performed on a triangle by simply sliding and rotating it in the plane. By contrast, it is necessary to pick the triangle up out of the plane and turn it over in order to accomplish an orientation reversing isometry.

In the next few sections we will study the decomposition into reflections in more detail.

12.3   ROTATIONS, TRANSLATIONS, AND GLIDE REFLECTIONS

We now explore the consequences of the theorems from the last section regarding existence and uniqueness of isometries. We will use Theorem 12.2.8 to construct additional examples of isometries of the plane. Corollary 12.2.11 promises that each of these isometries can be expressed as a composition of reflections. As we construct examples of isometries, we will analyze each type to see how it decomposes as a product of reflections. All of these results will eventually come together to prove a rather amazing theorem that completely describes and classifies all the rigid motions of the plane. That theorem will be proved in the next two sections.

We begin with the definition of a rotation. The rough idea is familiar: given an angle, the plane can be rotated through that angle, keeping the vertex of the angle fixed. It is necessary to be careful with the definition, however, because there are actually two different ways to rotate through a given angle; there is one rotation of the plane in the clockwise direction and a second rotation in the counterclockwise direction. Another technical point is that we need to include the possibility of a rotation through $180°$. We get around both those difficulties by thinking of the rotation as rotating one ray to another. Specifically, given any pair of rays $\overrightarrow{OA}$ and $\overrightarrow{OB}$ that share a common endpoint O, we will define a rotation that rotates $\overrightarrow{OA}$ to $\overrightarrow{OB}$ keeping O fixed. We will use the notation R_{AOB} for this isometry. Note that R_{AOB} and R_{BOA} are different transformations. In fact, $R_{BOA} = R_{AOB}^{-1}$ (Exercise 12.19).

EXAMPLE 12.3.1   The rotation with center O and angle $\angle AOB$

Let A, O, and B be three distinct points. The *rotation* R_{AOB} is defined as follows.

Case 1. $\overrightarrow{OA} = \overrightarrow{OB}$. In this case we define $R_{AOB} = \iota$ (the identity transformation).

Case 2. A, O, and B are noncollinear. Let A' be the point on $\overrightarrow{OB}$ such that $OA = OA'$. Use the Protractor and Ruler Postulates to locate a point B' such that B' is on the opposite side of $\overleftrightarrow{OB}$ from A, $\angle AOB \cong \angle BOB'$, and $OB = OB'$. Define R_{AOB} to be the unique isometry such that $R_{AOB}(O) = O$, $R_{AOB}(A) = A'$, and $R_{AOB}(B) = B'$. Such a transformation exists by Theorem 12.2.8. The point O is called the *center of rotation*.

Case 3. $\overrightarrow{OA}$ and $\overrightarrow{OB}$ are opposite rays. Again let A' be the point on $\overrightarrow{OB}$ such that $OA = OA'$. Choose a point C, not on $\overleftrightarrow{AB}$, such that $\overleftrightarrow{OC} \perp \overleftrightarrow{AB}$ and let C' be the point on the opposite side of $\overleftrightarrow{AB}$ such that O is the midpoint of $\overline{CC'}$. Define R_{AOB} to be the unique isometry such that $R_{AOB}(O) = O$, $R_{AOB}(A) = A'$, and $R_{AOB}(C) = C'$. Such a transformation exists by Theorem 12.2.8. In the special case in which $\overrightarrow{OA}$ and $\overrightarrow{OB}$ are opposite, the rotation R_{AOB} is called a *half-turn*. Again the point O is the center of rotation. The half-turn is well defined in the sense that

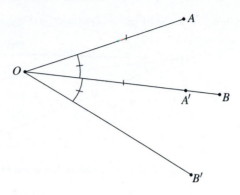

FIGURE 12.5: Definition of rotation in Case 2

the same isometry will result no matter which external point C is used (as long as $\overleftrightarrow{OC} \perp \overleftrightarrow{AB}$). ∎

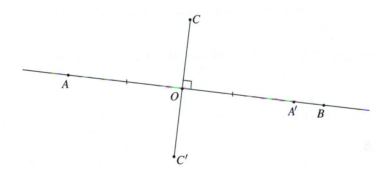

FIGURE 12.6: Definition of rotation in Case 3

You may have been expecting a definition of rotation that goes more like this: For each point P, define the point $P' = R_{AOB}(P)$ to be the point such that $\angle POP' \cong \angle AOB$ and $OP = OP'$. The problem with such a definition is that there are two points P' that satisfy the stated conditions, one on each side of $\overleftrightarrow{OP}$. Rather than trying to give the rule for choosing the correct point P' in every case, it is more convenient to give the rule for the two points A and B and then to rely on Theorem 12.2.8 to take care of the other points. Nonetheless it is true that $\angle POP' \cong \angle AOB$ for every P.

Theorem 12.3.2. *Let R_{AOB} be a rotation with center O. If A, O, and B are noncollinear, then $\angle POP' \cong \angle AOB$ for every point $P \neq O$.*

Sketch of proof. Observe, first, that the conclusion follows from the definition of R_{AOB} in case P lies on $\overrightarrow{OA}$ or $\overrightarrow{OB}$. If $\overrightarrow{OP}$ is between $\overrightarrow{OA}$ and $\overrightarrow{OB}$, then the conclusion

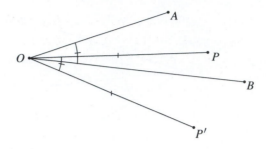

FIGURE 12.7: Alternative definition of rotation

follows from the fact that R_{AOB} preserves betweenness of rays and preserves angles. Let $B' = R_{AOB}(B)$ and let $B'' = R_{AOB}(B')$. Since $\overrightarrow{OB}$ is between rays $\overrightarrow{OA}$ and $\overrightarrow{OB'}$, the fact that R_{AOB} preserves betweenness of rays implies that $\overrightarrow{OB'}$ is between $\overrightarrow{OB}$ and $\overrightarrow{OB''}$. Thus the same proof can be used to show that the conclusion holds in case $\overrightarrow{OP}$ is between rays $\overrightarrow{OB}$ and $\overrightarrow{OB'}$. In this way we can continue to enlarge the set of points P for which the conclusion holds until we know that the conclusion holds for all points P. □

Corollary 12.3.3. *If $\overrightarrow{OA} \neq \overrightarrow{OB}$, then O is the only fixed point of R_{AOB}.*

Suppose $\overrightarrow{OA}$ and $\overrightarrow{OB}$ are opposite rays and $\ell = \overleftrightarrow{AB}$. The reflection ρ_ℓ and the half-turn R_{AOB} are alike in some ways. For example, both interchange the two sides of ℓ and both fix O. Another way in which they are alike is that both are examples of involutions. An *involution* is an isometry that is equal to its own inverse, so applying it twice results in the identity (see Exercise 12.3).

The reflection and the half-turn are different in that every point of ℓ is a fixed point of ρ_ℓ while R_{AOB} interchanges the rays $\overrightarrow{OA}$ and $\overrightarrow{OB}$. The half-turn preserves orientation while the reflection reverses orientation.

As a step in the direction of classifying all rigid motions of the plane we now prove that every rotation can be expressed as the composition of two reflections. This is the first of two theorems relating rotations and reflections. In the First Rotation Theorem we start with a rotation and produce a particular pair of lines such that reflection across those two lines results in the given rotation. The Second Rotation Theorem will describe a whole family of pairs of lines having the property that reflection across them produces the given rotation.

Theorem 12.3.4 (First Rotation Theorem). *An isometry f is a rotation with center O if and only if there exist two lines ℓ and m such that $O \in \ell \cap m$ and $f = \rho_m \circ \rho_\ell$.*

Proof. Assume, first, that $f = R_{AOB}$ is a rotation with center O (hypothesis). We will consider each of the cases in the definition of R_{AOB} separately. It is obvious that the identity transformation can be written as the composition of two reflections (in fact $\iota = \rho_\ell \circ \rho_\ell$ for any line ℓ), so the theorem holds in case $\overrightarrow{OA} = \overrightarrow{OB}$.

Now suppose $\overrightarrow{OA}$ and $\overrightarrow{OB}$ are opposite. Choose a point C such that $\overleftrightarrow{OC} \perp \overleftrightarrow{AB}$. Let $\ell = \overleftrightarrow{AB}$ and $m = \overleftrightarrow{OC}$. It is easy to check that R_{AOB} and $\rho_m \circ \rho_\ell$ map the three points A, O, and C to the same points, so $R_{AOB} = \rho_m \circ \rho_\ell$ (Corollary 12.2.9). Thus the theorem holds in the half-turn case as well and we need only consider the case in which A, O, and B are noncollinear.

Suppose that A, O, and B are noncollinear. Define $\ell = \overleftrightarrow{OA}$. Choose a point Q such that $\overrightarrow{OQ}$ is the bisector of $\angle AOB$, and define $m = \overleftrightarrow{OQ}$. We claim that $R_{AOB} = \rho_m \circ \rho_\ell$. Since $\rho_m \circ \rho_\ell$ is an isometry (Theorem 12.2.6), it suffices to show that the two transformations R_{AOB} and $\rho_m \circ \rho_\ell$ agree on the three noncollinear points O, A, and B (Corollary 12.2.9).

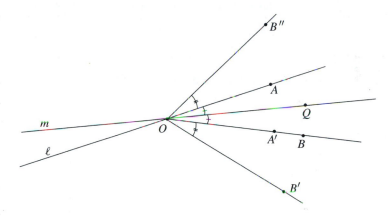

FIGURE 12.8: A rotation is the composition of two reflections

All three transformations fix O, so $R_{AOB}(O) = O = \rho_m \circ \rho_\ell(O)$. Since A lies on ℓ, A is a fixed point of ρ_ℓ. Furthermore, the choice of Q and m guarantee that $\rho_m(A) = A' = R_{AOB}(A)$. Thus $R_{AOB}(A) = \rho_m \circ \rho_\ell(A)$. Let $B' = R_{AOB}(B)$ and $B'' = \rho_\ell(B)$. Then $\angle BOA \cong \angle AOB'' \cong \angle BOB'$ and $\mu(\angle BOQ) = (1/2)\mu(\angle BOA)$. It is not hard to check that A is in the interior of $\angle B''OQ$ and B is in the interior of $\angle QOB'$, so $\angle B'OQ \cong \angle B''OQ$ (see Fig. 12.8). Therefore, $\rho_m(B'') = B'$ and $R_{AOB}(B) = \rho_m \circ \rho_\ell(B)$.

The proof of the converse is left as an exercise. (Exercise 12.17.)   □

It is a little hard to see the rotation in the proof above. In order to see it more clearly, choose a point P, on the opposite side of $\overleftrightarrow{OA}$ from B, such that $\mu(\angle POA) < (1/2)\mu(\angle AOB)$. We will follow P under the two reflections.

Define $P' = \rho_\ell(P)$ and $P'' = \rho_m(P')$. Since $\mu(\angle POA) = \mu(\angle AOP')$, P' lies in the interior of $\angle AOQ$. Therefore,

$$\mu(\angle AOP') + \mu(\angle P'OQ) = \mu(\angle AOQ)$$
$$= (1/2)\mu(\angle AOB).$$

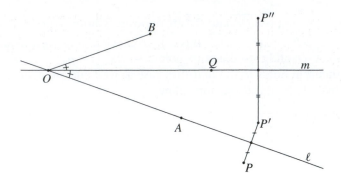

FIGURE 12.9: Following P under the two reflections

Hence

$$\mu(POP'') = \mu(\angle POP') + \mu(\angle P'OP'')$$
$$= 2\mu(\angle AOP') + 2\mu(\angle P'OQ) = \mu(\angle AOB).$$

Thus P is rotated to P'' through the angle $\mu(\angle AOB)$.

The decomposition of R_{AOB} into reflections is not unique. In fact, the next theorem tells us that we can specify either one of the lines to be any line whatsoever that passes through O.

Theorem 12.3.5 (Second Rotation Theorem). *If R_{AOB} is a rotation and n is any line with $O \in n$, then there exist lines s and t such that $R_{AOB} = \rho_s \circ \rho_n = \rho_n \circ \rho_t$.*

Proof. Choose a point $P \neq O$ on n and define $Q = R_{AOB}(P)$. It is not difficult to check that $R_{AOB} = R_{POQ}$ (Exercise 12.20). The proof of Theorem 12.3.4 shows that $R_{POQ} = \rho_s \circ \rho_n$, where s is the line determined by the bisector of $\angle POQ$. The proof of the other part is similar (Exercise 12.20). □

There is a second way to define half-turn that does not require Theorem 12.2.8. Let O be a point. For each point $P \neq O$, define $H_O(P) = P'$, where P' is chosen so that O is the midpoint of the segment $\overline{PP'}$. The next theorem shows that the transformation H_O just defined is the same as the half-turn determined by any pair of opposite rays that share endpoint O.

Theorem 12.3.6 (Half-Turn Theorem). *If $\overrightarrow{OA}$ and $\overrightarrow{OB}$ are opposite rays, then $R_{AOB} = H_O$. Furthermore, if n and s are any two lines such that $n \perp s$ and $O \in n \cap s$, then $H_O = \rho_s \circ \rho_n$.*

Proof. Exercise 12.21. □

One consequence of the Half-Turn Theorem is the fact that a half-turn depends only on the center of rotation and does not depend on the particular opposite rays used to define it. Notice the parallel between the new definition of half-turn and the definition of reflection given earlier in the chapter: In one case the center of rotation is the midpoint of every segment $\overline{PP'}$ and in the other case the line of reflection is the perpendicular bisector of every segment $\overline{PP'}$. In view of that relationship it is natural to think of a half-turn as being reflection in a point.

We now define another type of isometry, called a translation. Just as a rotation is associated with the composition of two reflections across intersecting lines, so a translation is associated with the composition of two reflections across parallel lines that share a common perpendicular.

EXAMPLE 12.3.7   The translation from A to B

Let A and B be two points. The *translation from A to B*, denoted T_{AB}, is defined as follows. If $A = B$ then T_{AB} is defined to be the identity. If $A \neq B$, let $\ell = \overleftrightarrow{AB}$ and let m be the line that is perpendicular to ℓ at A. Choose a point C that lies on m but not ℓ. Let C' be the point such that C and C' are on the same side of ℓ, $\overleftrightarrow{BC'} \perp \ell$ and $AC = BC'$. Let B' be the point on ℓ such that A and B' lie on opposite sides of $\overleftrightarrow{BC'}$ and $AB = BB'$. Then define T_{AB} to be the unique isometry such that $T_{AB}(A) = B$, $T_{AB}(B) = B'$, and $T_{AB}(C) = C'$.

There is exactly one isometry having the properties specified in the last paragraph by Theorem 12.2.8. It is easy to check that translation is well defined in the sense that the same isometry will result no matter which point C we use in the definition. ∎

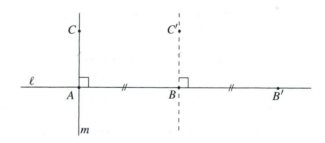

FIGURE 12.10: Definition of translation

Again this definition was probably not the one you were expecting. It seems more natural to define the translation of a point P by constructing a line through P that is parallel to $\overleftrightarrow{AB}$ and then locating a point P' on that line such that $PP' = AB$. One reason for taking the indirect approach is to avoid the technicality of choosing which of the two possible points P' is the right one. But this time there is also a much deeper reason for the indirect approach. In fact the kind of construction just described does not work in neutral geometry. In Euclidean geometry it is possible to define translation that way, but in hyperbolic geometry the function described does not preserve distances (see Exercise 12.24).

The next step in the direction of classifying all isometries of the plane is to prove that a translation is also the composition of two reflections, but this time across parallel lines.

Theorem 12.3.8 (First Translation Theorem). *An isometry f is a translation if and only if there exist lines ℓ and m that share a common perpendicular such that $f = \rho_\ell \circ \rho_m$.*

Proof. The idea of the proof is suggested by Fig. 12.11. The details are left as an exercise (Exercise 12.22). □

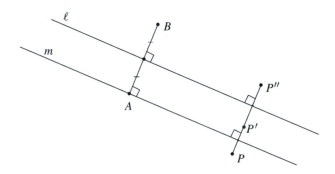

FIGURE 12.11: A translation is the composition of two reflections

Just as in the case of a rotation, it is possible to improve on the theorem by showing that we have a great deal of freedom in the choice of lines across which to reflect.

Theorem 12.3.9 (Second Translation Theorem). *Let A and B be two distinct points. For every line n that is perpendicular to $\overleftrightarrow{AB}$ there exists a line s, also perpendicular to $\overleftrightarrow{AB}$, such that $T_{AB} = \rho_s \circ \rho_n$.*

Proof. Exercise 12.23. □

The last several theorems show that every rotation and every translation can be written as the composition of two reflections. The following theorem summarizes what we have proved.

Theorem 12.3.10. *Let ℓ and m be two lines.*

1. *If $\ell = m$, then $\rho_\ell \circ \rho_m$ is the identity.*
2. *If $\ell \perp m$, then $\rho_\ell \circ \rho_m$ is a half-turn about the point of intersection.*
3. *If ℓ and m intersect in a point, then $\rho_\ell \circ \rho_m$ is a rotation about the point of intersection.*
4. *If ℓ and m are parallel and admit a common perpendicular, then $\rho_\ell \circ \rho_m$ is a translation.*

The only possibility not covered by the theorem is that in which ℓ and m are parallel but do not admit a common perpendicular. That situation can only occur in hyperbolic geometry, so we will postpone study of it until §12.5.

We want to define one more important type of isometry.

EXAMPLE 12.3.11   Glide reflection

Suppose A and B are two points on a line ℓ. The composite isometry $\rho_\ell \circ T_{AB}$ is called a *glide reflection*.

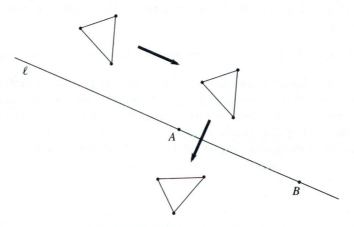

FIGURE 12.12: A glide reflection

In case $A = B$, the glide reflection is just an ordinary reflection in ℓ. If $A \neq B$, the glide reflection "glides" along ℓ and then reflects across ℓ. In case $A \neq B$, the two points A and B determine ℓ, so we will use the notation G_{AB} for the glide reflection. The two isometries involved could be performed in either order with the same effect. In other words, the fact that A and B lie on ℓ means that $G_{AB} = \rho_\ell \circ T_{AB} = T_{AB} \circ \rho_\ell$. Therefore either $\rho_\ell \circ T_{AB}$ or $T_{AB} \circ \rho_\ell$ could be taken as the definition of G_{AB}.   ■

At first glance a glide reflection looks like a very specialized kind of composite isometry, and you might wonder why we single out this particular type of isometry for study. But there is a good reason for it: In the next two sections we will prove the amazing and wonderful theorem that an isometry is a glide reflection if and only if it is the composition of exactly three reflections. This theorem is true in neutral geometry but we will prove it separately in Euclidean geometry so that the simplicity of the Euclidean proof will be evident.

The theorem is surprising because we are set up to expect a much longer list of possibilities for the composition of three isometries. Theorem 12.3.10 lists four different kinds of isometries that can result from reflecting across two lines and the paragraph after the theorem hints at even a fifth possibility. Thus we naturally expect the situation with three reflections to be even more complicated. Instead we have the unanticipated result that there is only one kind of isometry that can result from the composition of three reflections.

The fact that every composition of three reflections is a glide reflection will require a substantial proof. The converse, however, is easy to see. A translation is the composition of two reflections, so it is obvious that every glide reflection can be expressed as the composition of three reflections.

12.4 CLASSIFICATION OF EUCLIDEAN MOTIONS

In this section we will study isometries of the Euclidean plane. These isometries have a special name.

Definition 12.4.1. A *Euclidean motion* is an isometry of the Euclidean plane.

The objective of the section is to classify Euclidean motions. Specifically, we will prove that every Euclidean motion is either the identity, a reflection, a rotation, a translation, or a glide reflection. Since the identity is a special case of rotation and reflection is a special case of glide reflection, we can simplify this to say that every Euclidean motion is either a rotation, a translation, or a glide reflection.

The classification of rigid motions of the plane could be done in neutral geometry, but we will carry out the program in Euclidean and hyperbolic geometries separately. The main reason for treating the Euclidean case separately is that the classification is simpler in that setting and working through the simpler case first allows us to see the structure of the argument more clearly. A second reason for working out the Euclidean case first is so that the theorem and its proof can be appreciated by those who have not studied hyperbolic geometry.

Most of the hard work of the proof of the classification theorem has already been done in the last two sections. The one ingredient still needed is the theorem about glide reflections.

Theorem 12.4.2 (Glide Reflection Theorem). *An isometry is a glide reflection if and only if it can be written as the composition of three reflections.*

The Glide Reflection Theorem is a theorem in neutral geometry. The main point is that any composition of three reflections must be a glide reflection. The proof of this fact is divided into several special cases, each of which is relatively simple. We will state these special cases as separate lemmas. It should be noted that the proofs of the next four lemmas use only results of neutral geometry, so we can use the lemmas in the next section when we move on to the proof of the hyperbolic case of the theorem. There are fewer special cases in the Euclidean proof than there are in the hyperbolic proof.

Let us begin with the case in which the three lines are concurrent. In that case the composition of the three reflections is another reflection.

Lemma 12.4.3. *If ℓ, m, and n are three lines that are concurrent at P, then there exists a line s such that P lies on s and $\rho_\ell \circ \rho_m \circ \rho_n = \rho_s$.*

Proof. By Theorem 12.3.10, $\rho_\ell \circ \rho_m$ is a rotation. Hence we can apply the Second Rotation Theorem (Theorem 12.3.5) to conclude that there is a line s such that $\rho_\ell \circ \rho_m = \rho_s \circ \rho_n$. Therefore, $\rho_\ell \circ \rho_m \circ \rho_n = \rho_s \circ \rho_n \circ \rho_n = \rho_s$. □

A surprisingly similar proof works in case the three lines are mutually parallel.

Lemma 12.4.4. *If ℓ, m, and n are three lines that share a common perpendicular k, then there exists a line s such that $s \perp k$ and $\rho_\ell \circ \rho_m \circ \rho_n = \rho_s$.*

Proof. By Theorem 12.3.10, $\rho_\ell \circ \rho_m$ is a translation. Apply the Second Translation Theorem (Theorem 12.3.9) to conclude that there is a line s such that $s \perp k$ and $\rho_\ell \circ \rho_m = \rho_s \circ \rho_n$. Thus $\rho_\ell \circ \rho_m \circ \rho_n = \rho_s \circ \rho_n \circ \rho_n = \rho_s$. □

We come now to the case that should be thought of as the main special case. The process used in this proof is really the heart of the proof of the Glide Reflection Theorem.

Lemma 12.4.5. *If ℓ, m, and n are three lines such that ℓ and m intersect, then $\rho_\ell \circ \rho_m \circ \rho_n$ is a glide reflection.*

Proof. Let A be a point of $\ell \cap m$ (hypothesis) and define s to be the line such that A lies on s and $s \perp n$. By the Second Rotation Theorem there exists a line t such that $\rho_\ell \circ \rho_m = \rho_t \circ \rho_s$. Therefore, $\rho_\ell \circ \rho_m \circ \rho_n = \rho_t \circ \rho_s \circ \rho_n$.

Let B be the point at which s and n intersect. Drop a perpendicular k from B to t and let C be the foot of that perpendicular. Take h to be the line through B that is perpendicular to k. By the Half-Turn Theorem (Theorem 12.3.6), $\rho_s \circ \rho_n = \rho_h \circ \rho_k$. Therefore $\rho_t \circ \rho_s \circ \rho_n = \rho_t \circ \rho_h \circ \rho_k$. But $\rho_t \circ \rho_h$ is a translation along k. Specifically, if D is the point on k such that $D * C * B$ and $DC = CB$, then $\rho_t \circ \rho_h = T_{BD}$ and $k = \overleftrightarrow{BD}$. Therefore $\rho_\ell \circ \rho_m \circ \rho_n = \rho_t \circ \rho_h \circ \rho_k = T_{BD} \circ \rho_k = \rho_k \circ T_{BD}$ is a glide reflection. □

The process used in the last proof is called "moving the lines." It is a two-step process: First the lines at A are moved so that there is one right angle of intersection and then the lines at B are moved so that there is a second right angle of intersection. The two moves are illustrated in Fig. 12.13.

Notice that $D = C = B$ if and only if B lies on t. This could happen if $s = t$ or if $A = B$. Thus the composition will be a reflection if either $\ell = m$ or the three lines ℓ, m, and n are concurrent. In all other cases the glide reflection $\rho_k \circ T_{BD}$ will have a nontrivial translation part.

The same kind of proof can be used in case it is m and n that intersect.

Lemma 12.4.6. *If ℓ, m, and n are three lines such that m and n intersect, then $\rho_\ell \circ \rho_m \circ \rho_n$ is a glide reflection.*

Proof. Exercise 12.38. □

We are now ready to prove the Glide Reflection Theorem and then the Classification Theorem. *The remainder of this section is part of Euclidean geometry.*

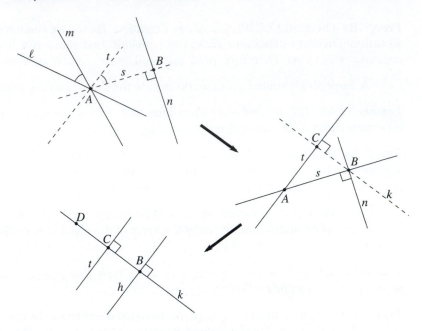

FIGURE 12.13: Moving the lines

Proof of Glide Reflection Theorem 12.4.2. It is obvious from the definition that every glide reflection can be expressed as the composition of three reflections. We will prove the converse. Let ℓ, m, and n be three lines (hypothesis). We must show that $\rho_\ell \circ \rho_m \circ \rho_n$ is always a glide reflection (which may be just a reflection, if the translation part is the identity).

Either $\ell \nparallel m$ or $\ell \parallel m$. If $\ell \nparallel m$, then the theorem follows from Lemma 12.4.5. We will show that the other case can be reduced to that in Lemma 12.4.6.

Suppose $\ell \parallel m$. Choose a point C on n. If it should happen that C lies on ℓ, take $s = \ell$; if C lies on m, take $s = m$. Otherwise take s to be the unique line through C that is parallel to both ℓ and m. (The line s exists by the Euclidean Parallel Postulate and Transitivity of Parallelism.) By Lemma 12.4.4, there exists a line t such that $\rho_t = \rho_\ell \circ \rho_m \circ \rho_s$. Thus

$$\rho_\ell \circ \rho_m \circ \rho_n = \rho_\ell \circ \rho_m \circ \rho_s \circ \rho_s \circ \rho_n = \rho_t \circ \rho_s \circ \rho_n.$$

Since C lies on both s and n, we can complete the proof by applying Lemma 12.4.6 to the composition $\rho_t \circ \rho_s \circ \rho_n$. $\square$

Theorem 12.4.7 (Classification of Euclidean Motions). *Every Euclidean motion is either the identity, a reflection, a rotation, a translation, or a glide reflection.*

Proof. Let T be a Euclidean motion. Then T can be written as the composition of zero, one, two, or three reflections (Corollary 12.2.11). If the number of reflections is zero, then T is the identity. If the number of reflections is one, then T is a reflection. If the number of reflections is two, then T is either the identity, a rotation, or a translation (Theorem 12.3.10). Finally, if the number of reflections is three, then T is a glide reflection (Theorem 12.4.2).   □

12.5   CLASSIFICATION OF HYPERBOLIC MOTIONS

This short section is part of hyperbolic geometry. In it we indicate the additions that must be made to the proofs in the previous section in order to classify all isometries of the hyperbolic plane. Such isometries are called *hyperbolic motions*. The proof of the classification theorem for hyperbolic motions follows the basic outline of the Euclidean proof. The hyperbolic proof is a little more complicated because there are two kinds of parallel lines to consider instead of one.

The fact that there is a second type of parallelism comes up in two separate places in the proof. First, there is an additional kind of isometry that can arise as the composition of two reflections since there is an additional possibility for the relationship between two lines. Second, there is an additional special case to consider in the proof of the Glide Reflection Theorem.

Definition 12.5.1. A *horolation* is an isometry that can be written as a composition $\rho_\ell \circ \rho_m$ in which ℓ and m are asymptotically parallel lines.

In the case of rotations and translations we started with a geometric description of the isometry and then proved that it could be written as the composition of two reflections. We are not in a position to give a comparable geometric description of a horolation. Such a description must wait until the next chapter when we can describe a horolation in terms of a specific model for hyperbolic geometry. At that time we will see that a horolation can be thought of as a rotation about a point at infinity.

We do have a theorem for horolations that is analogous to the second rotation and translation theorems.

Theorem 12.5.2 (Horolation Theorem). *Let ℓ and m be asymptotically parallel in the direction $\overrightarrow{AB}$. If n is any line such that ℓ and n are asymptotically parallel in the direction $\overrightarrow{AB}$, then there exists a line s such that ℓ and s are also asymptotically parallel in the direction $\overrightarrow{AB}$ and $\rho_m \circ \rho_\ell = \rho_s \circ \rho_n$.*

Proof. Exercise 12.39.   □

Since asymptotically parallel lines are the only pairs of lines not covered by Theorem 12.3.10, we can say that any isometry of the hyperbolic plane that can be written as the composition of two reflections must be the identity, a rotation, a translation, or a horolation. We now move on to compositions of three reflections. As in the Euclidean case, such a composition must be a glide reflection. We state the theorem again to emphasize the fact that it also holds in hyperbolic geometry.

Theorem 12.5.3 (Hyperbolic Glide Reflection Theorem). *A hyperbolic motion is a glide reflection if and only if it can be written as the composition of three reflections.*

Recall that the various special cases of the Glide Reflection Theorem that were proved as lemmas in the previous section were actually neutral results, so we can use them in the hyperbolic proof. There is one new special case to be considered separately.

Lemma 12.5.4. *If ℓ, m, and n are three lines that are all asymptotically parallel in the direction $\overrightarrow{AB}$, then there exists a line s such that s is also asymptotically parallel in the direction $\overrightarrow{AB}$ and $\rho_\ell \circ \rho_m \circ \rho_n = \rho_s$.*

Proof. Let $\rho_\ell \circ \rho_m \circ \rho_n = \rho_s$ be as in the statement of the lemma. By the Horolation Theorem, there exists a line s such that $\rho_m \circ \rho_\ell = \rho_s \circ \rho_n$. Therefore $\rho_\ell \circ \rho_m \circ \rho_n = \rho_s \circ \rho_n \circ \rho_n = \rho_s$. □

Proof of the Hyperbolic Glide Reflection Theorem. The proof is very similar to that in the Euclidean case. Again it is obvious from the definition that every glide reflection can be expressed as the composition of three reflections, so we focus on the converse. Let ℓ, m, and n be three lines (hypothesis). We will show that $\rho_\ell \circ \rho_m \circ \rho_n$ is a glide reflection.

Either $\ell \nparallel m$ or $\ell \parallel m$. If $\ell \nparallel m$, then the theorem follows from Lemma 12.4.5. As before, we will show that the second case can be reduced to that in Lemma 12.4.6.

Suppose $\ell \parallel m$. Then either ℓ and m admit a common perpendicular or they are asymptotically parallel (Theorem 8.4.18). Choose a point C on n. If ℓ and m admit a common perpendicular k, then choose s to be the line through C that is perpendicular to k. If ℓ and m are asymptotically parallel in the direction $\overrightarrow{AB}$, then choose s to be the line through C that is asymptotically parallel to ℓ in the direction $\overrightarrow{AB}$. (It could happen that C lies on ℓ; in that case take $s = \ell$.) By either Lemma 12.4.4 or Lemma 12.5.4, there exists a line t such that $\rho_\ell \circ \rho_m \circ \rho_s = \rho_t$. Thus $\rho_\ell \circ \rho_m \circ \rho_n = \rho_\ell \circ \rho_m \circ \rho_s \circ \rho_s \circ \rho_n = \rho_t \circ \rho_s \circ \rho_n$. Since C lies on both s and n, an application of Lemma 12.4.6 completes the proof. □

Theorem 12.5.5 (Classification of Hyperbolic Motions). *Every hyperbolic motion is one of the following: the identity, a reflection, a rotation, a translation, a horolation, or a glide reflection.*

Proof. The proof is almost exactly the same as in the Euclidean case. The only new feature is that there is one additional possibility for the composition of two reflections. □

12.6   A TRANSFORMATIONAL APPROACH TO THE FOUNDATIONS

Let us go back to the foundations of geometry and examine them from a transformational point of view. There are two places in Chapter 5 where transformations arise, at least implicitly. In both cases we avoided transformations when we could have used them.

The first point at which transformations arise implicitly is in the definition of congruence for triangles. We gave a static definition, saying that two triangles are congruent if there is a correspondence between their vertices so that corresponding parts of the triangle (edges and angles) are congruent. This is not the way we intuitively think of congruence. We usually think of it more dynamically, visualizing a motion that takes one triangle and moves it over to make it coincide with the other.

The second place where transformations arise implicitly is in the SAS Postulate. Here again it is natural to think, as Euclid did, of a dynamic process that takes one triangle and superimposes it on the other. We chose instead a static, passive approach that simply accepts the SAS triangle congruence condition as an axiom.

It was not until this chapter that we defined isometry and constructed the basic examples, such as reflections. A key ingredient in both the construction of a reflection and the proof that it is an isometry is the SAS Postulate. For example, SAS is required for the definition of a reflection in that it is used to construct the perpendiculars that determine P'.

We will now see that those relationships can be reversed. It is possible to accept transformation as the more basic concept and then to define congruence of triangles in terms of transformations and to prove SAS as a theorem using transformations. This transformational approach to the axioms has become standard in high school textbooks in recent years.

In the transformational approach to the axioms, we accept all the axioms from Chapter 5 except the Side-Angle-Side Postulate and we use all the definitions from Chapter 5 except that of congruence. In their places we put the new axiom stated below and the next three definitions.

If we do not take SAS as an axiom, we cannot prove the existence of perpendiculars and therefore we cannot define reflections. So we will take the existence of reflections as an axiom and we will use that axiom to prove SAS as a theorem. We also do not have the theorems which assert that distance-preserving transformations preserve collinearity and angle measure available to us, so we build those into the axiom as well. The definition of transformation remains the same: It is simply a function from the plane to the plane that is one-to-one and onto.

Axiom 12.6.1 (Reflection Postulate). *For every line ℓ there exists a transformation $\rho_\ell : \mathbb{P} \to \mathbb{P}$, called the reflection in ℓ, that satisfies the following conditions.*

1. *If P lies on ℓ, then P is a fixed point for ρ_ℓ.*
2. *If P lies in one of the half-planes determined by ℓ, then $\rho_\ell(P)$ lies in the opposite half-plane.*
3. *ρ_ℓ preserves collinearity.*
4. *ρ_ℓ preserves distance.*
5. *ρ_ℓ preserves angle measure.*

Definition 12.6.2. A *rigid motion* is a transformation of the plane that preserves collinearity, distance, and angle measure.[2]

Note that the Reflection Postulate implies that a reflection is a rigid motion. It is easy to check that any composition of reflections is also a rigid motion.

Definition 12.6.3. A *figure* is any subset of the plane.

Another way to define figure is to say that a figure is a set of points in the plane. Segments, angles, triangles, quadrilaterals, polygons, and circles are all examples of figures.

Definition 12.6.4. Two figures X and Y are *congruent* if there exists a rigid motion $f : \mathbb{P} \to \mathbb{P}$ such that $f(X) = Y$.

In particular, two triangles $\triangle ABC$ and $\triangle A'B'C'$ are congruent if there exists a rigid motion $f : \mathbb{P} \to \mathbb{P}$ such that $f(\triangle ABC) = \triangle A'B'C'$. The notation $\triangle ABC \cong \triangle A'B'C'$ should be interpreted to mean that there exists a rigid motion f such that $f(A) = A'$, $f(B) = B'$, and $f(C) = C'$. With this new definition of congruence, the assertion that corresponding parts of congruent triangles are congruent (CPCTC) becomes a theorem just as it is in high school geometry. Until now we have viewed CPCTC as the definition of congruence for triangles.

We will now prove Side-Angle-Side as a theorem, using the Reflection Postulate. This means that either Side-Angle-Side or the Reflection Postulate can be taken as an axiom for neutral geometry and then the other becomes a theorem. One way to view the following theorem is that it proves that SAS and the Reflection Postulate are logically equivalent—given the other axioms of neutral geometry. We already proved half that statement earlier in the chapter when we constructed reflections in neutral geometry. The next theorem is the other half of the statement. The proof is essentially the same as the proof of Theorem 12.2.8. We will go through all the same steps, but we must be careful to note that the only properties of reflections that are needed are those that are explicitly stated in the Reflection Postulate.

Theorem 12.6.5. *The Reflection Postulate implies the Side-Angle-Side triangle congruence condition.*

---

[2]We use the term *rigid motion* rather than isometry because we prefer to reserve the term isometry for the situation in which only distance-preserving is assumed. In UCSMP Geometry [71], an isometry is defined to be any transformation that is a composition of reflections. There is an "A-B-C-D Theorem" which asserts that every isometry preserves Angles, Betweenness, Collinearity, and Distance. Since an isometry is defined to be the composition of reflections and it is assumed in the Reflection Postulate that reflections preserve these properties, the A-B-C-D Theorem is really just asserting that if these properties are preserved by two transformations then they are preserved by the composition. We do not explicitly assume betweenness is preserved, either in this definition or in our formulation of the Reflection Postulate, because that follows from the other assumptions.

Before beginning the proof, it would be good to clarify just what may be used and what may not. We are accepting all the axioms of neutral geometry except SAS. Therefore, we may use any theorem that was stated and proved before SAS was introduced. There were not many such theorems, but two theorems that were stated and proved early were the existence of a perpendicular bisector for a segment and the existence of an angle bisector for an angle. Both those theorems were proved using only the Ruler and Protractor Postulates. The construction of a perpendicular through an external point, on the other hand, was based on SAS. Thus we may use the existence of perpendicular bisectors in the proof below, but we may not drop perpendiculars from external points.

The following observation is also useful in the proof. Let A, B and C be three noncollinear points and let A', B', and C' be their images under the reflection ρ_ℓ. The fact that ρ_ℓ preserves collinearity means that the image of any point on $\overleftrightarrow{AB}$ must lie on $\overleftrightarrow{A'B'}$. The fact that ρ_ℓ preserves distances implies that, in fact, the image of any point of $\overline{AB}$ is in $\overline{A'B'}$. Any point on $\overline{A'B'}$ is some distance from each of the endpoints, so it must be the image of the corresponding point on $\overline{AB}$. All those observations together imply that $\rho_\ell(\overline{AB}) = \overline{A'B'}$. The same is true of the other sides of $\triangle ABC$, so we see that $\rho_\ell(\triangle ABC) = \triangle A'B'C'$.

Proof of Theorem 12.6.5. Let $\triangle ABC$ and $\triangle DEF$ be two triangles such that $\overline{AB} \cong \overline{DE}$, $\angle BAC \cong \angle EDF$, and $\overline{AC} \cong \overline{DF}$ (hypothesis). We must prove that there is a rigid motion T such that $T(\triangle ABC) = \triangle DEF$. We will construct T as a composition of at most three reflections. By the observation immediately before the proof, it is enough to construct a T such that $T(A) = D$, $T(B) = E$, and $T(C) = F$.

Let ℓ be the perpendicular bisector of $\overline{AD}$ and let ρ_ℓ be the associated reflection (Reflection Postulate). Let G denote the point of intersection of ℓ and $\overline{AD}$. Since ρ_ℓ fixes G and preserves collinearity, the image of $\overrightarrow{GA}$ must be a ray emanating from G. Since ρ_ℓ interchanges the sides of ℓ and preserves angles, the image of $\overrightarrow{GA}$ must be $\overrightarrow{GD}$. Since ρ_ℓ preserves distances, $\rho_\ell(A) = D$. Define $B' = \rho_\ell(B)$ and $C' = \rho_\ell(C)$.

In case C', D, and F are noncollinear, let m be the bisector of $\angle C'DF$ and let ρ_m be the associated reflection. Then $\rho_m(D) = D$. The properties of ρ_m can be used as in the previous paragraph to show that $\rho_m(\overrightarrow{DC'}) = \overrightarrow{DF}$. It follows that $\rho_m(C') = F$. In the special case in which $\overrightarrow{DC'}$ and $\overrightarrow{DF}$ are opposite rays, define m to be the line that is perpendicular to $\overleftrightarrow{FC'}$ at D. Again $\rho_m(D) = D$ and $\rho_m(C') = F$. In case the rays $\overrightarrow{DC'}$ and $\overrightarrow{DF}$ are equal, it must be true that $C' = F$. By taking $g = \rho_m$ in the first two cases and $g = \iota$ in the third case, we construct a rigid motion g such that in every case $g(D) = D$ and $g(C') = F$.

Note that either $g(B') = E$ or $g(B')$ is the reflection of E across $\overleftrightarrow{DF}$. Define $n = \overleftrightarrow{DF}$ and define $f = \iota$ in case $\rho_m(B') = E$ and $f = \rho_n$ in the other case. Then $T = f \circ g \circ \rho_\ell$ is a rigid motion such that $T(A) = D$, $T(B) = E$, and $T(C) = F$. $\qquad\square$

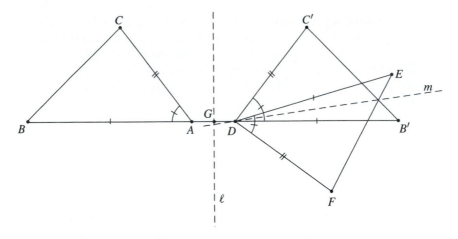

FIGURE 12.14: Proof of SAS

Many of the proofs given earlier in the course can be recast in transformational terms. The last proof is an example; it is a transformational proof of SAS. We provide two further illustrations by giving transformational proofs of two other familiar theorems. The reason for including these proofs is to further illuminate the transformational point of view and to make clear why many proofs in the high school textbooks are formulated the way they are.

The next proof is a transformational proof of the Isosceles Triangle Theorem. The transformational proof is essentially the same as the first proof that was given in Chapter 5. The only difference is a small change in emphasis; rather than using SAS to prove that the two subtriangles are congruent, the proof uses the fact that a reflection is a rigid motion and therefore maps a triangle onto a congruent triangle. It should finally be clear why that proof has a place in high school geometry textbooks even though it is longer and more complicated than the second proof given in Chapter 5. In the transformational approach, transformation is considered to be a more basic concept than SAS, which is derived from it.

Theorem 12.6.6 (Isosceles Triangle Theorem). *If $\triangle ABC$ is a triangle and $\overline{AB} \cong \overline{AC}$, then $\angle ABC \cong \angle ACB$.*

Transformational proof. By the Crossbar Theorem, the bisector of angle $\angle BAC$ must intersect $\overline{BC}$ in a point D. Let $\ell = \overleftrightarrow{AD}$. Then $\rho_\ell(A) = A$ and $\rho_\ell(D) = D$. Now the reflection preserves the angle $\angle BAD$, but reflects it to the other side of ℓ. The reflection also preserves distances, so $\rho_\ell(B) = C$. Since a reflection maps a triangle to a congruent triangle (definition of congruent), $\triangle ADB \cong \triangle ADC$ and so $\angle ABC \cong \triangle ACB$.   □

Our final example of a transformational proof comes from Euclidean geometry. We prove the Angle Sum Theorem. Since it is a proof in Euclidean geometry, we can make use of the Converse to the Alternate Interior Angles Theorem. We

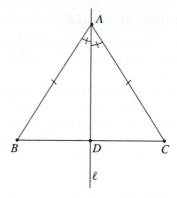

FIGURE 12.15: Transformational proof of the Isosceles Triangle Theorem

will assume that translation has already been defined (as the composition of two reflections).

Theorem 12.6.7 (Angle Sum Theorem). *For every triangle* $\triangle ABC$, $\sigma(\triangle ABC) = 180°$.

Transformational proof. Let $\triangle ABC$ be any triangle and let T_{AB} be the translation that maps A to B. Define $A' = T_{AB}(A)$, $B' = T_{AB}(B)$, and $C' = T_{AB}(C)$. Then $A' = B$ and A, B and B' are collinear (isometries preserve collinearity). Therefore

$$\mu(\angle ABC) + \mu(\angle CBC') + \mu(\angle C'BB') = 180°$$

(Protractor Postulate). But $\angle C'BB' \cong \angle CAB$ (isometries preserve angles) and so $\overleftrightarrow{AC} \parallel \overleftrightarrow{BC'}$ (Alternate Interior Angles Theorem). Therefore $\angle CBC' \cong \angle ACB$ (Converse to Alternate Interior Angles Theorem), and the proof is complete.   □

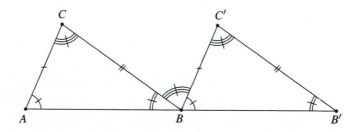

FIGURE 12.16: Transformational proof of the Angle Sum Theorem

Other examples of transformational proofs are included in the exercises.

12.7   EUCLIDEAN INVERSIONS IN CIRCLES

This section is part of Euclidean geometry. The proofs in the section rely on results about similar triangles from Section 7.4 and results about Euclidean circles from Section 10.4.

In the section we will construct and study a collection of transformations of the Euclidean plane called inversions in circles. These transformations provide examples of transformations that are very different from those studied earlier in the chapter. They are interesting and important in their own right, but they are also included here because they will play a key role in the construction of models in the next chapter.

Notation. In this section we will be looking at what happens to one circle when it is inverted in another circle. In order to avoid confusion and to make clear which circle is defining the inversion, we will consistently denote that circle by C. Other circles will continue to be called α, β, γ, and so on.

Definition 12.7.1. Let $C = C(O, r)$ be a circle. The *inversion in C* is the transformation $I_{O,r}$ defined as follows. For each $P \neq O$, let P' be the point on $\overrightarrow{OP}$ such that $(OP)(OP') = r^2$. Define $I_{O,r}(P) = P'$.

Notice that we have stretched the definition of transformation a bit by calling an inversion a transformation. In fact the inversion is not defined at O, so the domain of the function is not the entire plane as required in the definition of transformation. The inversion is also not onto because there is no point P such that $I_{O,r}(P) = O$. The usual way around these difficulties is to extend the plane by adding an additional point ∞ (called the *point at infinity*) to the plane and agreeing that the inversion interchanges O and ∞. The inversion then becomes a transformation of the extended plane.

Definition 12.7.2. The *inversive plane* is defined by $\mathbb{P}^* = \mathbb{P} \cup \{\infty\}$. Extend $I_{O,r}$ to a transformation of $\mathbb{P}^*$ by defining $I_{O,r}(O) = \infty$ and $I_{O,r}(\infty) = O$.

We defined $I_{O,r}(P)$ to be the point P' such that $(OP)(OP') = r^2$. While it is clear from the Ruler Postulate that there must be such a point, it is also possible to construct P' using straightedge and compass. The next two paragraphs tell how P' can be constructed.

Construction of P' in case P is inside C. Erect at P a perpendicular to $\overleftrightarrow{OP}$ and let T and U denote the two points at which the perpendicular intersects C. Draw tangents to C at T and U and let P' be the point at which those two tangents intersect $\overleftrightarrow{OP}$. The proof that $P' = I_{O,r}(P)$ is left as an exercise (Exercise 12.48).   □

Construction of P' in case P is outside C. Let M be the midpoint of $\overline{OP}$ and let α be the circle with center M and radius MP. Let T and U be the two points of intersection of C and α. Connect T and U with a segment and let P' be the point at which that segment crosses $\overleftrightarrow{OP}$. The proof that $P' = I_{O,r}(P)$ is left as an exercise (Exercise 12.49).   □

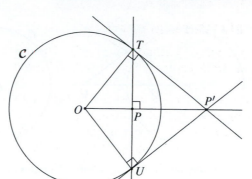

FIGURE 12.17: Construction of P' in case P is inside C

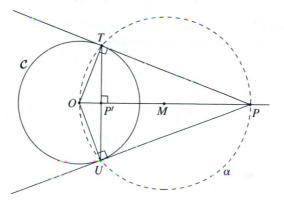

FIGURE 12.18: Construction of P' in case P is outside C

By moving to the inversive plane, we are able to consider $I_{O,r}$ to be a transformation. It is clear, however, that inversion does not preserve distances and so an inversion is definitely not an isometry. Despite the fact that distances are not preserved by inversions, there are other geometric properties that are preserved and most of the theorems in the remainder of the section are aimed at proving such results. The theorems in the remainder of the section are all based on the fact that certain triangles get transformed into similar triangles.

Theorem 12.7.3. *If $I_{O,r}$ is an inversion and P and Q are points that are not collinear with O, then $\triangle OPQ$ is similar to $\triangle OQ'P'$.*

Proof. By definition of inversion we have $(OP)(OP') = r^2 = (OQ)(OQ')$ and so $OP/OQ = OQ'/OP'$. Furthermore, $\overrightarrow{OP} = \overrightarrow{OP'}$ and $\overrightarrow{OQ} = \overrightarrow{OQ'}$, so $\angle POQ = \angle P'OQ'$. Therefore, $\triangle OPQ \sim \triangle OQ'P'$ by the SAS Similarity Criterion (Theorem 7.4.3). $\qquad\square$

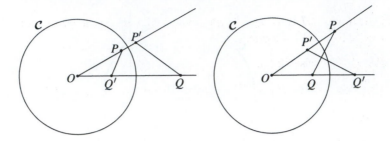

FIGURE 12.19: Two possible diagrams of Theorem 12.7.3

The next few theorems investigate what happens to circles and lines under inversions.

Theorem 12.7.4. *If $I_{O,r}$ is an inversion and ℓ is a line that does not contain O, then $I_{O,r}(\ell \cup \{\infty\})$ is a circle that contains O.*

Proof. Drop a perpendicular from O to ℓ and call the foot P. Let α be the circle with diameter $\overline{OP'}$. We will show that $\alpha = I_{O,r}(\ell \cup \{\infty\})$.

Let Q be a point on ℓ. If $Q = P$, then it is obvious that Q' lies on α. Assume $Q \neq P$. By Theorem 12.7.3, $\angle OQ'P' \cong \angle OPQ$ and so $\angle OQ'P'$ is a right angle. Hence Q' lies on α by Corollary 10.4.4. Since $I_{O,r}(\infty) = O$ we see that $I_{O,r}$ maps $\ell \cup \{\infty\}$ into α. In order to complete the proof we must show that every point on α is the image of some point of $\ell \cup \{\infty\}$.

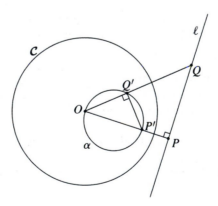

FIGURE 12.20: Line ℓ inverts to circle α

Let R be a point on α. If $R = O$, then $R = I_{O,r}(\infty)$. If $R \neq O$, then $\angle ORP'$ is a right angle (Corollary 10.4.2). Therefore $\angle OPR'$ is a right angle (Theorem 12.7.3) and so R' lies on ℓ. Hence $R = I_{O,r}(Q)$ where $Q = R'$. $\square$

Corollary 12.7.5. *If $I_{O,r}$ is an inversion and α is a circle such that $O \in \alpha$, then $I_{O,r}(\alpha - \{O\})$ is a line.*

Proof. Exercise 12.50.   □

Theorem 12.7.6. *If ℓ is a line and O lies on ℓ, then $I_{O,r}(\ell \cup \{\infty\}) = \ell \cup \{\infty\}$.*

Proof. Exercise 12.51.   □

Theorem 12.7.7. *If $I_{O,r}$ is an inversion and α is a circle such that O does not lie on α, then $I_{O,r}(\alpha)$ is a circle.*

Proof. Let A be the center of α and let s be the radius of α. If $A = O$, then it is easy to see that $I_{O,r}(\alpha)$ is the circle with center O and radius r^2/s. So we assume that $A \neq O$. In that case the line $\overleftrightarrow{OA}$ will intersect α in two points Q and R. Let β be the circle which has the segment $\overline{Q'R'}$ as diameter.

We claim that $\beta = I_{O,r}(\alpha)$. In order to prove this we show that a point P (different from Q or R) lies on α if and only if its inverse point P' lies on β. By Corollaries 10.4.2 and 10.4.4 it suffices to prove that $\angle R'P'Q'$ is a right angle if and only if $\angle QPR$ is a right angle. (See Fig. 12.21.)

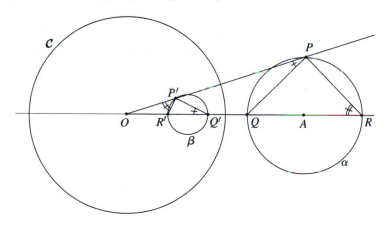

FIGURE 12.21: The image of α is β

Applying Theorem 12.7.3 gives $\angle OQ'P' \cong \angle OPQ$ and $\angle OP'R' \cong \angle ORP$. Since

$$\mu(\angle R'P'Q') = \sigma(\triangle OP'Q') - \mu(\angle OP'R') - \mu(\angle OQ'P') - \mu(\angle P'OQ')$$

and

$$\mu(\angle RPQ) = \sigma(\triangle OPR) - \mu(\angle ORP) - \mu(\angle OPQ) - \mu(\angle POR),$$

we see that $\mu(\angle R'P'Q') = \mu(\angle QPR)$. Thus $\angle R'P'Q'$ is a right angle if and only if $\angle QPR$ is.   □

The last few theorems can be summarized by saying that inversions preserve circles and lines. In other words, if γ is either a circle or a line, then $I_{O,r}(\gamma)$ is also a circle or a line. The remainder of this section will be devoted to an investigation of the sense in which inversions also preserve angle measure and distance.

Recall that the power of a point with respect to a circle was defined in §10.4. Here is a quick review of the definition. Let β be a circle and let O be a point that does not lie on β. Choose any line ℓ through O that intersects β. If ℓ intersects β at two points Q and R, the power of O is defined to be the product $(OQ)(OR)$. In case ℓ is tangent to β at P, the power of O is defined to be $(OP)^2$. (Refer to Fig. 10.29.) By Theorem 10.4.12, the power of O is well defined. We will use that fact in the proof of the next theorem.

Definition 12.7.8. Two circles α and β are *orthogonal* if they intersect at two points and their tangent lines are perpendicular at the points of intersection.

Theorem 12.7.9. *If the circle β is orthogonal to $C(O,r)$, then $I_{O,r}(\beta) = \beta$.*

The theorem does not assert that the points of β are fixed points for $I_{O,r}$. In fact the part of β that is inside C is mapped to the part that is outside C and vice versa.

Proof. Let P and T be the two points at which β intersects $C = C(O,r)$. Then $\overleftrightarrow{OP}$ and $\overleftrightarrow{OT}$ are tangent to β since the tangent and radius of a circle are perpendicular (Theorem 10.2.4). It follows that O is outside β (Theorem 10.2.5) and, in particular, O does not lie on β.

It is obvious that $I_{O,r}(P) = P$ and $I_{O,r}(T) = T$. If Q is any other point of β, then $\overleftrightarrow{OQ}$ is a secant line for β and intersects β in two points; call the second point R. Since the power of O with respect to β is well defined (Theorem 10.4.12), $(OQ)(OR) = (OP)^2 = r^2$. Hence $I_{O,r}(Q) = R$ and $I_{O,r}(R) = Q$. Therefore the image of every point on β is another point on β and every point on β is the image of some point of β. In other words, $I_{O,r}(\beta) = \beta$. □

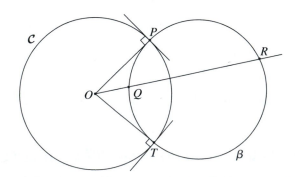

FIGURE 12.22: An orthogonal circle is invariant, Theorem 12.7.9

The next theorem gives a criterion that can be used to determine whether or not two circles are orthogonal.

Theorem 12.7.10. *Let $C = C(O, r)$ and β be two circles. If there exists a point Q on β such that $Q' = I_{O,r}(Q)$ also lies on β and $Q' \neq Q$, then C is orthogonal to β.*

Proof. Since Q and Q' are on opposite sides of C, the circle β must intersect C at two points P and T (Circular Continuity Principle, Theorem 10.5.4). Now P lies on C, so $(OP)^2 = r^2$. By definition of inversion, $(OQ)(OQ') = r^2$ as well. The power of O with respect to β is well defined (Theorem 10.4.12), so $\overleftrightarrow{OP}$ must be tangent to β. The fact that a radius of C is tangent to β implies that C is orthogonal to β (Theorem 10.2.4). ∎

Theorem 12.7.10 is stronger than a converse to Theorem 12.7.9, so we have the following corollary.

Corollary 12.7.11. *Let $C = C(O, r)$ and β be two circles. Then C is orthogonal to β if and only if $I_{O,r}(\beta) = \beta$.*

The next corollary will be important in the following chapter where we will want to construct circles that are orthogonal.

Corollary 12.7.12. *If $C = C(O, r)$ is a circle and R and S are two points inside C that do not lie on the same diameter of C, then there exists a unique circle γ such that R and S both lie on γ and γ is orthogonal to C.*

Proof. Since R and S do not lie on the same diameter, neither of them is equal to O. It also follows that R, S, and $S' = I_{O,r}(S)$ are not collinear. Let γ be the unique circle that passes through R, S, and S' (Theorem 10.3.4). It follows from Theorem 12.7.10 that γ is orthogonal to C. ∎

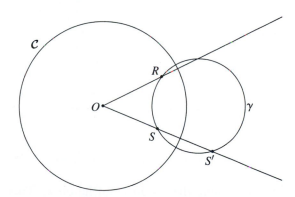

FIGURE 12.23: R and S lie on a circle γ that is orthogonal to C

Theorem 12.7.13. *Let $C = C(O, r)$ be a circle and let P be a point inside C that is different from O. For every line t such that P lies on t but O does not lie on t, there exists a unique circle α such that t is tangent to α at P and α is orthogonal to C.*

Proof. Let ℓ be the line such that P lies on ℓ and ℓ is perpendicular to t. There are two cases to consider: either O lies on ℓ or it does not.

Let us assume, first, that O lies on ℓ. Then $P' = I_{O,r}(P)$ lies on ℓ as well. Define M to be the midpoint of $\overline{PP'}$ and let α be the circle with center M and radius MP. By Theorem 12.7.10, α is orthogonal to C. It is clear from the construction that t is tangent to α at P.

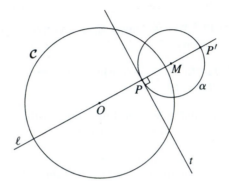

FIGURE 12.24: Case 1, O lies on ℓ.

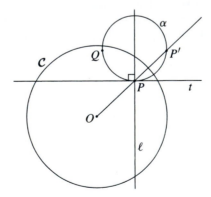

FIGURE 12.25: Case 2, O does not lie on ℓ.

Now let us assume that O does not lie on ℓ. Then $P' = I_{O,r}(P)$ does not lie on ℓ either. Define $Q = \rho_\ell(P')$ to be the reflection of P' across ℓ. Take α to be the unique circle that passes through the points P, P', and Q (Theorem 10.3.4). The fact that both P and P' lie on α implies that α is orthogonal to C (Theorem 12.7.10). The fact that both P' and Q lie on α implies that the center of α lies on ℓ

(Theorem 10.2.6). Since P lies on α and the center of α is on the line perpendicular to t at P, we see that t is tangent to α at P (Theorem 10.2.4).   □

Our next objective is to prove that inversions preserve angles between circles and lines. Before we can do that we must define what we mean by such angles.

Suppose γ is a circle with center at A and that P is a point on γ. Choose a second point C such C lies on γ and C is not antipodal to P. We will use $\gamma(P, C)$ to denote the *circular arc* on γ from P to C. Specifically,

$$\gamma(P, C) = \{P, C\} \cup \{Q \mid Q \text{ lies on } \gamma \text{ and } Q \text{ is in the interior of } \angle PAC\}.$$

The circular arc determines a tangent ray at P in the following way. Start with the line t that is tangent to γ at P and choose a point C^* on t such that C and C^* lie on the same side of $\overleftrightarrow{AP}$. We will say that the ray $\overrightarrow{PC^*}$ is tangent to the circular arc $\gamma(P, C)$ at P.

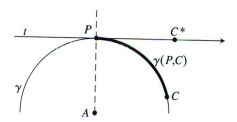

FIGURE 12.26: A circular arc and tangent ray

Definition 12.7.14. The *angle between the circular arc* $\gamma(P, C)$ *and the ray* $\overrightarrow{PQ}$ is defined to be $\angle C^* PQ$. If γ' is a second circle and D is a point on γ', then the *angle between the circular arcs* $\gamma(P, C)$ and $\gamma'(P, D)$ is defined to be the angle $\angle C^* PD^*$.

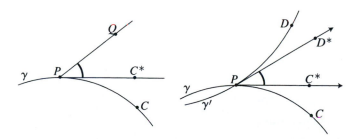

FIGURE 12.27: The angle between a ray and a circular arc; the angle between two circular arcs

We are now ready to prove that inversions preserve angles. We begin with two special cases.

Theorem 12.7.15. *If γ is a circle, P is a point on γ, $I_{O,r}$ is an inversion, and C is a point on γ that is not antipodal to P, then the angle between $\gamma(P, C)$ and $\overrightarrow{PP'}$ is congruent to the angle between $\gamma'(P', C')$ and $\overrightarrow{P'P}$.*

Proof. Let A be the center of γ. If $A = O$, then γ and $C = C(O, r)$ are either identical or disjoint. In the first case the conclusion holds vacuously. In the second case, $I_{O,r}(\gamma)$ is a concentric circle, so both angles are right angles.

Assume that $A \neq O$. Let R and S be the two points where $\overleftrightarrow{OA}$ intersects γ. By Theorem 12.7.7, $\gamma' = I_{O,r}(\gamma)$ is also a circle. In order to simplify the notation we will denote the center of γ' by B and the image of C under the inversion by D. We will show that $\angle PP'D^* \cong \angle P'PC^*$. (See Fig. 12.28.)

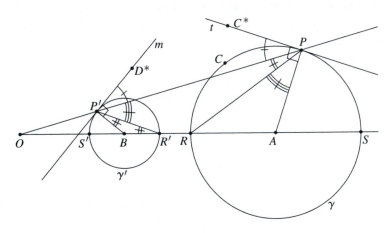

FIGURE 12.28: The angles $\angle PP'D^*$ and $\angle P'PC^*$ are congruent

Observe that $\angle BP'D^*$ and $\angle APC^*$ are both right angles (Theorem 10.2.4). Hence the proof will be complete if we show that $\angle BP'R' \cong \angle RPP'$ and $\angle R'P'P \cong \angle APR$. Now $\angle BR'P' \cong \angle RPP'$ by Theorem 12.7.3 and $\angle BR'P' \cong \angle BP'R'$ by the Isosceles Triangle Theorem, so $\angle BP'R' \cong RPP'$. The other congruence is proved similarly. □

The proof of the next theorem is similar.

Theorem 12.7.16. *If ℓ is a line that does not pass through O, $I_{O,r}$ is an inversion, γ' is the image of ℓ under $I_{O,r}$, and P and C are two distinct points on ℓ, then $\angle P'PC$ is congruent to the angle between $\gamma'(P', C')$ and $\overrightarrow{P'P}$.*

Proof. Exercise 12.52. □

The last two theorems allow us to prove that inversions preserve angles. The assertion that inversions preserve angles means that if we start with two curves, each of which is a line or a circle, then the angle between the curves at a point is congruent to the angle between the images of the curves at the image point. In order to avoid having to state lots of special cases, depending on whether each curve is a circle or a line and its image is a circle or a line, we will use the notation $\gamma(P, C)$ for arcs on lines as well as arcs on circles.

Theorem 12.7.17. *If each of γ_1 and γ_2 is either a line or a circle, P is a point that lies on both γ_1 and γ_2, C_1 and C_2 are points on γ_1 and γ_2, and $I_{O,r}$ is an inversion, then the angle between $\gamma_1(P, C_1)$ and $\gamma_2(P, C_2)$ is congruent to the angle between $\gamma_1'(P', C_1')$ and $\gamma_2'(P', C_2')$.*

Proof. This follows from Theorem 12.7.15 and subtraction—see Fig. 12.29.   □

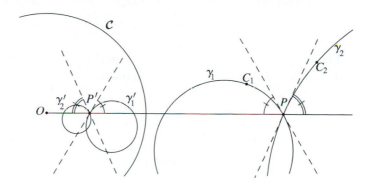

FIGURE 12.29: Theorem 12.7.17

Now that we know inversions preserve angle measure, we turn our attention to distance and investigate the sense in which inversions preserve distance. It is not the Euclidean distance itself that is preserved but rather a combination of distances called a cross-ratio.

Definition 12.7.18. *If A, B, P, Q are four distinct points, the* cross-ratio (AB, PQ) *is defined by*

$$[AB, PQ] = \frac{(AP)(BQ)}{(AQ)(BP)}.$$

Theorem 12.7.19. *If A, B, P, Q and O are all distinct and A', B', P', Q' are the images of A, B, P, Q under $I_{O,r}$, then*

$$[AB, PQ] = [A'B', P'Q'].$$

Proof. First observe that $AP/OA = A'P'/OP'$. In case O, A, and P are noncollinear, this follows from Theorem 12.7.3. Suppose O, A, P are collinear; let us say that P is closer to O than A is. The definition of inversion gives $(OA)(OA') = (OP)(OP')$ so $OP/OA = OA'/OP'$. Subtracting gives $1 - OP/OA = 1 - OA'/OP'$, so $AP/OA = A'P'/OP'$ in this case as well.

A similar proof shows that $AQ/OA = A'Q'/OQ'$. Combining the two equations gives

$$\frac{AP}{AQ} = \frac{AP}{OA} \cdot \frac{OA}{AQ} = \frac{A'P'}{OP'} \cdot \frac{OQ'}{A'Q'}.$$

It follows similarly that

$$\frac{BQ}{BP} = \frac{B'Q'}{OQ'} \cdot \frac{OP'}{B'P'}.$$

Finally, all the equations together show that

$$\frac{(AP)(BQ)}{(AQ)(BP)} = \frac{A'P'}{OP'} \cdot \frac{OQ'}{A'Q'} \cdot \frac{B'Q'}{OQ'} \cdot \frac{OP'}{B'P'} = \frac{(A'P')(B'Q')}{(A'Q')(B'P')}.$$

$\square$

This section contains many relatively technical theorems. The results are essential to the constructions in the next chapter, so it seems worthwhile to summarize them before moving on.

Summary. For each circle $C = C(O, r)$ there is a transformation $I_{O,r}$ of the inversive plane that has the following properties.

1. $I_{O,r}$ fixes each point of C and interchanges the inside and outside of C (definition).
2. $I_{O,r}$ maps a circle or a line to a circle or a line (Theorems 12.7.4–12.7.7).
3. $I_{O,r}$ maps a circle β to itself if and only if β is orthogonal to C (Theorem 12.7.9–Corollary 12.7.11).
4. For every pair of points P and Q inside C that do not lie on a diameter there is a unique circle γ such that P and Q lie on γ and γ is orthogonal to C (Corollary 12.7.12).
5. For every point $P \neq O$ inside C and for every line t that passes through P but not through O there exists a unique circle α such that t is tangent to α at P and α is orthogonal to C (Theorem 12.7.13).
6. $I_{O,r}$ preserves angles (Theorem 12.7.17).
7. $I_{O,r}$ preserves cross-ratios (Theorem 12.7.19).

EXERCISES

12.1. Prove that every function $f : \mathbb{P} \to \mathbb{P}$ that preserves distances is one-to-one and onto.

12.2. Prove that a reflection is an isometry (see Definition, page 286). Use only theorems of neutral geometry to prove this. In your proof you will need to consider several cases, depending on whether the two given points are on the same side of ℓ, opposite sides of ℓ, or one lies on ℓ. Think about which theorems of neutral geometry you need to apply in your proof. In what ways does the proof (indirectly) depend on Side-Angle-Side?

12.3. Prove that reflections and half-turns are involutions. That is, prove that if ρ_ℓ is a reflection in ℓ, then $\rho_\ell \circ \rho_\ell = \iota$ and that $\rho_\ell^{-1} = \rho_\ell$. Formulate and prove the analogous equations for half-turns.

12.4. Let S be a set of points in the plane. A line ℓ is called a *line of symmetry* for S if $\rho_\ell(S) = S$. Find all lines of symmetry for each of the following.
  (a) An isosceles triangle.
  (b) An equilateral triangle.
  (c) A regular polygon.
  (d) A circle.

12.5. Dilations and collinearity.
  (a) Prove that every dilation of the Euclidean plane preserves collinearity.
  (b) Does a dilation of the hyperbolic plane preserve collinearity? Justify your answer.

12.6. Prove that an isometry preserves betweenness of points (Part 2 of Theorem 12.2.7).

12.7. Prove that an isometry preserves segments (Part 3 of Theorem 12.2.7).

12.8. Prove that an isometry preserves lines (Part 4 of Theorem 12.2.7).

12.9. Prove that an isometry preserves betweenness of rays (Part 5 of Theorem 12.2.7).

12.10. Prove that an isometry preserves angles (Part 6 of Theorem 12.2.7).

12.11. Prove that an isometry preserves triangles (Part 7 of Theorem 12.2.7).

12.12. Prove that an isometry preserves circles (Part 8 of Theorem 12.2.7).

12.13. Prove that an isometry preserves areas (Part 9 of Theorem 12.2.7).

12.14. Find examples of Euclidean isometries f and g such that $f \circ g \neq g \circ f$.

12.15. Assume the Cartesian model for Euclidean geometry. Find the images of the points $(0,0)$, $(0,2)$, and $(2,1)$ under the following isometries.
  (a) The reflection across the line with equation $x = y$.
  (b) The reflection across the line $x + y = 5$.
  (c) The half-turn about $(-1, 1)$.
  (d) The rotation through $45°$ in the counterclockwise direction about the point $(2, 2)$.
  (e) The glide reflection G_{AB}, where $A = (-2, 0)$ and $B = (0, 1)$.

12.16. Prove: If A, O, and B are noncollinear, then O is the only fixed point of the rotation R_{AOB}.

12.17. Complete the proof of the First Rotation Theorem (Theorem 12.3.4) by showing that any composition of two reflections across intersecting lines is a rotation.

12.18. Let S be a set of points in the plane. We say that S has *rotational symmetry* if there exist points A, O, and B such that $R_{AOB}(S) = S$. Find all rotational symmetries of each of the following sets in Euclidean geometry.
  (a) An isosceles triangle.
  (b) A rectangle.
  (c) A circle.
  (d) A regular n-sided polygon.

12.19. Prove that $R_{AOB}^{-1} = R_{BOA}$.

12.20. Check that $R_{POQ} = R_{AOB}$ in the proof of the Second Rotation Theorem (Theorem 12.3.5). Complete the proof of the theorem by proving that the line t exists.

12.21. Prove the Half-Turn Theorem (Theorem 12.3.6).

12.22. Prove the First Translation Theorem (Theorem 12.3.8).

12.23. Prove the Second Translation Theorem (Theorem 12.3.9).

12.24. Fix two distinct points A and B. The *pseudotranslation* S_{AB} is defined as follows. Let P be an arbitrary point in the plane. If P lies on $\overleftrightarrow{AB}$, then define P to be the point on $\overleftrightarrow{AB}$ such that $PP' = AB$ and the rays $\overrightarrow{AB}$ and $\overrightarrow{PP'}$ intersect in a ray. (The last condition ensures that P' is on the correct side of P.) In case P does not lie on $\overleftrightarrow{AB}$, construct the line $\overleftrightarrow{AP}$ and choose a point Q such that $A * P * Q$. Construct a ray $\overrightarrow{PR}$ such that $\angle QPR \cong \angle PAB$ and R lies on the same side of $\overleftrightarrow{AP}$ as B. Define $P' = S_{AB}(P)$ to be the point on $\overrightarrow{PR}$ such that $PP' = AB$. (See Fig. 12.30.)

(a) Prove that $S_{AB} = T_{AB}$ in Euclidean geometry.

(b) Prove that S_{AB} is not an isometry in hyperbolic geometry (except in the special case $A = B$) and conclude that $S_{AB} \neq T_{AB}$.

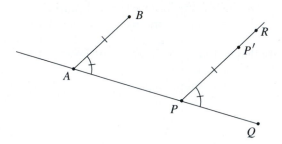

FIGURE 12.30: Definition of pseudotranslation

12.25. Use Part (a) of Exercise 12.24 to prove that the composition of two Euclidean translations is another translation. Prove that Euclidean translations commute $(T_{CD} \circ T_{AB} = T_{AB} \circ T_{CD})$.

12.26. Let A and B be two distinct points, let ℓ be the line that is perpendicular to $\overleftrightarrow{AB}$ at A, and let m be the line that is perpendicular to $\overleftrightarrow{AB}$ at B. Prove that $H_B \circ H_A = \rho_m \circ \rho_\ell$. Conclude that the composition of two half-turns is a translation.

12.27. This exercise leads to a second proof that the composition of two Euclidean translations is a translation. (The first is in Exercise 12.25.) Let S and T be two Euclidean translations. By the First Translation Theorem there exist parallel lines ℓ and m and parallel lines n and k such that $S = \rho_m \circ \rho_\ell$ and $T = \rho_k \circ \rho_n$.

(a) Prove that $T \circ S$ is a translation in case either $m = n$ or $m \parallel n$.

(b) Suppose m and n intersect in one point P. Let A be the foot of the perpendicular from P to ℓ and let B be the foot of the perpendicular from P to k. Let $t = \overleftrightarrow{PA}$ and let $u = \overleftrightarrow{PB}$.

  (i) Prove that $t \perp m$ and $u \perp n$.
  (ii) Prove that $\rho_n \circ \rho_m = \rho_u \circ \rho_t$.
  (iii) Prove that $T \circ S = H_B \circ H_A$.
  (iv) Prove that $T \circ S$ is a translation.

12.28. Let ℓ and m be two lines such that ℓ and m intersect in one point. Prove that $\rho_\ell \circ \rho_m \neq \rho_m \circ \rho_\ell$ unless $\ell \perp m$. Prove that $\rho_\ell \circ \rho_m = (\rho_m \circ \rho_\ell)^{-1}$.

12.29. Prove that the composition of two rotations with the same center is a rotation.

12.30. Prove that the composition of two Euclidean rotations is either a rotation or a translation. Under what conditions will the composition of two Euclidean rotations be a translation?

12.31. Prove that the composition of a Euclidean translation and a Euclidean rotation is a rotation.

12.32. This is an exercise in Euclidean geometry. It shows that *orientation preserving* is well defined.

    (a) Use Exercises 12.25 through 12.31 to show that the composition of any four Euclidean reflections can be rewritten as a composition of two reflections.

    (b) Prove that the identity transformation cannot be written as the composition of an odd number of reflections.

    (c) Prove that no composition of an even number of reflections can equal a composition of an odd number of reflections.

12.33. Assume $A \neq B$. What is the inverse of the glide reflection G_{AB}? Is G_{AB} an involution?

12.34. Suppose R_{AOB} and $R_{A'O'B'}$ are two rotations. Under what conditions could R_{AOB} and $R_{A'O'B'}$ commute? Justify your answer.

12.35. Let S be a set of points in the plane. A *symmetry* of S is an isometry T such that $T(S) = S$. Find all symmetries of each of the following sets in Euclidean geometry.

    (a) An isosceles triangle.

    (b) A rectangle.

    (c) A square.

    (d) A regular pentagon.

    (e) A regular hexagon.

    (f) A circle.

    (g) A regular n-sided polygon.

12.36. Suppose T is a rigid motion of the Euclidean plane that has no fixed points. What is T? Suppose T has exactly one fixed point; what is T? Suppose the set of fixed points for T is one line; what is T? Prove that your answers are correct.

12.37. A line ℓ is an *invariant line* for the isometry T if $T(\ell) = \ell$. (This is not the same as saying that the points of ℓ are fixed points.) If T has exactly one invariant line, what is T? (Be careful—the answer is different in Euclidean and hyperbolic geometries.)

12.38. Prove Lemma 12.4.6.

12.39. Prove the Horolation Theorem (Theorem 12.5.2).

12.40. Give a transformational proof of the Vertical Angles Theorem.

12.41. Give a transformational proof of the Euclidean theorem which asserts that the diagonals of a parallelogram bisect each other (Theorem 7.2.10).

12.42. Let ℓ, m, and n be three concurrent lines. Prove that if $\rho_\ell(m) = n$, then ℓ is the bisector of one of the angles between m and n.

12.43. Let $\triangle ABC$ be a triangle, let ℓ be the bisector of the interior angle at A, let n be the bisector of the interior angle at B, let O be the point of intersection of ℓ and n, and let m be the line through O that is perpendicular to $\overleftrightarrow{AB}$. Prove that $\rho_n \circ \rho_m \circ \rho_\ell(\overleftrightarrow{AC}) = \overleftrightarrow{BC}$. Use that result, Exercise 12.42, and Lemma 12.4.3 to give a transformational proof of the fact that the three interior angle bisectors of $\triangle ABC$ are concurrent.

12.44. Use the same idea as in the last exercise to give a transformational proof of the fact that the bisectors of two exterior angles and the remote interior angle of a triangle are concurrent. (Recall that the point of concurrency is the center of one of the exscribed circles.)

12.45. Let $\triangle ABC$ be a triangle, let ℓ be the perpendicular bisector of $\overline{AB}$, let n be the perpendicular bisector of $\overline{AC}$, let D be the point at which ℓ and n intersect, and let $m = \overleftrightarrow{AD}$. Prove that $\rho_n \circ \rho_m \circ \rho_\ell(B) = C$. Use this observation and Lemma 12.4.3 to give a transformational proof of the fact that the three perpendicular bisectors of a Euclidean triangle are concurrent. Explain where the Euclidean Parallel Postulate was used in the proof.

12.46. Analyze the proof of Lemma 12.4.5 and use it to prove the following result: *Suppose ℓ, m, and n are three lines such that $\ell \cap m \neq \emptyset$. If $\rho_n \circ \rho_m \circ \rho_\ell$ has a fixed point, then ℓ, m, and n are concurrent.*

12.47. Use Exercise 12.46 to give a transformational proof of the concurrency of the three perpendicular bisectors of the sides of a Euclidean triangle. Where is the Euclidean Parallel Postulate used in the proof?

12.48. Let $\mathcal{C} = C(O, r)$ be a circle and let P be a point inside $\mathcal{C}$. Prove that $I_{O,r}(P)$ is the point P' constructed on page 310.

12.49. Let $\mathcal{C} = C(O, r)$ be a circle and let P be a point outside $\mathcal{C}$. Prove that $I_{O,r}(P)$ is the point P' constructed on page 310.

12.50. Prove Corollary 12.7.5.

12.51. Prove Theorem 12.7.6.

12.52. Prove the second case of Theorem 12.7.15.

12.53. Let $I_{O,r}$ and $I_{O,s}$ be inversions in circles that share the same center point. Prove that $I_{O,r} \circ I_{O,s}$ is a dilation with center O. What is the proportionality constant of the dilation?

12.54. Assume the Cartesian model of Euclidean geometry and let $O = (0, 0)$.
 (a) Find $I_{O,2}((1, 1))$.
 (b) Find $I_{O,2}((2, 3))$.
 (c) Let ℓ be the vertical line through $(1,0)$. Sketch $I_{O,2}(\ell)$. Find an equation for the curve $I_{O,2}(\ell)$.
 (d) γ be the circle of radius 1 with center at $(1,0)$. Sketch $I_{O,2}(\gamma)$. Find an equation for the curve $I_{O,2}(\gamma)$.
 (e) Prove that $\mathcal{C} = \{(x, y) | x^2 + y^2 = 4\}$ is orthogonal to $\beta = \{(x, y) | (x - 2)^2 + (y - 2)^2 = 4\}$.
 (f) Find an equation for the circle that is orthogonal to $\mathcal{C} = \{(x, y) | x^2 + y^2 = 4\}$ and contains the points $(1, 2)$ and $(-1, 2)$. Sketch the circle.

12.55. A mechanical inverter. In the nineteenth century there was great interest in the problem of constructing mechanical linkages that perform geometric transformations. A mechanical linkage consists of a number of rigid bars with joints at which the linkage can freely bend. Each bar in the mechanical linkage remains straight and of constant length; the angle between bars is allowed to vary. Choose positive numbers x, y and r such that $r^2 = x^2 - y^2$. Prove that the mechanical linkage shown in Fig. 12.31 performs the inversion $I_{O,r}$; that is, if P and P' are moved around in the plane, it is always the case that $P' = I_{O,r}(P)$.[3]

---

[3]This mechanical inverter is known as the *Peaucellier cell* after the French Naval engineer Captain Charles-Nicolas Peaucelleir who invented it in 1873.

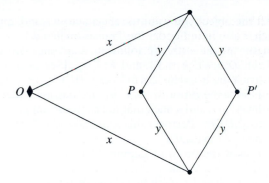

FIGURE 12.31: The angle at each • may change; the entire linkage may rotate about the point marked ♦ and the angle at ♦ may change

The reason this particular linkage was of such great interest historically is that it can be used to convert circular motion to straight line motion. Prove that if an additional bar is added to confine P to a circle (see Fig. 12.32), then P' will trace out a straight line. Will it trace out the entire line?

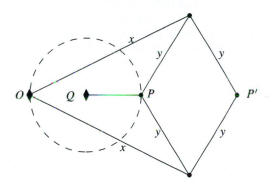

FIGURE 12.32: An extra bar has been added that confines P to a circle

TECHNOLOGY EXERCISES

T12.1. Let A, O, and B be the three points in the Cartesian plane with coordinates $(4, 3), (2, 3)$, and $(3, 5)$. [To construct points in Sketchpad that have specified Cartesian coordinates, use the "Plot Points..." command, which is found under the Graph menu.]

(a) Follow the proof of Theorem 12.3.4 to find two lines ℓ and m such that $R_{AOB} = \rho_m \circ \rho_\ell$.

(b) Construct the lines ℓ and m and apply the two reflections to several triangles in order to verify that $R_{AOB} = \rho_m \circ \rho_\ell$.

[To reflect across a line ℓ in Sketchpad, select ℓ and then mark it as the line of reflection using the "Mark Mirror" command under the Transform menu. Then select the object you want to reflect across ℓ and apply the "Reflect" command, which is also found under the Transform menu.]

T12.2. Plot the points with the following Cartesian coordinates: $A(0,3)$, $B(-5,3)$, $C(1,1)$, $D(5,4)$, $E(5,9)$, and $E(7,3)$. [See previous exercise for instructions.] Construct the triangles $\triangle ABC$ and $\triangle DEF$.

(a) Follow the proof of the existence part of Theorem 12.2.8 to construct three lines ℓ, m, and n such that the composition $\rho_n \circ \rho_m \circ \rho_\ell$ transforms $\triangle ABC$ to $\triangle DEF$. Perform this sequence of reflections and verify that $\triangle ABC$ is mapped to $\triangle DEF$.

(b) Follow the proof of Lemma 12.4.5 to move the lines ℓ, m, and n to lines k, s, and t such that $s \perp k$, $t \perp k$, and $\rho_k \circ \rho_t \circ \rho_s = \rho_n \circ \rho_m \circ \rho_\ell$. Find two points P and Q such that the glide reflection G_{PQ} maps $\triangle ABC$ to $\triangle DEF$.

(c) Use Sketchpad's translation and reflection commands to verify that this glide reflection carries $\triangle ABC$ to $\triangle DEF$.

T12.3. Make tools that implement the two constructions of the image of a point under an inversion in a circle $\mathcal{C} = C(O, r)$. (See page 310).

(a) Construct a line ℓ that is disjoint from $\mathcal{C}$ and construct a point P on ℓ. Use one of the tools you constructed to find the inverse P'. Find the curve traced out by P' as P moves along ℓ.

(b) Do the same thing in case ℓ is a secant line for $\mathcal{C}$ but does not pass through the center O of $\mathcal{C}$. You will have to use one of the tools for points of ℓ that are outside of $\mathcal{C}$ and the other tool for points P that are inside $\mathcal{C}$.

(c) Construct a circle β that is disjoint from $\mathcal{C}$ and construct a point P on β. Use one of the tools you constructed to find the inverse P'. Find the curve traced out by P' as P moves along β.

(d) Do the same thing for a circle β that intersects $\mathcal{C}$.

T12.4. While inversion in a circle is not at all the same as a dilation, Sketchpad's dilation command can still be used to make one tool that inverts all points, both those inside $\mathcal{C}(O, r)$ and those outside. Fix O and r and let $I_{(O,r)}$ be inversion in $\mathcal{C}(O, r)$. Verify that

$$OP' = \frac{r^2}{(OP)^2} OP.$$

(a) Make a tool that inverts a point P. The tool should accept as givens the center O, a point Q on the circle of inversion, and the point P. It should produce the point P' as a result. The point P' should be the result of applying a dilation to P. The center of the dilation is O and the ratio is $r^2/(OP)^2$. Be sure to use the calculate and measure commands to specify the ratio so that it will adjust appropriately when the givens change.

(b) Use the tool from (a) to make a tool that inverts lines. The tool should accept as givens the center O, a point Q on the circle of inversion, and two points A and B on the line. It should produce the line $\overleftrightarrow{AB}$ and the inverse of $\overleftrightarrow{AB}$ as results. The inverse of $\overleftrightarrow{AB}$ should be a circle, so you will probably want to make use of your circumscribed circle tool to construct the necessary circle.

(c) Use the tool from (a) to make a tool that inverts circles that do not pass through O. The tool should accept as givens the center O, a point Q on the circle of inversion, and three points A, B, and C on the circle. It should produce the circle ABC and the inverse of that circle as results.

CHAPTER 13

Models

13.1 THE SIGNIFICANCE OF MODELS FOR HYPERBOLIC GEOMETRY

In this chapter we finally construct models for plane geometry. In particular, we will construct one model for neutral geometry in which the Euclidean Parallel Postulate holds and another in which it does not. In the terminology of Chapter 2, the existence of such models proves that the Euclidean Parallel Postulate and the Hyperbolic Parallel Postulate are both *independent* of the neutral axioms. This means that neither of them can be proved or disproved from the neutral axioms alone. Thus the efforts of the many geometers who tried for centuries to supply a proof for Euclid's Fifth Postulate were bound to fail.

The chapter begins with a quick review of the Cartesian model for Euclidean geometry; in this model *point* is interpreted as an ordered pair of real numbers. After that review we will construct a basic model, called the Poincaré disk model, for hyperbolic geometry. It is one of several models for hyperbolic geometry that were discovered by Henri Poincaré (1854–1912). The Poincaré disk model is constructed within Euclidean geometry, using inversions in Euclidean circles (§12.7) as the main technical tool. So what we really prove in the chapter is that we can use the real numbers and standard properties of the real numbers to construct a model for Euclidean geometry and that inside any model for Euclidean geometry we can construct a model for hyperbolic geometry.

One interesting aspect of this state of affairs is how wrong Saccheri turned out to be. His idea was to "vindicate" Euclid by proving that the addition of the Hyperbolic Parallel Postulate to the postulates of neutral geometry would lead to a logical contradiction. However, contained within any model for Euclidean geometry there is a model for hyperbolic geometry. So any logical contradiction in hyperbolic geometry would imply a logical contradiction in Euclidean geometry. Exposing such a logical contradiction in Euclidean geometry would hardly have vindicated Euclid! Even if Saccheri had succeeded in proving that the Euclidean

Parallel Postulate is a consequence of the neutral postulates, he would still not have vindicated Euclid—he would merely have proved that the Euclidean Parallel Postulate is a redundant postulate. The true vindication of Euclid is the fact that the parallel postulate is independent of the remaining postulates. There were many other unstated assumptions that Euclid allowed into his proofs, but the fact that he identified this one as the key property that should be explicitly stated as an assumption is a testament to his profound insight.

Later in the chapter we will carefully check that the Poincaré disk model satisfies the various postulates of hyperbolic geometry. This is the real point of the "discovery of non-Euclidean geometry." Many accounts of non-Euclidean geometry simply describe points and lines on the sphere or some other surface, and leave the impression that that is the substance of the discovery of non-Euclidean geometry. If that were all that had been discovered, the discovery would not have the significance that is generally attached to it. It is obvious that the ancients were aware of non-Euclidean geometry in the sense that they knew about geometry on the sphere and could easily have drawn something like the Klein disk model (Example 2.5.9) for hyperbolic geometry. But they would not have been impressed with this as a demonstration of the independence of Euclid's Fifth Postulate. The reason is that these models, as described, do not reflect the infinite, unbounded character of the plane that is implicit in Euclid's Second Postulate. The heart of the construction of models in this chapter is the careful definition of a distance function that satisfies the Ruler Postulate and is consistent with the measurement of angles in the sense that SAS is satisfied.

There are two additional properties of general axiomatic systems that should be mentioned in this context. One property that the axioms of both Euclidean and hyperbolic geometries share is that they are *categorical*. This means that there is essentially only one model for each geometry in the sense that any two models are isomorphic. We will not prove that the two axiomatic systems are categorical, but we will exhibit isomorphisms between the various models for hyperbolic geometry that we construct.

A final property that we would want our axiomatic systems to possess is consistency. Recall that a system of axioms is said to be *consistent* if no logical contradiction can be derived from the axioms. The results of this chapter prove a type of relative consistency: *If Euclidean geometry is consistent, then hyperbolic geometry is consistent.* The reason we can only claim to prove consistency in this relative sense is that our construction of a model for hyperbolic geometry is based on the assumption that there is a model for Euclidean geometry and the construction of a model for Euclidean geometry is based on assumptions about the real numbers.

In fact we cannot absolutely prove that any branch of mathematics is consistent. While that statement strikes most people as surprising the first time they hear it, it is actually in harmony with one of the basic themes of this course: You can't prove everything, you must start by making some foundational assumptions and build from there. In mathematics we keep pushing our assumptions back in the direction of simpler, more intuitive objects and properties. But those assumptions cannot be pushed back indefinitely and mathematicians have generally agreed that all of mathematics should ultimately be built on set theory.

One of the fundamental theorems of mathematical logic, which was proved by Kurt Gödel (1906–1978), says roughly that any axiomatic system that is consistent and complex enough to include arithmetic and to allow the formulation of the question of whether or not it is consistent will be too weak to furnish an internal proof of its own consistency. Thus set theory cannot be used to prove that set theory is consistent, but the consistency of set theory must be assumed. If the consistency of set theory is assumed, then the real numbers can be constructed, the consistency of Euclidean geometry can be proved, and so can the consistency of hyperbolic geometry.

The stereotypical view of mathematics is that it is the one area of life in which we have absolute certainty about our results because we prove them using pure logic. In fact that is not true at all. Our use of mathematics rests on the *assumption* that mathematics is consistent. Everything in our intuition and our experience confirms that this is the case, but ultimately it is something we accept without proof. This should not be viewed as a problem with the foundations of mathematics. The very thing that many outsiders dislike about mathematics is the fact that they see it as too cold and too certain; the fact that there is actually some significant mystery surrounding the foundations of mathematics makes it clear that mathematics is a human activity with definite limitations.

Our claims regarding consistency must therefore be conditional. In the next section we will discuss the Cartesian model for Euclidean geometry. We do not claim to be proving that there is a model for Euclidean geometry; rather, we are only making the claim that if there are models for set theory and the real numbers, then there is a model for Euclidean geometry. Perhaps that is the most surprising aspect of the entire chapter. The one thing that geometers of previous ages, as well as most students of geometry today, have taken for granted is the presumed fact that there is a model for Euclidean geometry. Ironically that is the one thing we cannot prove.

13.2 THE CARTESIAN MODEL FOR EUCLIDEAN GEOMETRY

In this section we briefly review the Cartesian plane model for Euclidean geometry. As indicated in the previous section, we will assume the existence of the set of real numbers $\mathbb{R}$ satisfying the standard properties (listed in Chapter 4). The purpose of this brief review is just to point out that, even in this familiar context, it is still necessary to define distance and angle measure and that it is not immediately obvious that the familiar postulates are satisfied.

The interpretation

A *point* in the Cartesian plane is an ordered pair of real numbers, (x, y). A *line* is a particular type of set of points. For each triple of real numbers a, b, c with a and b not both 0, there is a line ℓ defined by $\ell = \{(x, y) \mid ax + by + c = 0\}$. Distance is defined by the usual formula. Let $A = (x_1, y_1)$ and let $B = (x_2, y_2)$. Define the *distance from A to B* by

$$AB = \sqrt{(x_1 - x_2)^2 + (y_1 - y_2)^2}.$$

Suppose ℓ is the line determined by the equation $ax + by + c = 0$. We define the two *half-planes determined by* ℓ as follows:

$$H_1 = \{(x, y) \mid ax + by + c > 0\} \text{ and } H_2 = \{(x, y) \mid ax + by + c < 0\}.$$

The final undefined term is *angle measure*. This is the most difficult of the undefined terms to interpret rigorously. We will just give a rough outline of how it can be done. The problem is to assign a real number to an angle. There are two different ways in which to go about this. The naïve way is to extend one of the rays to a line so that the angle is contained in a 180° angle. Then the method of successive bisection can be used to find a binary representation of the angle measure. This method is intuitively right because it is essentially what we do when we use a physical protractor to measure an angle, but it is still a little unsatisfying because it involves an infinite process. It can be used to find any number of bits in the binary expansion, but it does not find the real number exactly.

Another way to proceed is to use calculus. Recall that the inverse tangent function can be defined as follows:

$$\arctan x = \int_0^x \frac{1}{1 + t^2}\, dt.$$

Given the equations for two nonvertical lines in the Cartesian plane, we can write them in the form $y = m_1 x + b_1$ and $y = m_2 x + b_2$. If $m_1 m_2 = -1$, then the lines are perpendicular. Otherwise, the smaller angle between them has measure

$$\left(\frac{180}{\pi}\right) \left| \arctan\left(\frac{m_1 - m_2}{1 + m_1 m_2}\right) \right|.$$

The measure of the larger angle between the two lines is obtained by subtracting the value above from 180°.

The postulates

We must check that the various postulates are satisfied in the interpretation above. The Existence Postulate is obvious. To verify the Incidence Postulate, start with two points $A = (x_1, y_1)$ and $B = (x_2, y_2)$. If $x_1 = x_2$, then the line $x = x_1$ is the unique line containing them. Otherwise the unique line is given by the familiar point-slope formula

$$y - y_1 = \left(\frac{y_2 - y_1}{x_2 - x_1}\right)(x - x_1).$$

In Chapter 5 (Example 5.4.12) we defined coordinate functions for the Cartesian model. In effect we were proving there that the Cartesian model satisfies the Ruler Postulate. It is relatively easy to check the Plane Separation Postulate. The key observation is that the coordinate functions are continuous functions, so the Intermediate Value Theorem can be used to show that a segment that begins in one half-plane and ends in the other must cross the line separating them. The verification of the Protractor Postulate is a bit messy, so we omit it.

It is not difficult to prove that the Euclidean Parallel Postulate is satisfied. Any line is either vertical and has an equation of the form $x = c$, or it is not vertical and has an equation that can be written in the form $y = mx + b$. (In the latter case we say that the line has *slope m*.) Here the key observation is that two distinct lines are parallel if and only if either both are vertical or both are nonvertical and have the same slope. This can be verified algebraically. Now consider a line ℓ and an external point P. There is exactly one vertical line through P and there is exactly one line through P of slope m for every $m \in \mathbb{R}$ (by the point-slope formula). Thus there is exactly one line through P that is parallel to ℓ.

13.3   THE POINCARÉ DISK MODEL FOR HYPERBOLIC GEOMETRY

For the remainder of this chapter, the existence of a model for Euclidean geometry will be assumed. Within that Euclidean model we will build a hyperbolic one. Thus an understanding of the proofs in this section requires a fairly high level of thinking:[1] We will be using intricate constructions in Euclidean geometry to prove things about hyperbolic geometry.

One simple way to understand what we are doing is to think in terms of maps. We want to construct a precise, detailed map of hyperbolic space. (Not a "map" in the sense in which the term is usually used in mathematics—as a synonym for function—but a "map" in the sense of geography.) We are all familiar with maps of the world. The surface of the earth is roughly spherical, but a map represents the surface of the earth on a flat plane. A map that represents a small portion of the earth's surface can be quite accurate (except for scale), but a map that shows the entire surface of the earth must distort some relationships in a significant way. Often the polar regions are greatly stretched out so that the map shows them as much larger than they really are in relation to other parts of the earth. In the same way our map of the hyperbolic plane will be drawn in the Euclidean plane and will inevitably distort some aspects of hyperbolic geometry. So you should not expect to see diagrams of the hyperbolic plane in which all lines look straight and all distances are measured in the usual way. Instead we will give a precise interpretation of each undefined term and verify that with these interpretations the postulates are true statements. We will do this without reference to whether or not the objects we construct look like the familiar objects that have the same names.

The interpretation

Fix a circle γ in the Euclidean plane. For example, we could use the Cartesian model for the Euclidean plane and take γ to be the circle of radius 1 centered at the origin. A *point* in the Poincaré disk model is a Euclidean point inside γ.

The term *line* is interpreted in two different ways in the model. One kind of line is an open diameter of γ. Specifically, start with a Euclidean line ℓ that passes through the center of γ. The associated Poincaré line consists of all the points on ℓ that are inside γ. The second type of Poincaré line starts with a Euclidean circle β

---

[1]See the Van Hiele model, Appendix D.

that is orthogonal to γ. The associated Poincaré line consists of the set of all points on β that lie inside γ. The two kinds of lines are illustrated in Figure 13.1.

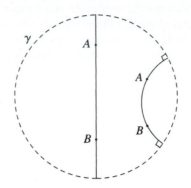

FIGURE 13.1: Two kinds of Poincaré lines

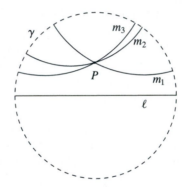

FIGURE 13.2: Parallel lines in the Poincaré disk model

We will now interpret *half-plane* in the Poincaré disk model. Let m be a line in the model. If m is a line of the first kind, then it is determined by a Euclidean line ℓ that passes through the center of γ. There are two Euclidean half-planes determined by ℓ. The Poincaré half-planes determined by m are defined to be the intersections of the Euclidean half-planes with the interior of γ. In case m is determined by the Euclidean circle β, we define the half-planes determined by m to be the intersection of the interior and exterior of β with the interior of γ.

We really should not define angle measure until after we have defined distance (because the definition of angle requires the definition of ray which requires the definition of betweenness which requires the definition of distance), but we want to define angle measure first since it is much simpler in this model. For points on either type of Poincaré line there is a natural notion of betweenness that is inherited from Euclidean geometry. We will assume this definition of betweenness and so rays are defined in the Poincaré disk model. Suppose we are given two Poincaré rays with endpoint A. Each ray is part of a Euclidean circle or line, so we can define the angle between them exactly as we did in the last chapter. In other words, two Poincaré rays determine two Euclidean tangent rays and we define the measure of the angle between the Poincaré rays to be the measure of the Euclidean angle between the tangent rays—see Fig. 13.3.

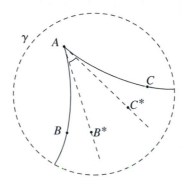

FIGURE 13.3: The measure of the Poincaré angle $\angle BAC$ is defined to be the Euclidean measure of the Euclidean angle $\angle B^* A C^*$

A model, such as this one, with the property that angles are faithfully represented is called a *conformal* model. Of the four models we will construct, two are conformal and two are not. The advantage of a conformal model is that the diagrams we draw accurately portray (to our Euclidean eyes) at least some of the geometric relationships. Distances are still seriously distorted, but at least the angles are correct.

Now that we know how to interpret angle measure and line, we can begin to draw diagrams in the Poincaré disk model. The next several figures show what some of the geometric objects we have studied would look like in this model.

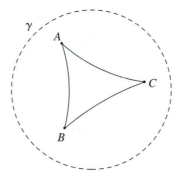

FIGURE 13.4: A triangle in the Poincaré disk model

The final undefined term that must be interpreted is *distance*. Obviously distance between points in the model cannot be interpreted in the usual Euclidean way. Let A and B be two points in the model. If A and B lie on a diameter of γ,

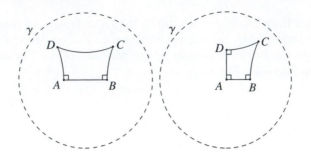

FIGURE 13.5: A Saccheri quadrilateral and a Lambert quadrilateral

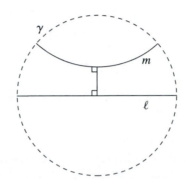

FIGURE 13.6: A common Perpendicular segment

then define P and Q to be the endpoints of the diameter. If A and B do not lie on a diameter, then there is a unique Euclidean circle β that contains the two points and is perpendicular to γ (Corollary 12.7.12). Let P and Q be the two points at which β intersects γ. (See Fig. 13.8.)

Note. Be sure to notice that P and Q are points on γ and are therefore *not* points in the Poincaré disk model of hyperbolic geometry. Even though they are not points in the model we are constructing, they are useful in defining the distance function for that model.

Definition 13.3.1. Define the *distance* from A to B by

$$d(A, B) = |\ln([AB, PQ])|,$$

where $[AB, PQ]$ is the cross ratio.

Recall that $[AB, PQ]$ is defined by

$$[AB, PQ] = \frac{(AP)(BQ)}{(AQ)(BP)}$$

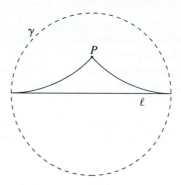

FIGURE 13.7: Limiting parallel rays

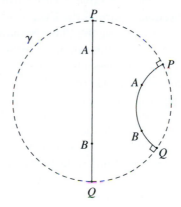

FIGURE 13.8: The definition of Poincaré distance

and so

$$[BA, PQ] = \frac{1}{[AB, PQ]} = [AB, QP].$$

It follows that d is symmetric in the sense that

$$d(B, A) = |\ln([BA, PQ])| = |-\ln([AB, PQ])| = d(A, B)$$

and that the definition is independent of the order in which P and Q are listed. This calculation also makes clear the need for the absolute value in the definition of d.

The postulates

We will prove that all the postulates of hyperbolic geometry are satisfied. The Existence Postulate is obvious. The fact that the Hyperbolic Parallel Postulate holds is also obvious (see Figure 13.2).

In order to verify the Incidence Postulate, start with two points in the model. The two points either lie on a diameter of γ or they do not. If they do, then the diameter is the unique line that contains both of them. (There is no Poincaré line of the second type that contains two points on a diameter of γ because the center point of γ lies outside any circle that is perpendicular to γ—see Exercise 13.2.) If the two points do not lie on a diameter of γ, then there is a unique Poincaré line of the second type that contains them by Corollary 12.7.12. Thus the Incidence Postulate holds.

To verify the Ruler Postulate, fix a Poincaré line m and choose a point A on m. As above, the Poincaré line m determines two Euclidean points P and Q on γ. Define $f : m \to \mathbb{R}$ by $f(X) = \ln([AX, PQ])$. It is not too difficult to verify that f is a coordinate function for m; that is, f is a one-to-one correspondence that has the properties required in the Ruler Postulate (Exercise 13.3). In particular, standard properties of the natural logarithm function can be used to show that $d(B, C) = |f(B) - f(C)|$ and that f varies from $-\infty$ to $+\infty$.

The fact that a circle separates the Euclidean plane into an inside and an outside makes it relatively easy to prove the Plane Separation Postulate (Exercise 13.5).

Since the model is conformal, most parts of the Protractor Postulate are automatic. The one part that needs to be checked out is Part 3. But we have already proved the necessary result in Theorem 12.7.13.

The final neutral postulate is Side-Angle-Side . By Theorem 12.6.5, it is sufficient to verify the Reflection Postulate in place of SAS. Let ℓ be a Poincaré-hyperbolic line. Either ℓ is a diameter of γ or ℓ is part of a Euclidean circle that is perpendicular to γ. In either case we can find a hyperbolic reflection associated with ℓ. In the first case, the hyperbolic reflection is the same as the Euclidean reflection across ℓ. It is easy to check that this reflection preserves all the relevant hyperbolic data. In the second case, ℓ is part of a Euclidean circle β. Let I be the Euclidean inversion in β. Observe first that I maps γ to itself (Theorem 12.7.9) and so I maps the inside of γ to the inside of γ. Thus I determines a transformation of the Poincaré disk. In order to distinguish this transformation from I, we will name it ρ. By Theorems 12.7.17 and 12.7.19, ρ satisfies the conditions in the Reflection Postulate.

Note. The two transformations ρ and I are the same in the sense that $\rho(P) = I(P)$ for every point P in the Poincaré disk. The difference between I and ρ is that I has the entire Euclidean inversive plane as its domain and range while ρ has just the Poincaré disk as its domain and range.

13.4 OTHER MODELS FOR HYPERBOLIC GEOMETRY

The Beltrami-Klein disk model

The second model for hyperbolic geometry is the Klein disk model (Example 2.5.9). This model is also sometimes called the Beltrami-Klein disk model for hyperbolic geometry.[2] Again we fix a circle γ in the Euclidean plane. A *point* in the Klein disk model is a Euclidean point that is inside γ. A *line* in the model is the portion of a Euclidean line that lies inside γ.

---

[2]Eugenio Beltrami constructed the first detailed model of hyperbolic geometry in 1868.

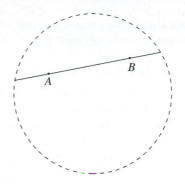

FIGURE 13.9: A line in the Beltrami-Klein disk model

We do not directly interpret distance and angle measure in this model; instead we exhibit a transformation h that maps the Klein disk to the Poincaré disk in such a way that lines are mapped to lines. The hyperbolic distance between two points in the Klein disk is then interpreted to be the distance between their images in the Poincaré disk under h. In a similar way the measure of the angle between two rays in the Klein disk is interpreted to be the measure of the angle between the images of the rays under h. This model is not conformal.

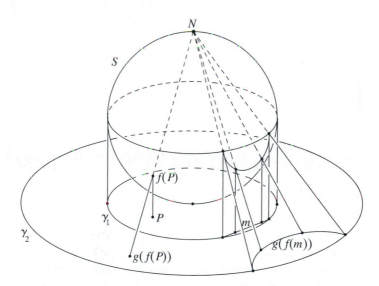

FIGURE 13.10: An isomorphism from the Klein disk to the Poincaré disk

Fig. 13.10 suggests the definition of h. More precisely, think of the Euclidean plane as a horizontal plane in three-dimensional space. Let S be a sphere that is tangent to the Euclidean plane. Define γ_1 to be the circle that is directly below the equator of S. We will define the points of the Klein disk to be the points inside γ_1.

Let f be the transformation of the Klein disk to the lower hemisphere of S that simply moves a point straight up until it hits the sphere. This is one half of the transformation h.

The second half of h is defined to be projection from N, the north pole of S. For each point $P \neq N$ on S, define $g(P)$ to be the point at which the ray $\overrightarrow{NP}$ intersects the Euclidean plane. The transformation g is called *stereographic projection*. It maps all of the sphere except the north pole to the Euclidean plane. Let γ_2 be the image of the equator of S under stereographic projection. We will assume that the Poincaré disk model was defined using the circle γ_2.

Consider Fig. 13.10. Notice that the point P is first mapped straight up to $f(P)$ and then is projected radially out from N to the point $g(f(P))$. Since this process is reversible, it defines a one-to-one transformation of the points inside γ_1 to the points inside γ_2. Thus $h = g \circ f$ is a transformation of the Klein disk to the Poincaré disk.

We need to see that the transformation maps a line in the Klein model to a line in the Poincaré model. Let m be a line in the Klein model; then m is a chord for γ_1. There is a vertical plane through m and f maps m to the intersection of that plane with the sphere S. Thus $f(m)$ is a semicircle on S that meets the equator of S at right angles. When this circle is projected radially out from N, it is mapped to a circle that meets γ_2 at right angles. Hence $g(f(m))$ is a line in the Poincaré model. The details of the proof that $g(f(m))$ is a Poincaré line are left as an exercise (Exercise 13.12).

The Klein disk model has the advantage that it is very simple to describe the points and lines of the model. The disadvantage is the fact that it is not conformal. Neither distances nor angle measures are faithfully represented in the model. In fact it is quite complicated and difficult to give direct definitions of either angle measure or distance in this model. For that reason we define them indirectly using h. In effect we simply define the transformation h to be an isomorphism.

The Poincaré half-plane model

Our third model for hyperbolic geometry is also due to Poincaré. Begin by choosing a fixed line λ in the Euclidean plane and one of the two half planes associated with it. A *point* in the half-plane model is one of the Euclidean points in the fixed half-plane. Again *line* is interpreted in two different ways. One kind of line consists of the part of a Euclidean line perpendicular to λ that lies in the half-plane. The second type of line consists of all the points of a Euclidean circle orthogonal to λ that lie in the half-plane. The two types of lines are illustrated in Fig. 13.11.

Fig. 13.12 shows how to define an isomorphism from the Poincaré half-plane to the Poincaré disk. This time we think of the Euclidean plane as being a vertical plane in three-dimensional space, we take λ to be a horizontal line, and we take points in the Poincaré half-plane to be points that are below λ. Let S be a sphere that is tangent to the plane at a point of λ and let M be the point of S that is opposite to the point of tangency. Define k to be stereographic projection from M. Under k, the lower hemisphere of S is projected radially from the point M onto the lower half of the plane. The transformation from the Poincaré disk to the half-plane model

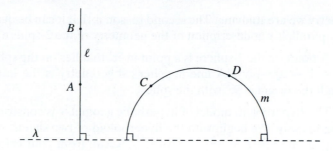

FIGURE 13.11: Lines ℓ and m in the Poincaré half-plane

is $h' = k \circ g^{-1}$, where g is still the stereographic projection that takes the lower hemisphere of S to the Poincaré disk.

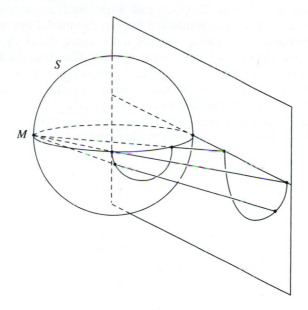

FIGURE 13.12: An isomorphism from the sphere to the Poincaré half-plane

Once again the *distance* between two points in the Poincaré half-plane is defined to be the distance between the corresponding points under the inverse of this transformation and the *measure* of an angle is defined to be the measure of the corresponding angle in the Poincaré disk. It is not too difficult to see that h' preserves angles, so this Poincaré model is also conformal.

The hyperboloid model

The last model is particularly nice for a couple of reasons. One is that it involves hyperbolas and goes a long way toward explaining the name *hyperbolic* for the geometry we are studying. The second reason is that it can be described in a way that nicely parallels our description of the geometry of the 2-sphere (Example 2.5.8).

A point on the 2-sphere is a point in $\mathbb{R}^3$ that lies on the sphere whose equation is $z^2 = 1 - x^2 - y^2$. A line in spherical geometry is the intersection of a plane through the origin in $\mathbb{R}^3$ with the sphere.

The hyperboloid model of hyperbolic geometry is constructed in a completely analogous way. We begin with the hyperboloid of two sheets, which is the graph of the equation $z^2 = 1 + x^2 + y^2$ in $\mathbb{R}^3$. (Recall from your calculus course that the hyperboloid of two sheets is the surface that results from revolving the hyperbola $z^2 - x^2 = 1$ about the z-axis.) A *point* in the hyperboloid model is a point in $\mathbb{R}^3$ that lies on the upper sheet ($z > 0$) of this hyperboloid. A *line* in the hyperboloid model is the intersection of a plane through the origin in $\mathbb{R}^3$ with the upper sheet of the hyperboloid.

Again we define distance and angle measure indirectly by exhibiting an isomorphism that takes points and lines in this model to points and lines in one of the previous models. Define the Klein disk using the circle γ that lies in the plane $z = 1$ and has radius 1 and center (0,0,1). The function from the Klein disk to the hyperboloid model is simply radial projection from the origin. Fig. 13.13 shows a point P and a line ℓ in the Klein disk along with their images P' and ℓ' in the hyperboloid model. The line ℓ in the Klein disk is the intersection of the plane Π with the interior of γ and the line ℓ' in the hyperboloid model is the intersection of the same plane with the sheet of the hyperboloid.

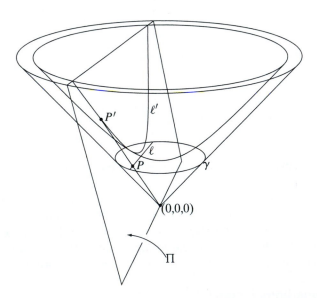

FIGURE 13.13: An isomorphism from the Klein disk to the hyperboloid model

13.5 MODELS FOR ELLIPTIC GEOMETRY

Before leaving the topic of models we should think briefly about models for non-Euclidean geometries other than hyperbolic geometry. The 2-sphere (Example 2.5.8) is an example of a continuous geometry that satisfies the Elliptic Parallel Postulate. The geometry of the sphere is called *spherical geometry*. Spherical geometry satisfies all the postulates of neutral geometry except the Incidence Postulate and the Ruler Postulate. It differs from neutral geometry in that two points do not necessarily determine a unique straight line and lines cannot be extended to be arbitrarily long.

It is possible to modify this model to obtain an elliptic geometry that satisfies the Incidence Axiom as well. In order to do that, simply identify antipodal points on the sphere and think of each pair of antipodal points on the sphere as representing one point. The resulting space is called the *projective plane* and the geometry of the projective plane is called *elliptic geometry*. In elliptic geometry, *point* is interpreted to mean a pair of antipodal points on the sphere. Since each line through the origin in three-dimensional space intersects the sphere in a pair of antipodal points, an equivalent way of interpreting point is to say that a *point* is a line through the origin in three-dimensional space. Elliptic geometry satisfies the Elliptic Parallel Postulate and all of the postulates of neutral geometry except for the Ruler Postulate and the Plane Separation Postulate. This model improves on the sphere in that it satisfies the Incidence Postulate, but the price that was paid is that it no longer satisfies the Plane Separation Postulate (a line has only one "side").

This last model helps to clarify the role of the Ruler Postulate in some of our earlier work. The real reason that parallel lines do not exist on the sphere (and triangles can have angle sum that exceeds 180°) is not because the Incidence Postulate fails but rather because the lines on a sphere close up and cannot be extended indefinitely.

SUGGESTED READING

Chapters 7 and 8 of *The Non-Euclidean Revolution*, [70].

EXERCISES

13.1. Explain the formula in the text for angle measure in the Cartesian model.

13.2. Suppose γ and β are Euclidean circles that are perpendicular and intersect at the points P and Q. Prove that the two tangent lines to β at P and Q intersect at the center of γ. Conclude that the center of γ must lie outside β. (This fact was used in the proof that the Poincaré disk model satisfies the Incidence Postulate.)

13.3. Prove that the function $f : m \to \mathbb{R}$ for the Poincaré disk model, defined on page 336, is a coordinate function. Prove that the Ruler Postulate is satisfied in the Poincaré disk model.

13.4. Suppose the distance function for the Poincaré disk model is defined by $d(A, B) = |\log_{10}(AB, PQ)|$, using the logarithm with base 10 in place of the natural logarithm. Would the Ruler Postulate still be satisfied?

13.5. Verify that the Poincaré disk model satisfies the Plane Separation Postulate.

13.6. Suppose the Poincaré disk model is defined using the Cartesian model for Euclidean geometry and taking γ to be the circle of radius 2 centered at $(0,0)$. Calculate the Poincaré distances $d(A, B)$ between the following pairs of points.
   (a) $A = (0,0)$, $B = (1,0)$.
   (b) $A = (-1,0)$, $B = (1,0)$.
   (c) $A = (1,1)$, $B = (1,-1)$.
   (d) $A = (1,1)$, $B = (1,0)$.

13.7. Let O be the center of the Euclidean circle γ that is used to define the Poincaré disk model. If P is a point in the Poincaré disk such that the Poincaré distance from O to P is r_0, find a formula for the Euclidean distance OP.

13.8. Let O be the center of the Euclidean circle γ that is used to define the Poincaré disk model. If P is a point in the Poincaré disk such that the Euclidean distance from O to P is r_1, find a formula for the Poincaré distance $d(O, P)$.

13.9. Let A be a point in the Poincaré disk and let $\mathcal{C}$ be the hyperbolic circle of radius r centered at A. Thus $\mathcal{C} = \{B \mid d(A, B) = r\}$. Prove that $\mathcal{C}$ is also a Euclidean circle. What is the Euclidean radius of $\mathcal{C}$? Prove that the Euclidean center of $\mathcal{C}$ is A if and only if A is the center of the circle γ that is used in the definition of the model.

13.10. One way to define an absolute standard of length in hyperbolic geometry is to use the distance x such that the measure of the angle of parallelism is $45°$. Calculate that distance x in the Poincaré disk model. Fig. 13.14 should help you get started.

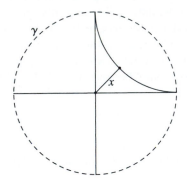

FIGURE 13.14: The distance x satisfies $\kappa(x) = 45°$

13.11. Review the definition of *horolation* on page 303. Draw two asymptotically parallel lines in the Poincaré disk model. Explain how the horolation determined by the two lines can be interpreted as a rotation of the Poincaré disk about a point at infinity.

13.12. Prove that the transformation h from the Klein disk model to the Poincaré disk model maps Klein lines to Poincaré lines.

13.13. Let O be the origin in the Cartesian plane, let N be the point $(0, r)$ for some positive number r, and let γ be the circle that has $\overline{ON}$ as diameter. For each point $P \in \gamma \smallsetminus \{N\}$, define P' to be point at which the ray $\overrightarrow{NP}$ intersects the x-axis. Prove that $P' = I_{N,r}(P)$ (see Fig. 13.15). (The transformation from P to P' is stereographic projection of $\gamma \smallsetminus \{N\}$ to the x-axis, so this exercise shows that stereographic projection in a circle is the same as inversion in a circle of twice the radius.)

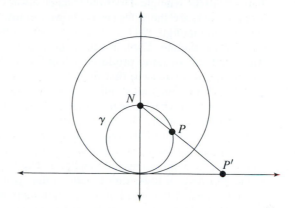

FIGURE 13.15: Stereographic projection from N equals inversion in $C(N, r)$

13.14. Let ℓ be a line in the Klein disk model that passes through the center point O of γ and let m be a second line in the Klein disk model. Prove that ℓ and m are perpendicular in the Klein disk model if and only if they are perpendicular in the Euclidean plane.

13.15. The Poincaré upper half-plane model for hyperbolic geometry is the half-plane model that results from starting with the x-axis in the Cartesian model and using the half-plane above it (points with positive y coordinate) in the definition of the model. Draw figures that illustrate the following in the Poincaré upper half-plane.
   (a) Multiple parallel lines through an external point.
   (b) A triangle.
   (c) A Lambert quadrilateral.
   (d) A Saccheri quadrilateral.
   (e) Parallel lines that admit a common perpendicular.
   (f) Limiting parallel rays.

TECHNOLOGY EXERCISES

T13.1. Start with a circle γ that is to be the circle used in the construction of the Poincaré disk model. The circle is specified by its center O and a point R on the circle. Make a tool (or macro) that finds the Poincaré line through two points A and B in the Poincaré disk determined by γ. Your tool should accept the four points O, R, A, and B as givens and could produce a Euclidean circle as result. (It would be difficult to trim off the part of the circle that extends outside γ, so just leave it there.) You may assume that A and B do not lie on a diameter of γ. The construction you need for this is implicit in the proof (in this chapter) that two points in the Poincaré disk determine a unique line and the construction (in the previous chapter) of the inverse of a point that lies inside γ. Move A and B around inside γ and observe how the line changes. What happens when you move the two points so that they lie on a diameter of γ?

T13.2. Use the tool you created in the preceding exercise to construct a triangle in the Poincaré disk model. Move the vertices around inside the disk and watch what happens to the triangle. Does the triangle always look like the one in Figure 13.4? In particular, are the sides always bowed in towards the inside of the triangle as they are in that figure? Draw a sketch of the different shapes that are possible for a Poincaré triangle. Explain where the vertices must be located relative to each other and O in order to produce each different shape.

T13.3. Make a tool (or macro) that finds the Poincaré distance between two points A and B. Again your tool should accept four points O, R, A, and B as givens and should produce the Poincaré distance $d(A, B)$ as result. This time the result should be a number. To calculate the number $d(A, B)$ you must first use the tool in Exercise T13.1 to construct the Poincaré line through A and B, locate the points P and Q used in the definition of d, and then use the the Calculate tool under the Measure menu in Sketchpad. The calculate tool includes the absolute value and natural logarithm functions that you need. Move B around inside γ while keeping A fixed and observe what happens to the value of $d(A, B)$. What happens as B approaches A? What happens as B approaches γ? Does your tool give an accurate result in both cases? If not, can you explain why not?

T13.4. Use the tools you just created in the last three exercises to verify the Median Concurrence Theorem in the Poincaré disk model. Specifically, you should start with three arbitrary points A, B, and C in the Poincaré disk, construct the Poincaré lines through the vertices, find the Poincaré midpoints of the sides, construct the Poincaré medians for the triangle, and verify that the medians are concurrent. The trickiest part is finding the Poincaré midpoints. To find the midpoint of $\overline{AB}$, mark a point F on the Poincaré line through A and B, measure the Poincaré distance $d(A, F)$, and move F along the line until $d(A, F)$ is half of $d(A, B)$. All this will of course be approximate, but the tool you made should be accurate enough for the construction to work. Try this with at least two triangles of different shapes.

CHAPTER 14

Polygonal Models and the Geometry of Space

In this final chapter of the book we address two questions that have been implicit in the discussion for much of the course. First, what would a hyperbolic plane (in which both angle measure and distance are accurately represented) really be like? How, in particular, could the lines in such a plane be "straight" and still exhibit many of the properties we associate with curved lines? Second, what is the relationship between the geometries studied in this course and the geometry of the real world? Does non-Euclidean geometry have any practical significance or usefulness?

Fortunately we can gain insight into the answers to both questions at the same time. In order to do so, we will study curved spaces; we will think of space itself as being curved rather than the lines in it. After we have become familiar with curved surfaces and the way in which curvature relates to geometry, we will consider the geometry of the physical universe. Our study will lead us right to the edge of what is known about geometry and into the realm of current research.

The style of this chapter is quite different from that of the rest of the book. We will abandon the axiomatic method that we have used until now and turn instead to the method of analogy. Rather than proving theorems or stating axioms, we will try instead to gain insight into certain relationships by the use of analogies and comparisons.

We turn away from the axiomatic method because the construction of the models in Chapter 13 represent the end of the road we started down when we laid out the axioms for neutral geometry at the beginning of the course. The models show conclusively that the Hyperbolic Parallel Postulate is just as consistent with the axioms of neutral geometry as is the Euclidean Parallel Postulate. This answers the fundamental question about the foundations that was left unresolved by the study of the various geometries in the early chapters of the book and is the logical culmination of the axiomatic study of the foundations of geometry.

While the models of Chapter 13 complete our formal understanding of the foundations, they do not seem to help us to appreciate the usefulness and practical significance of non-Euclidean geometry. The models are less than completely satisfying because they look like toy models to us; the lines in the Poincaré disk model appear to us to be curved and, from our perspective at least, appear to be only finitely long. The lines do not look like good models for the "lines" we see around us in the physical world. Thus we could be left with the impression that the significance of the models is just to answer an abstract question in logic that was raised by mathematicians of an earlier age.

In this chapter we will reach a better intuitive understanding of non-Euclidean geometry by building a new collection of models for the hyperbolic plane. These are not models in the technical sense in which we defined the term in Chapter 2 but instead are models in the everyday sense that they help us to visualize what the hyperbolic plane is like. The new models are polygonal surfaces that we can construct out of paper and whose geometry closely approximates that of the hyperbolic plane. In these models both angle measure and distance are faithfully represented, at least approximately. On them we can draw lines that really are straight but nonetheless bound triangles with angle sums less than 180°. This helps us to form a mental picture of what the hyperbolic plane is like. Once we understand these curved surfaces, we can imagine analogous structures in three dimensions and can therefore imagine new possibilities for the geometry of the universe.

14.1 CURVED SURFACES

The word *space* refers to any set of points in which there is a concept of distance. The main point of this chapter is that the geometry of a space is related to the curvature of that space. The curvature of a space is something that is most easily detected by an observer who is viewing the space from the outside. For example, the curvature of the surface of the earth is most obvious to someone who views the earth from outer space. Those of us who are confined to the surface of the earth can see indirect indications that the earth is round, but that fact is most readily apparent to us when we look at photographs taken from a spacecraft that is well above the surface of the earth.

The two spaces that we particularly want to understand are the hyperbolic plane and the physical universe. Unfortunately we are not able to take an outsider's view of either one of those spaces. In the case of the hyperbolic plane this is because there is no way to accurately represent the entire hyperbolic plane in Euclidean 3-space. In the case of the physical universe, we cannot achieve the outsider view because we are three-dimensional beings who live inside the space. Only a four-dimensional creature would have the ability to view our universe from the outside.

Those obstacles are not going to stop us from doing our best to understand either of the spaces of interest or from understanding them in terms of curvature. In this section we will take a look at the curvature of the spaces that we can see from the outside. In the following sections we will apply the lessons we learn in the simple setting to help us understand the more difficult examples.

The place to begin is with the study of curves and surfaces in three-dimensional Euclidean space $\mathbb{R}^3$. Such objects are studied in multivariate calculus courses, so they are at least somewhat familiar to all mathematics students. Recall that a curve is one-dimensional and that a surface is two-dimensional. More precisely, a curve is described by a function of one variable and a surface is described by a function of two variables. Thus a curve is defined by a function of the form $f : [a, b] \to \mathbb{R}^3$. It will usually be assumed that f is differentiable and that the image of f is smooth enough so that it has a unique tangent line at each point.

FIGURE 14.1: A curve with tangent and normal vectors

In calculus you learned to compute the *curvature* of a curve. One way to define the curvature is to change the parametrization f so that the length of the tangent vector is always 1. Once that has been done, the acceleration vector will be perpendicular to the tangent vector and the curvature is defined to be the length of the acceleration vector. The intuitive idea is that we travel along the curve at constant unit speed, so the only acceleration we experience is due to change in direction. The length of the acceleration vector represents the amount of that change, so it is a good measure of the amount of curving. The larger the curvature, the sharper the turning in the curve. Two standard examples studied in calculus are the line and the circle. A line has curvature 0. A circle of radius r has curvature $1/r$; thus a large circle has small curvature and a small circle has large curvature.

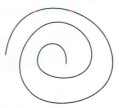

FIGURE 14.2: This curve has much greater curvature than does the curve in the preceding figure

The definition reviewed in the preceding paragraph is about as far as most calculus courses go into the subject of differential geometry. We need to go one step further and study the curvature of a surface. Calculus courses include some discussion of surfaces, but they do not usually include a definition of the curvature of a surface. In fact it is not at all obvious how to define one number that describes the way in which a two-dimensional surface curves, and it took the great Gauss to solve that problem in a satisfactory way. Our next objective is to describe Gauss's definition of curvature for surfaces in 3-space.

A *surface* is defined to be the image of a function of two variables.[1] For now we will restrict to the case in which the range of the function is a subset of three-dimensional Cartesian space $\mathbb{R}^3$, so a surface is described by a function of the form $f : [a, b] \times [c, d] \to \mathbb{R}^3$. The function f should have enough derivatives and be smooth enough so that there is a normal vector and a tangent plane at each point of the image. The function should also be one-to-one, except that points on the boundary of the rectangle $[a, b] \times [c, d]$ may be identified. The surface is the image of f and the function f is called a *parametrization* of the surface. Here are three standard examples.

EXAMPLE 14.1.1   A sphere

Define $f : [0, 2\pi] \times [\pi, 2\pi] \to \mathbb{R}^3$ by $f(u, v) = (\cos u \sin v, \sin u \sin v, \cos v)$. This function parametrizes the 2-sphere $\mathbb{S}^2$. The domain of f is a rectangle. The entire top of the rectangle is mapped to the north pole and the entire bottom of the rectangle is mapped to the south pole. The horizontal lines in the domain are mapped to parallels of latitude and the vertical lines in the domain are mapped to longitudes of the sphere. Fig. 14.3 was created in *Mathematica* using this parametrization. ∎

FIGURE 14.3: A sphere

---

[1]More generally, a surface is a space having the property that for every point in the space, the part of the space that surrounds the point is the image of a continuous one-to-one function of two variables. Such a function, whose image covers part of the surface, is called a *coordinate patch*.

EXAMPLE 14.1.2   A torus

A doughnut-shaped surface is called a *torus*. It can be parametrized by the function $f : [0, 2\pi] \times [0, 2\pi] \to \mathbb{R}^3$, whose equation is $f(u, v) = ((\cos u)(2 + \cos v), (\sin u)(2 + \cos v), \sin v)$. Fig. 14.4 was created using this parametrization. ∎

FIGURE 14.4: A torus

EXAMPLE 14.1.3   A plane

Another example of a surface that should not be overlooked is a plane. Fig 14.5 shows the portion of a plane that lies in the first octant. A plane cannot be covered by a single coordinate patch $f : [a, b] \times [c, d] \to \mathbb{R}^3$ but the area surrounding any point on the plane can be covered by such a patch, so we also consider a plane to be a surface. ∎

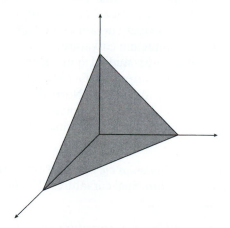

FIGURE 14.5: A plane

As a step toward the definition of curvature for a surface we define a curvature in each tangent direction. Pick a point x on a surface S. Let **n** be a normal vector at the point and let **u** be a tangent vector at that point. The plane Π spanned by **n** and **u** will intersect the surface in a curve $\Gamma(\mathbf{u})$. The curve $\Gamma(\mathbf{u})$ has a curvature, which we will denote by $k(\mathbf{u})$, called the *normal curvature* in the direction **u**. In a plane it is possible to distinguish between curving to the right and curving to the left, so the normal curvature can be either positive or negative. The sign is determined by the choice of normal vector. If the opposite normal vector is used, the curvature will change sign.

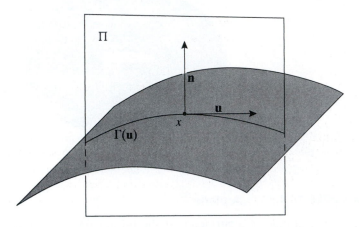

FIGURE 14.6: The normal curvature $k(\mathbf{u})$ of the surface is the curvature of $\Gamma(\mathbf{u})$

If the surface is smooth, the normal curvature will vary continuously as **u** changes. As a result, there will be one direction in which the curvature is greatest and one direction in which it is smallest. These directions are called the *principal directions* at x and the curvatures in those directions are called the *principal curvatures*. The two principal curvatures at x are denoted by k_1 (the maximum curvature) and k_2 (the minimum curvature).

Normal curvature was studied quite extensively by Leonhard Euler, who discovered the somewhat surprising fact that the principal curvatures are always achieved in perpendicular directions. Normal curvature is still an essentially one-dimensional concept and does not completely describe the curvature of a surface. It was Gauss who combined the two principal curvatures into one number that measures the two-dimensional curvature.

Definition 14.1.4. The *Gaussian curvature* of the surface S at the point x is defined to be the product of the principal curvatures. The Gaussian curvature is denoted by K, so the formula is $K = k_1 k_2$.

The definition of Gaussian curvature uses vectors and planes in $\mathbb{R}^3$. The way the definition is presented, it seems to depend not just on the surface itself, but on the way in which the surface is situated in $\mathbb{R}^3$. In this context it is helpful to distinguish between properties of the surface that are *intrinsic* and properties that are *extrinsic*. An intrinsic property of the surface is a property that depends only on relationships that are internal to the surface itself and does not depend on relationships with anything outside the surface or on the particular parametrization that is being used to describe the surface. Objects such as normal vectors and tangent planes are extrinsic. They are associated with the surface, but they are extrinsic in that the only way to measure them involves constructions that go outside the surface.

Even though Gaussian curvature appears from its definition to be extrinsic, Gauss discovered the amazing fact that it is actually an intrinsic property of the surface. Gauss was so impressed with this discovery that he called it *Theorema Egregium* (Notable Theorem), and it still goes by that Latin name today. The *Theorema Egregium* asserts that Gaussian curvature is an intrinsic property of a surface.

The *Theorema Egregium* implies that Gaussian curvature is something that can, in principle, be detected and measured by a two-dimensional creature whose entire existence is confined to the surface and who has no information about anything outside the surface. Of course the two-dimensional creature would have to do this calculation in some indirect way because it does not have access to normal vectors and the other objects that are part of the definition of Gaussian curvature. The indirect methods that can be used include measuring the circumference of a circle of radius r, measuring the defect of a triangle, and watching what happens when a line is translated around a closed curve. In the next section we will do some of those calculations for ourselves.

It is time to look at some examples of Gaussian curvature.

EXAMPLE 14.1.5   Curvature of a plane

The Gaussian curvature of a plane is 0. At each point of the plane, all the curves $\Gamma(\mathbf{u})$ are straight lines. Since all the normal curvatures $k(\mathbf{u})$ are 0, the principal curvatures are both 0 and their product is 0. ■

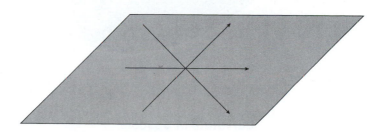

FIGURE 14.7: The Gaussian curvature of a plane is zero

EXAMPLE 14.1.6   Curvature of a sphere

The Gaussian curvature of a sphere of radius r is $1/r^2$. At each point x on the sphere, all of the curves $\Gamma(\mathbf{u})$ are great circles on the sphere and are therefore circles of radius r. Thus $k(\mathbf{u})$ is $1/r$ for every $\mathbf{u}$; the principal curvatures are both $1/r$ and $K = 1/r^2$. ■

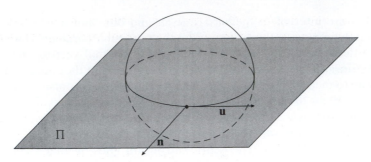

FIGURE 14.8: The Gaussian curvature of a sphere of radius r is $1/r^2$; any plane Π that contains the normal vector **n** will pass through the origin and intersect the sphere in a great circle

The two examples we just looked at are both very special in that all the normal curvatures are the same at every point and in every direction. Of course this will not always be the case. In fact, it is possible for some normal curvatures to be positive and some to be negative. A saddle-shaped surface, for example, has negative Gaussian curvature at every point but the numerical value of the curvature varies from point to point.

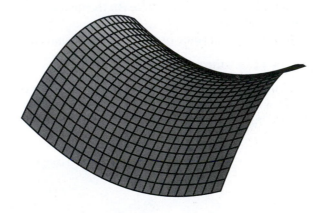

FIGURE 14.9: A saddle

EXAMPLE 14.1.7 Curvature of a saddle

The saddle shown in Fig. 14.9 has equation $z = x^2 - y^2$. The figure was drawn by *Mathematica* using the parametrization $f : [-2, 2] \times [-2, 2] \to \mathbb{R}^3$, where $f(x, y) = (x, y, x^2 - y^2)$.

In Fig. 14.10, four different curves $\Gamma(\mathbf{u})$ at the point x are shown in bold. There are two directions on the saddle in which the normal curvature is 0. In all other directions the curvature is nonzero. In some directions the curve $\Gamma(\mathbf{u})$ is on one side of the tangent plane and in other directions it is on the opposite side. Some normal curvatures are positive and some are negative; therefore the Gaussian curvature of the saddle is negative. The Gaussian curvature of this particular saddle is equal to -4 at the origin (see Exercise 14.4) and monotonically approaches 0 as the distance from the origin increases. ■

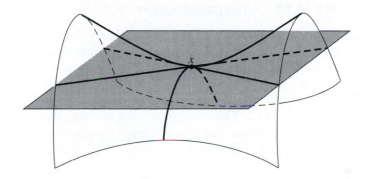

FIGURE 14.10: The Gaussian curvature of a saddle is negative

In the next example the Gaussian curvature varies even more significantly from one point to another.

EXAMPLE 14.1.8   Curvature of a torus

The torus has points at which the Gaussian curvature is positive, negative, and zero. Fig. 14.11 shows three sample points; curves in the principal directions at the three points are indicated in bold.

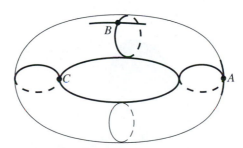

FIGURE 14.11: The Gaussian curvature of a torus varies from point to point; it is positive at A, zero at B, and negative at C

There is a circle of points on the top of the torus at height $z = 1$ and another circle of points at the bottom of the torus at height $z = -1$. The torus has Gaussian curvature 0 at each point on those two circles. The two circles separate the torus into two pieces. At each point of the piece that is close to the z-axis, the curvature is negative; at each point of the piece that is farther from the z-axis, the curvature is positive. ■

Now let us return to negative curvature. The Gaussian curvature of the saddle is negative at every point but changes from one point to another. It is not easy to find examples of surfaces of constant negative curvature. Probably the simplest example is the pseudosphere.

EXAMPLE 14.1.9   **The pseudosphere**

The *pseudosphere* is the surface generated by revolving a *tractrix* about its axis. A tractrix is a planar curve that has a special physical significance. Suppose a heavy weight is placed at the point $(0, 1)$ in the (x, y)-plane and connected by a rope of length 1 to a tractor located at the origin. If the tractor drives along the positive x-axis dragging the weight behind it, the weight will trace out a curve like that shown in Figure 14.12. This curve is called a tractrix. The geometric property that distinguishes a tractrix is that every tangent line intersects the x-axis a distance 1 away from the point of tangency.

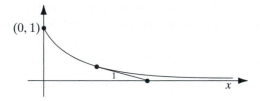

FIGURE 14.12: The tractrix

The pseudosphere is defined to be the surface obtained by revolving the tractrix about the x-axis (see Fig. 14.13). ∎

FIGURE 14.13: The pseudosphere

It is fairly clear that the pseudosphere has negative Gaussian curvature because the tractrix curves away from the x-axis while the circles of revolution curve toward it. The Gaussian curvature is the same at every point of the pseudosphere. We will not prove this, but it should seem intuitively plausible. The tractrix curves less and less as the x-coordinate increases while the circles of revolution get smaller and smaller. Thus the maximum normal curvature increases in absolute value while the minimum normal curvature decreases in absolute value. The two changes exactly balance each other and the Gaussian curvature, which is the product of the principal curvatures, is constant on the pseudosphere.

Now that we have a good supply of examples of surfaces, it is time to think about what it should mean for a line to be "straight" on a curved surface. To avoid confusion, we will not give a technical meaning to the term straight itself, but will use the term *geodesic* instead. (This is the standard terminology.) Roughly speaking, a geodesic is a curve on a surface that is as straight as it can be, given that it is confined to the surface. There are two ways in which to make this definition of geodesic more precise and rigorous, each of which captures one aspect of what we normally think of as straightness.

One way to define geodesic is to say that the geodesic segment between two nearby points on a surface is the shortest path in the surface joining the two points.[2] This is the more straightforward way to formulate the definition and clearly identifies one of the properties we normally associate with straight lines. The familiar colloquial version of the definition goes like this: "A straight line is the shortest distance between two points." Of course a mathematics student always shudders when he or she hears the definition put this way since it asserts the equality of two different kinds of objects (a line and a distance), but the meaning is clear.

The second way to define geodesic is to say that a geodesic is a curve that goes straight ahead on the surface. We are still considering smooth surfaces that are situated in $\mathbb{R}^3$, so the curve has a velocity vector and an acceleration vector. Since the surface itself is curved, we expect the acceleration vector to be nonzero. It will have a component that is tangent to the surface as well as a component that is normal to the surface. A more precise way to make the second definition of geodesic is to say that a geodesic is a curve in the surface whose acceleration vector is always normal (perpendicular) to the surface. This means that the curve is intrinsically straight in the sense that it would appear to a two-dimensional inhabitant of the surface to be straight. It would only look curved from an extrinsic point of view.

The second definition of geodesic can be explained by analogy with a familiar situation. An airplane flying straight ahead at a constant altitude will appear to its occupants to be flying in a straight line. In fact it really is flying along a straight path in the sense that the pilot is not turning the plane to the left or the right. However, it is not flying in a Euclidean straight line, but rather in a circular path in a surface. The surface to which the airplane is confined is a sphere at a constant height above the surface of the earth and the geodesic is a great circle in that sphere.

A surface is said to be *smooth* if it has a unique tangent plane at every point and that tangent plane varies continuously from point to point. The two definitions of geodesic are equivalent on smooth surfaces. More specifically, a curve on a smooth surface is a geodesic according to the second definition if and only if any

---

[2] A *path* is the image of a continuous function defined on a closed interval.

short piece of it is a geodesic segment according to the first definition. This is not obvious, but it is a theorem in differential geometry. So we can use the two definitions interchangeably (provided the surface is smooth).

EXAMPLE 14.1.10 Geodesics on a torus

The curves a and c shown on the torus in Fig. 14.14 are geodesics. The curve b is not a geodesic because its acceleration vector is in the plane tangent to the torus. ∎

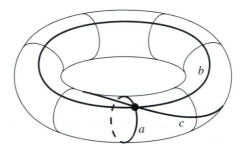

FIGURE 14.14: The curves a and c are geodesics; b is not

The geodesics on a surface are the curves that play the role of lines on that surface. We can imagine doing geometry on any surface, once the geodesics on that surface have been identified. The *points* in the geometry are the points on the surface and the *lines* are the geodesics.

The important point is that the geometry of a surface is determined by its Gaussian curvature, at least in the small. Positive Gaussian curvature is associated with spherical geometry, zero Gaussian curvature is associated with Euclidean geometry, and negative Gaussian curvature is associated with hyperbolic geometry.

Fig. 14.15 illustrates this last point by showing examples of geodesics on three of the surfaces we have studied. In each case there is a triangle whose sides are geodesic segments on the surface. Observe that the geodesic triangles on the positively curved surface have angle sum greater than 180° while those on the negatively curved surface have angle sum less than 180°. Triangles on the flat surface are "just right" in that they have angle sum exactly equal to 180°.

The word *nearby* in the distance-minimizing definition of geodesic should not be overlooked. If two points are close to each other, the shortest path joining them is a geodesic segment. But a given segment of a geodesic curve may not be the shortest path joining two points. For example, if A and B are two nearby points on the sphere, they determine a unique great circle. The entire great circle is a geodesic. The two points divide the great circle into two circular arcs. One of the two circular arcs is the shortest path connecting the two points; the other is much longer. In other words, the shortest path joining two nearby points is a geodesic segment, but a segment of a geodesic curve is not necessarily the shortest path joining its two endpoints (see Fig. 14.16).

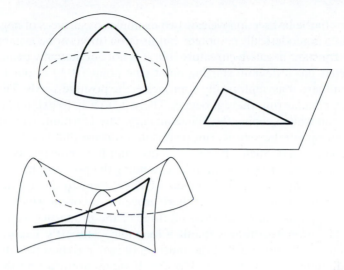

FIGURE 14.15: Geodesic triangles on the sphere, the plane, and the saddle

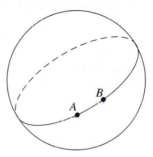

FIGURE 14.16: There is a long geodesic connecting *A* and *B* as well as a short one

14.2   APPROXIMATE MODELS FOR THE HYPERBOLIC PLANE

We can use the terminology of the previous section to describe the hyperbolic plane in a new way: the hyperbolic plane is a surface of constant negative Gaussian curvature that extends infinitely far in every direction. The fact that the curvature of a hyperbolic plane should be negative can be inferred from Fig. 14.15 and our knowledge of properties of hyperbolic geometry. The curvature must be constant because the Side-Angle-Side Postulate requires that triangles can be moved around in the hyperbolic plane without distortion. The Ruler Postulate implies that every geodesic must extend infinitely far. A surface on which geodesics can be extended without limit is called *geodesically complete*.

If we want to see a hyperbolic plane, therefore, we must locate a surface that possesses all three of the properties that were identified in the previous paragraph: negative curvature, constant curvature, and geodesically complete. As we will see, finding such a surface is $\mathbb{R}^3$ is easier said than done. But we can at least approximate one.

In the last section we saw two examples of surfaces of negative curvature. The saddle is geodesically complete, but its curvature is not constant. The pseudosphere has constant negative curvature, but it is not geodesically complete. Recall that the tractrix has a definite starting point (the point $(0, 1)$ in our description) and that from there it asymptotically approaches the positive x-axis. Thus the pseudosphere has a circular rim where it begins. There is no way to extend the pseudosphere past this rim while keeping the Gaussian curvature constant. Any attempt to extend the pseudosphere beyond its rim results in a surface that becomes wavy, like the brim of a floppy hat. Such a floppy brim does not have constant curvature, so we cannot construct the surface we want by extending the pseudosphere.

It is not just difficult to do what we want to do; there is a technical sense in which it is impossible. Hilbert proved a theorem which says that there is no smooth surface in $\mathbb{R}^3$ that has all three of the properties we need. Fortunately for us, Hilbert's Theorem contains a loophole. Hilbert's Theorem requires that the surface be so smooth that the function of two variables that describes it locally has continuous derivatives of high order. If we do not insist on this much smoothness, the construction can be carried out; it is possible to construct something like the hyperbolic plane in $\mathbb{R}^3$ but the surface will be forced to be more and more wavy as it spreads out. A complete hyperbolic plane would be hopelessly floppy and unwieldy, but we can still construct large pieces of one and see enough of it to gain a good understanding of what it is like.

We will look at constructions that produce approximations to portions of the hyperbolic plane. The surfaces we construct will differ from the actual hyperbolic plane in two ways. First, they will be polygonal rather than smooth. The reason we use polygonal pieces is simply so that we can make them for ourselves out of pieces of paper. A piece of paper is flat and (by Gauss's theorem) cannot be bent into the shape of a surface of negative curvature. So we will construct surfaces that approximate constant negative curvature in the sense that they have small flat polygonal pieces, but these pieces fit together in such a way that there is negative curvature concentrated at the vertices.

The second way in which these surfaces differ from the hyperbolic plane is that they are finite in extent. This limitation is a result of the simple fact that we will construct the surface out of a finite number of pieces of paper. Even though it is physically impossible to construct a complete hyperbolic plane, there is no definite limit to the constructions we will use. We can build sufficiently large pieces of the hyperbolic plane so that we can see for ourselves how floppy the surface is forced to become. This will give insight into why it is so difficult to represent a complete hyperbolic plane in $\mathbb{R}^3$ and why we have so much trouble visualizing hyperbolic geometry.

The idea of constructing polygonal models of this type apparently originated with William Thurston (b. 1946). He described such constructions in his book [69], which was first published in 1997. Since that time the idea has been developed by other authors; see [31] and [73], for example.

Let us begin with the familiar tiling of the Euclidean plane by equilateral triangles that is shown in Fig. 14.17. This pattern can obviously be extended to cover the entire plane. Around each vertex in the pattern there are six equilateral triangles that fit together to form a flat piece (curvature zero) of the plane. We will modify the pattern in two different ways, first to produce a surface of positive curvature and then to produce one of negative curvature.

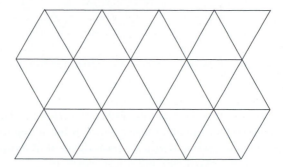

FIGURE 14.17: Tiling the plane with equilateral triangles

In order to have positive curvature at a vertex we need to force the surface to bend away from us. This can be done by removing one of the triangles at a vertex and pulling the remaining ones together into a positively curved surface. The result is a surface on which there are five equilateral triangles surrounding each vertex. If this pattern is continued, a closed surface of positive curvature results. It is the familiar icosahedron (Fig. 14.18). In Construction Project CP14.2 at the end of the chapter you will be asked to construct a paper model of the icosahedron. When you do so, you will find that the positive curvature forces the surface to close up and that the pattern of five equilateral triangles around a vertex cannot be continued indefinitely the way the pattern of six triangles can be.

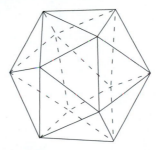

FIGURE 14.18: The icosahedron

Now we will make a surface of negative curvature. In order to be negatively curved, the surface must curve towards us in one direction and curve away from us in another direction. We can accomplish this by inserting an extra triangle at each vertex. In the new pattern each vertex is surrounded by seven equilateral triangles.

You should begin by constructing a paper model of the triangles surrounding one vertex. Start with the Euclidean pattern of six equilateral triangles surrounding

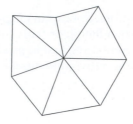

FIGURE 14.19: Seven equilateral triangles surrounding one vertex

the vertex, cut a slit, and insert one additional triangle. This is our model for a small piece of the hyperbolic plane. Observe how it has a negative curvature even though it is made up of flat pieces. If you set it on the desk it naturally assumes a position in which some parts curve down and others curve up.

In Construction Project CP14.3 at the end of the chapter you will be asked to continue this construction to produce a model for a large portion of the hyperbolic plane. This is not meant to be a thought experiment! You should get out some paper, a pair of scissors, and tape and make one of these surfaces for yourself. Only by doing this can you truly appreciate the sense in which the model becomes floppy. In the project you will be asked to perform some simple experiments on the surface. The purpose of these experiments is to verify that the geometry of the surface is approximately hyperbolic. You will also find that the curvature of this surface is so great that it soon becomes unwieldy. This will help you to appreciate the second model, which will be described below.

FIGURE 14.20: A small part of a triangular model of the hyperbolic plane

One important observation is that the pattern of seven triangles around a vertex can be continued indefinitely. By contrast, the pattern of five triangles around a vertex results in a surface that has positive curvature and is forced to close up. The pattern of seven triangles produces a negatively curved surface. It can be enlarged indefinitely, but it takes more and more triangles to enlarge the radius of the plane by any significant amount. There is no definite limit to how big the model can be, but it soon becomes impractical to continue to enlarge it. This gives real insight into what the hyperbolic plane is like. In the small it is not too different from the Euclidean plane, but as it spreads out it soon begins to include an unimaginably large amount of area. As a result, the model is forced to fold back on itself. By making such observations we begin to understand why it is so difficult for us to visualize a hyperbolic plane.

Remark. We are applying the term *curvature* to polygonal surfaces even though we have only defined Gaussian curvature for smooth surfaces. Curvature of a polygonal surface should be understood in an approximate sense. Smoothing out the sharp corners of the polygonal surface results in a smooth surface that approximates the polygonal one. It is the curvature of this smooth approximation that you should think of. While this interpretation gives the right intuitive understanding, it is worth noting that it is also quite easy to give a precise definition of what is meant by curvature of a polygonal surface—see Exercise 14.5.

As you have undoubtedly observed by now, the weakness of the triangular model for the hyperbolic plane is that it curves too much and soon becomes difficult to work with. There are many ways to vary this first simple construction in order to produce surfaces that approximate the hyperbolic plane but do not have quite as much curvature and therefore do not become so unwieldy so quickly. We will describe just one.

This time we start with the familiar hexagonal tiling of the Euclidean plane (Fig. 14.21). In this pattern each vertex is surrounded by three regular hexagons. The three regular Euclidean hexagons fit together to form a flat surface.

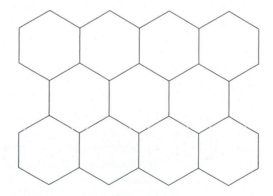

FIGURE 14.21: Tiling the plane with regular hexagons

In order to introduce positive curvature we replace one of the hexagons at each vertex with a pentagon. The new pattern is that each vertex is surrounded by one regular pentagon and two regular hexagons. The length of a side of the pentagon should equal the length of the side of the hexagons. In Construction Project CP14.4 you will be asked to construct such a surface for yourself. You will find that once again the positive curvature forces the surface to close up on itself. The construction produces the familiar soccer ball shape (Fig. 14.22).

FIGURE 14.22: A spherical soccer ball

Now let us go back to the hexagonal tiling of the Euclidean plane and modify it a different way in order to produce a polygonal surface with negative curvature. This time we replace one hexagon at each vertex with a heptagon. The new pattern is that each vertex is surrounded by two regular hexagons and one regular heptagon. The resulting surface is negatively curved, so it does not close up on itself. Instead the pattern can be continued indefinitely and results in a second model for the hyperbolic plane. A paper model of a small part of it is shown in Fig. 14.23. Because of the analogy with the ordinary soccer ball we will call this model the *hyperbolic soccer ball* (even though it is about as useless as anything could be for playing an actual game of soccer).

In Construction Project CP14.5 you will be asked to build a hyperbolic soccer ball for yourself. You should make one that is at least 10 polygons across. This is quite a bit larger than the hyperbolic soccer ball illustrated in Fig. 14.23. Constructing this model will take a lot of polygons because this surface too will become wavy and floppy as it grows. But that is characteristic of any model of the hyperbolic plane in which distances are accurately portrayed and therefore is part of the point of the construction.

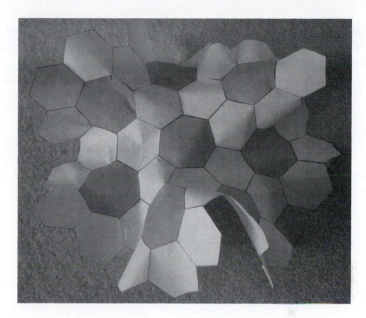

FIGURE 14.23: Part of a hyperbolic soccer ball

By constructing and studying this surface you will see for yourself something of what the hyperbolic plane is like. You can use a straightedge to draw line segments joining two points. You will see that each line is straight when it is drawn with a straightedge. But if you draw three line segments and then let the surface assume its natural position, you will see that the straight lines assume a curved shape relative to each other. The result is that triangles on the hyperbolic soccer ball have positive defect. Other experiments in the project will allow you to verify that the hyperbolic soccer ball satisfies the Hyperbolic Parallel Postulate and estimate the angle of parallelism. You will also plot area versus defect for triangles and plot circumference and area versus radius for circles.

The real purpose of the activities in this section is to develop an intuitive understanding of the geometry we studied abstractly in earlier parts of the book. While you are working through the projects, be sure to keep this in mind and reflect on the ways in which theorems from earlier sections are illustrated on the models.

14.3 GEOMETRIC SURFACES

The second goal of the chapter is to understand the relationship between the geometry we have studied in the course and the geometry of the physical universe. Before we can fully appreciate that relationship, we need to look at another aspect of the geometry of a surface.

Gaussian curvature is local in the sense that it measures the way in which a surface is curving in the vicinity of a point. As we have seen, the curvature at a point determines the geometry in the region near to that point. There is also a global aspect to the geometry of a surface. For each surface there is an overall, or global, geometry of complete geodesics on the surface. This overall geometry is determined by a global curvature that averages out all the local curvatures.

In this section we will study the global geometry of a surface. To understand it, we need to view surfaces somewhat differently.

The torus is a good example with which to begin. One way to construct a torus is by taking a square and gluing top to bottom and left to right as indicated in Fig. 14.24. This is exactly what happens in many computer games in which anything that goes off one edge of a rectangular screen emerges on the other side. The game-board for such a game is really a virtual torus. In order to make a physical torus this way, we need to use a square that is made of some kind of flexible material that can be bent and stretched into the shape of the torus.

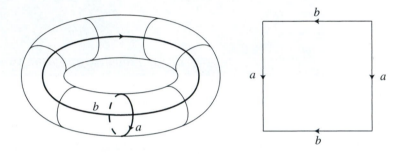

FIGURE 14.24: The torus is constructed by gluing the two *a* curves together and gluing the two *b* curves together

This construction cannot be done with a paper square. Paper cannot be stretched, so it is impossible to take a paper square and tape up the edges to form a torus without putting some serious creases in the paper. It is Gauss's *Theorema Egregium* that prevents it: The piece of paper has Gaussian curvature 0 while the torus has varying Gaussian curvature, so they are intrinsically different surfaces.

We can define the torus abstractly by starting with the square and abstractly identifying the bottom with the top and the left with the right. That is done by simply declaring each point on the left edge to be identical with the corresponding point on the right edge and each point on the top edge to be identical with the corresponding point on the bottom edge. This abstract identification does not require bending the paper at all. The abstract version of the torus defined this way is called the *flat torus*.

The observation that the torus can be constructed by identifying points on the boundary of a square is important because it shows that the torus possesses a geometry of constant curvature and that that constant is 0. Let us call part of the surface surrounding a point a *neighborhood* of the point. We will examine the neighborhoods of points on the flat torus. There are three different kinds. A neighborhood of any point *A* in the interior of the square is identical to a neighborhood of a point in the plane. A point *B* on the top of the square is identified with a point on the bottom of the square, so a neighborhood of *B* consists of two semicircles glued together. All four corner points of the square are identified to one point C in the flat torus. A neighborhood of C in the flat torus consists of four quarter-circles glued together. In each case the neighborhood is flat and this justifies the name "flat" torus.

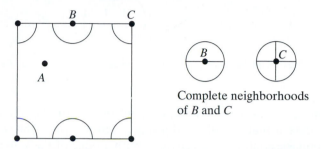

Complete neighborhoods of B and C

FIGURE 14.25: Neighborhoods of points in the flat torus

The flat torus is not the same as the ordinary torus that we studied earlier. For one thing, the flat torus has zero curvature at every point while the ordinary torus has variable Gaussian curvature. There is no isometry (distance-preserving transformation) from one to the other because the existence of such an isometry would contradict the *Theorema Egregium*. On the other hand, the ordinary torus and the flat torus are closely related. To transform the flat torus into the ordinary torus we must bend and stretch the square into a shape that allows the necessary gluing to be done in $\mathbb{R}^3$. This can be accomplished with a continuous one-to-one deformation, so most of the structure of the torus is preserved by the transformation. A transformation that is continuous and has a continuous inverse is called a *topological* transformation. Using that terminology we can explain the relationship between the two versions of the torus by saying that the flat torus and the ordinary torus have the same topology but different geometries.

The flat torus is an abstract object that cannot be realized as a surface in $\mathbb{R}^3$. It can be realized as a surface in a higher-dimensional Euclidean space, but we will not make use of that observation. One of the necessary prerequisites for understanding the geometry of the universe is a willingness to start thinking of curvature as an intrinsic property of a space that can be understood and studied without reference to some larger space that contains it. So we will study the flat torus as an abstract object that exists independently of any surrounding space.

The fact that the natural overall geometry for the torus is in some sense Euclidean is dimly reflected in the geometry of the ordinary torus. In particular, the geodesics on the torus satisfy the Euclidean Parallel Postulate. This is illustrated in Fig. 14.26. The point P is an external point for the geodesic ℓ and there is exactly one geodesic through P that is parallel to ℓ.

There is a much deeper connection between the geometry of the flat torus and that of the Euclidean plane. Imagine an inhabitant of the torus looking at her surroundings. When she looks in some direction, she will not see the edge of the square, but will see right across it. In other words, a geodesic on the flat torus will go

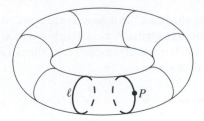

FIGURE 14.26: Geodesics on the torus satisfy the Euclidean Parallel Postulate

straight across an edge of the square and emerge on the other side of the edge. We can represent that visually by drawing another copy of the square beyond the first. The geodesic continues and crosses another edge of the square; again we represent this by drawing a third copy of the square. The result is a picture like Fig. 14.27.

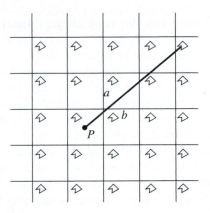

FIGURE 14.27: What an inhabitant of the flat torus sees

The way to understand Figure 14.27 is to imagine the torus being constructed from the plane by rolling it up like a carpet. Rather than just using one copy of the square we tile the entire plane with copies of the square. Each of those squares corresponds to the entire flat torus. In order to accomplish that correspondence, first roll the plane up side-to-side to identify the *a* curves and then roll it up top-to-bottom to identify the *b* curves. The first step can be done with an actual physical carpet, but Gauss's theorem prevents the second step from being done without distorting the local geometry of the carpet. That is why the rolling-up process results in a flat torus rather than an ordinary torus.

Looking at the process from the other point of view, a geodesic on the flat torus unrolls to a straight Euclidean line. The geodesics on the flat torus behave just like lines in the Euclidean plane because they can be unrolled to actually become straight lines in the Euclidean plane. For this reason we will say that the geometry of the flat torus is Euclidean.

An inhabitant of the torus who looks around will see what appears to be the Euclidean plane. But she will suspect that this is not the usual Euclidean plane because any object that she sees will actually be visible in lots of different directions. There will be one copy of the object in each of the infinitely many squares. More specifically, an observer at point P in Fig. 14.27 will see infinitely many copies of an object. One geodesic from P to the object is shown in bold. In the unrolled version there is one geodesic from the observer to each copy of the object. When the plane is rolled up to form the torus, these become different geodesics that connect the observer to the same object.

Now let us try to understand the geometry of another surface, the double torus, in a similar way. The *double torus* is the surface of a doughnut with two holes. If the double torus is cut open along the curves a, b, c, and d indicated in Fig. 14.28, the surface can be flattened out into an octagon. Reversing this process, the double torus can be constructed from the octagon by making the identifications indicated in Fig. 14.28. To be specific, the double torus is constructed from the octagon by gluing together the two edges marked a, and then the two edges marked b, and so on. The directions of the arrows should be matched when the the sides are glued together.

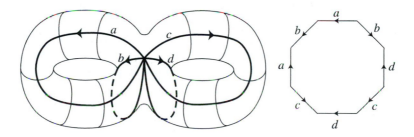

FIGURE 14.28: The double torus is constructed by gluing up the edges of an octagon in pairs

It is not possible to make a double torus in $\mathbb{R}^3$ without bending and stretching the octagon, but we can try to abstractly define a flat double torus in the same way we defined a flat torus. When the edges of the octagon are identified, a point on the interior of one edge is identified with a point in the interior of the corresponding edge. A neighborhood of the new point consists of two semicircles glued together, so again the neighborhood is flat. But there is a problem at the corner. When the edges are identified, all the vertices of the octagon are identified to one point. A neighborhood of a vertex of the octagon is a 135° slice of a circle. When eight of these are glued together there is a total angle of 1080° surrounding the vertex. Thus the abstract double torus we are defining is not flat but has negative curvature, all of it concentrated at one point.[3]

In order to avoid having a wrinkle in the surface at the vertex, we need to use an octagon that has eight angles of measure $360/8 = 45°$. Obviously no such octagon is to be found in Euclidean geometry, but such octagons do exist in

---

[3]See Exercise 14.5 for a definition of the curvature of a polygonal surface.

hyperbolic geometry. Fig. 14.29 shows a regular octagon in the Poincaré disk model of hyperbolic geometry, each interior angle of which has measure $45°$.

FIGURE 14.29: A hyperbolic octagon with eight $45°$ angles

The natural way to build a double torus is to start with this hyperbolic octagon and perform the abstract identifications on it. We abstractly define a double torus that has constant curvature by using a hyperbolic octagon. In this sense the overall curvature of the double torus is negative and the natural geometry on the double torus is hyperbolic geometry.

Again the fact that the natural geometry is hyperbolic is reflected even in the complete geodesics of the ordinary double torus. Fig. 14.30 illustrates the fact that the geodesics on the double torus satisfy the Hyperbolic Parallel Postulate: P is an external point for ℓ and both m and n are parallel to ℓ (in the sense that they do not intersect ℓ).

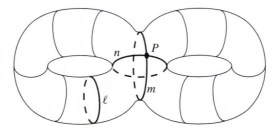

FIGURE 14.30: Geodesics on the double torus satisfy Hyperbolic Parallel Postulate

The abstract double torus can be unrolled just like the flat torus, but this time it unrolls into a hyperbolic carpet. (One lesson you should have learned in the last section is that a hyperbolic carpet would be extremely unwieldy.) Fig. 14.31 shows what this would look like in the Poincaré disk model. Just as the Euclidean plane can be tiled with square tiles, the hyperbolic plane can be tiled with regular octagonal tiles. The various octagonal tiles shown in the figure don't look the same to us, but in the Poincaré disk model they are all congruent. In this sense the natural geometry of the double torus is hyperbolic. The geodesics on the abstract double torus are exactly like lines in the hyperbolic plane in the sense that the double torus can be unrolled to a hyperbolic plane in which the unrolled geodesics really are hyperbolic lines.

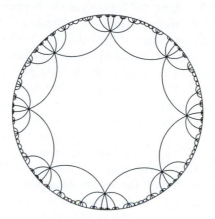

FIGURE 14.31: The double torus unrolls to a hyperbolic plane tiled by regular octagons

The flat torus and the abstract double torus are examples of geometric surfaces. A *geometric surface* is one that has constant curvature and a natural geometry associated with it. We have looked at only two examples, but the same kinds of techniques we used to understand them can be applied to other surfaces. It can be shown that every closed surface has its own overall curvature and its own natural geometry in the sense that it is topologically equivalent to a geometric surface.

Every geometric surface has one of the three standard geometries: spherical, Euclidean, or hyperbolic. One surprise is that those with hyperbolic geometries predominate. To be specific, the surfaces in $\mathbb{R}^3$ that close up on themselves can be arranged in a sequence: sphere, torus, double-torus, three-holed torus, four-holed torus, ... The first surface in the sequence has positive curvature, the second is flat, and all the rest have negative overall curvature. Just one of these surfaces has spherical geometry, one has Euclidean geometry, and infinitely many have hyperbolic geometry.

14.4 THE GEOMETRY OF THE UNIVERSE

In order to understand the relationship between the axiomatic geometry we have studied in this book and the geometry of the physical universe, it helps to be aware of one last historical development regarding the foundations of geometry. As was mentioned in Chapter 8, Bernhard Riemann developed a comprehensive theory of geometry that encompasses both Euclidean and non-Euclidean geometries. Riemann's theory provides the right framework within which to understand the geometry of the universe.

Riemann was just beginning his career at the University of Göttingen at the same time that Gauss was approaching the end of his career there. The last step in

Riemann's appointment to the position of lecturer was the delivery of an inaugural lecture. Riemann submitted a list of three topics from which the faculty was to choose one. Gauss chose the topic "On the hypotheses that lie at the foundations of geometry," even though there are indications that Riemann was much better prepared to speak on either of the other two topics. After seven weeks of intense preparation, Riemann delivered his lecture on June 10, 1854.

The lecture had two parts. In the first part Riemann described the class of spaces that provide the natural setting in which to do geometry. In the second part of the lecture Riemann explained how it would be possible to do geometry on such spaces. The key ingredient is a quadratic function that is defined at each point of the space. This function measures the local deviation from Euclidean geometry by specifying the formula that must replace the familiar $a^2 + b^2$ of the Pythagorean Theorem. The length of a path can be computed by integrating this quadratic function. Once the length of a path is defined, the distance between two points is defined to be the length of the shortest path connecting them. The distance-minimizing idea is also used to define geodesics. The geodesics play the role of lines, so everything is in place to do geometry.

The high-dimensional spaces described by Riemann are now called *manifolds*. A manifold should be thought of as a high-dimensional generalization of surface, so *surface* and *two-dimensional manifold* mean exactly the same thing. In general, an n-dimensional manifold is a space having the property that the neighborhood of any point can be covered with an n-dimensional patch. More specifically, a manifold is a space M such that for each point x in M there is a continuous one-to-one function $f : [a_1, b_1] \times [a_2, b_2] \times \cdots \times [a_n, b_n] \rightarrow M$ such that the image of f is a neighborhood of x. The function f is a generalization of the coordinate functions that we studied in connection with the Ruler Postulate and so it is also called a coordinate function. Ordinarily the coordinate functions are required to be compatible where their images overlap so that it is possible to move smoothly from one patch to another. High-dimensional manifolds can be curved. The information about the curvature at each point is encoded by the quadratic function mentioned in the previous paragraph. Manifolds with such structures are called *Riemannian manifolds*.

The theory of general relativity that was developed by Albert Einstein in the early twentieth century is built on Riemannian geometry. According to the theory of relativity, space and time cannot be separated and should be considered as one. The combined spacetime is a four-dimensional Riemannian manifold. A point in this spacetime manifold is an *event*, a position in time and space. Even though space and time cannot be separated, it is possible to identify a three-dimensional slice of spacetime that we can intuitively think of as the three-dimensional physical universe in which we live. One way to do this is to consider all the points in spacetime that are a certain fixed number of seconds after the big bang. This is a three-dimensional manifold and it constitutes our physical universe.

Just the assertion that the universe is a three-dimensional manifold already gives new insight. When we look around us we see what appears to be $\mathbb{R}^3$. Ordinarily we assume that what we see locally extends infinitely far in every direction; in other words, we usually operate on the assumption that the universe is $\mathbb{R}^3$ because the part of it that we can see looks like $\mathbb{R}^3$. But now we can realize that this is analogous to assuming that the earth is flat just because the part we see in our vicinity looks flat. Thinking of the universe as a three-dimensional manifold opens up the possibility that it might have interesting large-scale structure that we have not previously dreamed of.

The theory of relativity also implies that the universe is curved. In fact that curvature is the explanation for gravitation: a massive object actually bends or curves the space around it and that is why other objects are attracted to it. Thus the curvature of the universe varies from place to place, just like the curvature of a surface in $\mathbb{R}^3$. In particular, the curvature is greater near the sun than it is near the earth.

A photon traveling through the universe will follow a geodesic. Thus a ray of light is in reality curved, even though it looks straight to us. The reason it looks straight to us is that we human beings are confined to our three-dimensional universe and can only have an intrinsic view of the light ray. As far as we know, the universe is all there is to space and certainly the only view of geodesics that is available to us is the insider view. The geodesic is intrinsically straight, so it is straight as far as we can tell.

From our intrinsic point of view we can only gather indirect evidence for the curvature of the universe. It is interesting to speculate about the possibility that our universe might be contained within a higher-dimensional manifold and what the universe might look like from the outside if we were able to acquire four-dimensional bodies and see things from the outside. But that is just speculation.[4]

One way to verify experimentally that the universe is curved is via a gravitational lens. If a really massive object, like a star or a galaxy, gets between us and a light source, the massive object can actually bend the space through which the light rays must travel. Thus there can be two different geodesics connecting the observer to the light source and so an object will sometimes be visible in two different directions (see Fig. 14.32). Many examples of this phenomenon have been detected in the universe.

A diagram such as Fig. 14.32 makes it clear that the geodesics in a curved universe do not satisfy the Incidence Postulate: There can be many different geodesics connecting two points. On the other hand, if two points are close together there will be a unique shortest geodesic segment connecting them. One of Riemann's points in his lecture was that the axioms of geometry can only hold locally and that we should not expect them to be true on a global scale.

The question of whether Euclidean or hyperbolic geometry is the correct one for the space in which we live now seems quite naïve. The quick answer is that the geometry of the universe is neither Euclidean nor hyperbolic. The actual geometry of space varies from one location to another and may be spherical at some places and hyperbolic at other locations. At the same time, the curvature of the universe

---

[4]Such speculation does serve the purpose of helping us to develop our intuition and is good practice at reasoning by analogy.

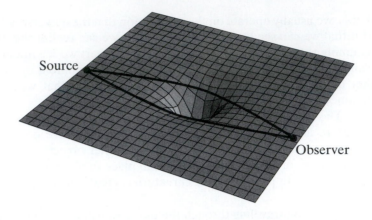

FIGURE 14.32: A gravitational lens

is rather small in comparison with the sizes that we deal with in everyday life. For that reason, Euclidean geometry continues to be the geometry we use for everyday purposes. Just as Newtonian mechanics suffices for almost everything we do in daily life, so Euclidean geometry will suffice for most applications. It is important to note, however, that Euclidean geometry does not suffice for all applications. When we deal with things on a very large (or very small) scale, we must account for deviations from Euclidean geometry. Just as there are times when Newtonian mechanics fails and the theory of relativity is needed, so there are situations in which non-Euclidean geometries are appropriate.

While the kind of curvature we have been discussing is only significant on scales that are quite large compared with distances we encounter in everyday life, it is still local in the sense that it varies from point to point. In addition to this local curvature, the three-dimensional universe apparently has an overall curvature and a natural geometry associated with it. This situation is analogous to that of surfaces in $\mathbb{R}^3$, which also have an overall curvature and a natural geometry associated with them. Another useful analogy is with the two-dimensional surface of the earth. The earth's local curvature varies enormously from one point to another; for example, the curvature is much greater in Colorado than it is in Kansas. At the same time, the surface of the earth has a global shape or curvature: It is spherical. In much the same way, mass causes small bumps and dents in the universe but there is still an overall, global shape that transcends these local variations in curvature.

Perhaps the question about which geometry is the correct geometry for the real world is not so naïve and misplaced after all. We can ask whether or not the universe is a geometric three-dimensional manifold that has an overall curvature and a natural geometry associated with it. If so, what is that geometry?

This is where we approach the frontier of what is actually known. At the time this book is being written, it is not known with certainty that every three-dimensional manifold has natural geometries associated with it. Nor is it known what the overall shape or curvature of the universe actually is. In that sense we do not yet know whether the geometry of the universe is spherical, Euclidean, or hyperbolic.

The idea that there might be a natural geometry associated with a three-dimensional manifold is due to William Thurston. (This is now the second major idea in the chapter that is attributed to Thurston.) His groundbreaking work on three-dimensional manifolds in the late 1970s and the 1980s revolutionized that subject and led to a revival of interest in hyperbolic geometry, especially in three dimensions.

The situation is, of course, more complicated in three dimensions than it is in two. There is not a single geometry associated with each three-dimensional manifold, but instead the manifold can be cut up into pieces, each of which has a characteristic geometry. There are a total of eight different three-dimensional geometries. Thurston's Geometrization Conjecture states that every closed three-dimensional manifold can be cut up into a finite number of pieces, each of which has one of the eight geometries as its natural geometry. It also gives a precise sense in which those with hyperbolic geometry predominate (just as they do in two dimensions).

Thurston formulated this conjecture, accumulated a great deal of evidence in its favor, and proved it in special cases. Recently the Russian mathematician Grigori Perelman announced that he has solved Thurston's Geometrization Conjecture in its entirety. As the time this book is being written (the summer of 2004), Perelman has posted proofs of some parts of the conjecture on the World Wide Web for other mathematicians to check. There is a growing consensus among professional mathematicians that the proof is well thought out and likely to be correct, but the entire proof has not yet been verified.

In view of Thurston's conjecture, it seems reasonable to assume that the three-dimensional universe has a natural geometry associated with it. There are eight different three-dimensional geometries, but five of them can be ruled out as possible geometries for the universe. They can be eliminated because the universe appears to be both homogeneous and isotropic. This assumption is usually called the *cosmological principle*. To say that the universe is *homogeneous* simply means that it is (roughly) the same at every point. On a large scale at least, the universe looks approximately the same from every point. *Isotropic* means that the universe looks roughly the same in every direction.

We do not know for certain that the universe satisfies the cosmological principle, but all indications are that it does. Furthermore, there is no reason to think that any particular location in the universe is special or that any direction is special. In the absence of any reason to think that some locations are different from others it must be assumed that they are all the same. Five of the eight three-dimensional geometries are nonisotropic, so the cosmological principle allows us to rule them out as possible geometries for the universe. The three remaining geometries correspond naturally to the three two-dimensional geometries. The fact that the universe is homogeneous also allows us to rule out the possibility that the universe might divide up into pieces with different geometries.

So we can assume that the universe has one of the three standard geometries and that the geometry is determined by the overall curvature. But which of the three is it? That is not known. You should be able to understand why it is difficult to determine: We have to find some intrinsic, indirect way to measure the curvature and this is not easy to do. It is difficult to measure even the relatively small local curvature; detecting the great overall curvature of the universe is still beyond the reach of current technology. But it is something that will very likely be accomplished in our lifetimes.

There are professionals who analyze data from satellites and try to find indirect evidence that would indicate whether the curvature is positive, negative, or zero. So far they have not been able to give a conclusive answer, but they are close. While this book was being written, a group of scientists announced that they had found the answer [45]. They claimed that the universe is a three-dimensional manifold that is formed by identifying opposite faces of a solid dodecahedron. This three-dimensional manifold is called the Poincaré dodecahedral space.

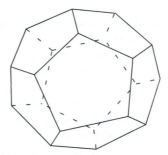

FIGURE 14.33: A dodecahedron

The Poincaré dodecahedral space is formed in a way that is completely analogous to the construction of the torus and double torus shown in Fig. 14.24 and Fig. 14.28. Start with the solid three-dimensional object whose boundary is the dodecahedron shown in Fig. 14.33. Then glue each face of the dodecahedron to the opposite face, but with a twist. The result is a three-dimensional manifold that has positive curvature and that unwraps to a three-dimensional version of the sphere that is tiled by a finite number of copies of the dodecahedron. A more detailed description of this manifold may be found on pages 220–224 of the book [73] by Jeffrey Weeks, who is one of the mathematicians involved in this effort.

If the claims of Weeks and his group are correct, the overall curvature of the universe is positive and the universe has spherical geometry. One implication is that the universe is finite and that geodesics must close up. At the present time the claim is in doubt. So we must admit that we still do not know the answer to the question of which geometry is the correct geometry for the real world. We can at least be fairly certain that it is one of the three that we know about.

14.5 CONCLUSION

Our study of the foundations of geometry has brought us all the way from the mathematics of the ancient Greeks to present day mathematical and cosmological research. It is amazing that the technical issue Euclid identified 2300 years ago when he stated his fifth postulate could have motivated so much work in geometry and inspired so many new ideas over the intervening years. It is even more amazing that these ideas continue to be vital and are the subject of intense interest even today. The mathematical community has still not exhausted their implications.

Not only does this vindicate Euclid's choice of postulate, but it also shows the value of the abstract, axiomatic method that the ancient Greeks developed. Without the freedom to experiment with alternative axiom systems, mathematicians would not have come up with the great ideas that we have studied. It is only the freedom to explore other worlds that allows us to imagine the possibilities for our own. The story of non-Euclidean geometry is just one of many examples from the history of mathematics in which ideas that were explored initially as purely abstract mathematics have turned out to have powerful practical applications.

It should be clear from the discussion of the geometry of the universe in the preceding section that abstract geometries, and hyperbolic geometry in particular, have useful practical applications. Albert Einstein, for example, freely acknowledged that he would not have been able to develop his theory of general relativity if he had not been acquainted with the ideas of hyperbolic geometry and Riemannian geometry. It is possible, therefore, to justify the pursuit of such abstract mathematical ideas on purely utilitarian grounds.

But it should also be clear that those who developed the ideas we have studied were not just motivated by the hope of practical applications for their work. Geometry is a beautiful subject that captivates many people who enjoy thinking about it and contributing to it simply because they love the subject. It is hoped that by now every reader of this book has come to understand and appreciate some of the aesthetic qualities of the subject. In the end, the question of why a subject that is so abstract and is often pursued for such impractical reasons should have such profound practical applications remains; it is a mystery that our study of the foundations of geometry has not solved. It is the author's opinion that this mystery adds to the appeal of the subject.

14.6 FURTHER STUDY

This chapter has covered a lot of ground. It attempts to develop an intuitive understanding and to tie together many different ideas. It does so by giving only a brief introduction to most of the topics discussed. For the benefit of those who want to explore the ideas in greater depth, suggestions for further study are included here.

Flatland: A Romance of Many Dimensions by Edwin A. Abbott, first published in 1884, remains the classic introduction to the use of reasoning by analogy about higher-dimensional spaces. It is written in the form of a novel that is a joy to read—the experience is highly recommended. *Flatland* is also an example of Victorian satire. This is a genre of writing that is no longer common and it can easily be misunderstood by modern readers. If you do read the book, be sure you also read an explanation of the satire. The Introduction to the Princeton University Press edition [1], written by Thomas Banchoff, is a great place to start.

This chapter only scratches the surface of differential geometry. A standard undergraduate text in the subject is *Elementary Differential Geometry* by Barrett O'Neill [56]. A more intuitive introduction to the subject may be found in Chapter IV of *Geometry and the Imagination* by Hilbert and Cohn-Vossen [33]. There is a nice, accessible explanation of some of the ideas in this chapter in Volume II of the monumental work on differential geometry by Michael Spivak [65]. In particular, Spivak includes a translation and an explication of the seminal papers of both Gauss and Riemann.

Thurston's book *Three-Dimensional Geometry and Topology* [69] contains a description of the polygonal models of the hyperbolic plane. It also describes geometric surfaces and geometric three-dimensional manifolds as well as Thurston's Geometrization Conjecture. Perelman's papers have not yet appeared in print but they can be found on the World Wide Web.

A great introduction to the geometry of surfaces and three-dimensional manifolds is *The Shape of Space* by Jeffrey Weeks [73]. This book is written in a way that promotes intuitive understanding of the concepts without demanding a lot of technical background or prerequisites. It includes a description of the method that is being used to determine the geometry of the universe from the satellite data.

The announcement by Weeks and others that evidence points to the universe being the Poincaré dodecahedral space appeared in the journal *Nature* in 2003 [45]. The article [74] appeared while this chapter was being written; it contains the latest information on the status of that claim.

The article *The geometry of the universe* by Roger Penrose [57] contains an exposition of the relationship between geometry and the theory of general relativity. In particular, Penrose describes the three two-dimensional geometries and relates them to the geometry of the universe.

In this chapter we have discussed only the large-scale geometry of the universe. There is also interesting geometry at the other end of the size spectrum. While the theory of general relativity does a good job of describing the large-scale structure of the universe, it does not accurately describe what happens at small scales. Quantum theory provides the best description of small-scale phenomena, but it conflicts with relativity theory. Albert Einstein spent the last years of his life searching for one unified field theory that would describe physical phenomena at both large and small scales. Einstein did not succeed in this effort, but theories have recently been proposed that would provide a comprehensive description. From our point of view, the most interesting thing about these new theories is the fact that once again the explanation of physical phenomena is framed in terms of geometry. The theories are called superstring theories because they model elementary particles as one-dimensional "strings." The leading expert in this field is Edward Witten (b. 1951) of the Institute for Advanced Study. The book *The Elegant Universe* by Brian Greene [28] gives a nice exposition of superstring theory for the novice.

There are many sources of information about the philosophy of mathematics. Chapter 8 of the book by Greenberg [27] addresses questions in the philosophy of mathematics from the perspective of the discovery of non-Euclidean geometry. Greenberg is quite negative about the state of the philosophy of mathematics and characterizes it as a "mess." Those who do not share Greenberg's pessimistic view can still learn from the chapter because it raises many interesting questions and because it points to many other sources of good information. A newer approach to the philosophy of mathematics is contained in the recent book by Corfield [15]. He attempts to account for such aspects of mathematics as intuition and the method of analogy that have often been ignored in traditional treatments of the philosophy of mathematics.

SUGGESTED READING

Parts IV and V of *Euclid's Window*, [50].

EXERCISES

14.1. Use trigonometry to show that the area of a regular Euclidean n-gon of side length L is given by the formula

$$A = \frac{nL^2}{4} \cot\left(\frac{\pi}{n}\right).$$

(This formula will be used in Construction Project CP14.7.)

14.2. Give a geometric description of the parametrization of the torus in Example 14.1.2. Your description should be similar to the description of the function f in Example 14.1.1. It should identify the image of each horizontal and vertical line segment in the domain of f.

14.3. Use the fact that the curvature of a circle of radius r is $1/r$ to find the numerical value of the Gaussian curvature of the torus at the points A and C in Figure 14.11.

14.4. Use the calculus formula for curvature of a curve to compute the normal curvature of the saddle of Example 14.1.7 at $(0,0,0)$ in each of the two coordinate directions. Assume that the coordinate directions are the principal directions and compute the Gaussian curvature of the saddle at $(0,0,0)$.

14.5. (Polygonal Curvature) A *polygonal surface S* is a surface that is written as the union of convex Euclidean polygonal regions in a particular way. It is required that two polygonal regions on the surface intersect in either an edge or a vertex or they do not intersect at all. All the surfaces studied in §14.2 are examples of polygonal surfaces.

The *polygonal curvature* of S is defined to be 0 at each point that is in the interior of a polygonal region or in the interior of an edge. Each vertex v of S is a vertex of a finite number of polygonal regions and each such polygonal region has an interior angle with v as vertex. Let $s(v)$ be the sum of the radian measures of all the interior angles that have v as a vertex. Define the polygonal curvature of S at v to be $k(v) = 2\pi - s(v)$.

For example, suppose S is the Euclidean plane expressed as the union of equilateral triangles as in Fig. 14.17. Each vertex is a vertex of six equilateral triangles and each interior angle of the equilateral triangle is $\pi/3$ radians. Therefore $k(v) = 2\pi - 6 \cdot (\pi/3) = 0$. This is the expected answer since the plane is flat and has curvature 0.

(a) Explain why it is reasonable to define the curvature to be 0 at each point of an edge even though two polygonal faces might meet at an angle along that edge. Your explanation should draw an analogy between polygonal curvature and Gaussian curvature.

(b) Find the polygonal curvature at each vertex of the following surfaces.

   (i) The icosahedron (Figure 14.18).
   (ii) The triangular hyperbolic plane (Figure 14.19).
   (iii) The hexagonal tiling of the Euclidean plane (Figure 14.21).
   (iv) The spherical soccer ball (Figure 14.22).
   (v) The hyperbolic soccer ball (Figure 14.23).
   (vi) The dedecahedron (Figure 14.33).

14.6. Let C be a right circular cylinder in $\mathbb{R}^3$. For example, C might be the graph of the equation $x^2 + y^2 = 1$.

(a) Show that the Gaussian curvature of the cylinder is 0 at each point.

(b) The cylinder can be constructed by taking a piece of paper and rolling it up. Perform the following experiment to verify that the cylinder and a flat piece of paper have the same intrinsic geometry: Mark two points A and B on the piece of paper and use a ruler to draw the straight line segment from A to B. Roll the paper up into a cylinder and stretch a piece of string from A to B. Pull the string tight so that the length from A to B is minimized; the string will then represent a geodesic from A to B. Observe that the taut string exactly follows the straight line segment you drew on the flat paper.

14.7. Construct a cone out of a piece of paper as follows: Start with a circle of some unspecified radius. Cut out a circular sector with central angle θ and tape the two radial edges of the circular sector together (see Figure 14.34). The point on the cone that corresponds to the center of the circle is called the *cone point*. (The cone point is the sharp point of the cone.) The angle θ is the *cone angle*. In this exercise the cone angle is assumed to be less than $360°$.

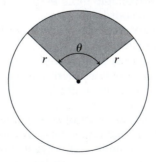

FIGURE 14.34: Construction of a cone with cone angle θ

On the paper cone you can draw geodesics as follows: given two points A and B, flatten the cone into a plane in such a way that the two points are on the same side of the flattened cone. Use a ruler to draw a straight line segment connecting the points on the flattened cone, and then restore the cone to its original shape.

(a) Find an equation in Cartesian coordinates that describes the cone of cone angle θ as a subset of $\mathbb{R}^3$.

(b) If the cone point is removed from the cone, the remaining surface is a smooth manifold. Show that the Gaussian curvature of this smooth manifold is 0 at every point. Explain how this is consistent with the Theorema Egregium.

(c) Can you find two points on the cone such that shortest path joining them passes through the cone point? Explain.

(d) Construct a geodesic triangle on the cone such that the cone point is outside the triangle. Use a protractor to measure the angles of the triangle and compare the angle sum with $180°$. Explain the results in terms of curvature.

(e) Construct a geodesic triangle on the cone such that the cone point is inside the triangle. Use a protractor to measure the angles of the triangle and compare the angle sum with $180°$. Find a relationship between the angle sum of the triangle and the cone angle θ. Does the angle sum depend in any way on the radius r of the circle used in the construction of the cone?

(f) Prove that the formula you proposed in part (e) is valid for any cone angle θ and any geodesic triangle on the cone, provided the cone point is on the inside of the triangle.

14.8. The *mean curvature* of a smooth surface at a point is the mean, or average, of the principal curvatures. The mean curvature is denoted by H, so the formula is $H = (1/2)(k_1 + k_2)$.

(a) Find the mean curvature of each of the following surfaces.

    (i) A plane.

    (ii) A sphere of radius r.

    (iii) A right circular cylinder of radius r.

    (iv) A saddle (at the saddle point).

    (v) The torus at points A and C (Figure 14.11).

(b) Is mean curvature an intrinsic property of the surface? Explain.

14.9. Let S be the surface that consists of all the points of $\mathbb{R}^2$ that lie inside the circle of radius 2 centered at the origin. Remove the origin to form a surface S'. The surface S' is called a *punctured disk*.

(a) Show that for every positive number ϵ there is a path in S' from $(-1, 0)$ to $(1, 0)$ whose length is less than $2 + \epsilon$.

(b) Show that there is no path of length 2 in S' that connects $(-1, 0)$ to $(1, 0)$.

(c) Conclude from (a) and (b) that there is no shortest path in the punctured disk joining the two points.

(d) Show that there are points in S' that are arbitrarily close together but for which there is no shortest path in S' joining them.

(e) Explain how the result in (d) is compatible with the assertion in the chapter that nearby points on a smooth surface are the endpoints of a shortest geodesic.

14.10. Let S be the surface of a rectangular box that has dimensions 2×2 at each end and 2×4 on the top, bottom, and sides.

(a) Find the shortest path on this surface from the center of one end to the center of the other end.

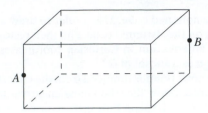

FIGURE 14.35: The rectangular box

(b) Find the shortest path on this surface from one corner to the opposite corner.

(c) Find the shortest path on this surface from the point A halfway up one edge to the point B halfway up the opposite edge (see Figure 14.35).

14.11. Is the ordinary torus in $\mathbb{R}^3$ homogeneous? Explain.

14.12. The construction of the soccer ball models began with the hexagonal tiling of the Euclidean plane (Figure 14.21). This tiling has three hexagons at each vertex. Suppose that instead of replacing just one hexagon with a pentagon we replace all three hexagons with pentagons. The result is a positively curved polygonal surface that has three regular pentagons at each vertex.

(a) Identify this surface with one of the figures in the chapter.

(b) How many pentagons are in the surface?

(c) What is the curvature at each vertex? Is the curvature greater or smaller than that of the ordinary soccer ball?

(d) Verify that this surface is dual to the icosahedron in the sense that there is one vertex in the surface for each face of the icosahedron, there is one face of the surface for each vertex of the icosahedron, and there is one edge in the surface for each edge in the icosahedron.

CONSTRUCTION PROJECTS

CP14.1. A paper model of the pseudosphere. An *annulus* (plural *annuli*) is the region in the Euclidean plane bounded by two concentric circles. In this project you will use paper annuli to construct a model that approximates the pseudosphere.

On page 384 there is a diagram of an annulus. Use a copy machine to enlarge this annulus to full page size and make five copies of the enlarged annulus. Cut out the copies of the annulus. Start with one annulus that will form the rim of the pseudosphere. Take a second annulus, cut across it, and then tape the outer boundary of the second annulus to the inner boundary of the first. The outer boundary is longer than the inner boundary, so there will be a small piece of the second annulus left over; cut it off. Now take the third annulus, cut across it, and tape the outer boundary of the third annulus to the inner boundary of the second. This time there will be even more left over; again cut it off and discard it. Continue this process until you have at least eight rings. Some of the discarded pieces of early rings are long enough to use as later rings.

Draw a geodesic triangle on the faux pseudosphere you constructed and measure its defect. To construct the geodesic joining two points on the paper pseudosphere, pull the surface tight between the the two points and place a ruler on top to hold that part of the surface straight while you draw the line. You can make a roughly equilateral triangle by selecting two points near the rim but less than half way around the rim and a third point up near the point of the pseudosphere. Observe how each geodesic is straight when you draw it, but the three geodesics assume a curved shape relative to each other that gives the triangle its defect.

CP14.2. A paper model of the icosahedron. In the template section at the end of this chapter you will find a page of equilateral triangles. Use a copy machine to enlarge the diagram to full page size and make copies of it so that you have a supply of triangles for this exercise and the next one. If you can, copy the triangles onto card stock so that you have stiffer material to work with. (But paper works too.)

Construct an icosahedron by taping together paper triangles. Part of the challenge here is to find an efficient way to do this. You should not cut all the triangles apart and then tape them back together again. If you begin by taping triangles together in the flat pattern shown in Figure 14.36, you can assemble the whole icosahedron at once. Put a crease along each edge separating triangles if you want to produce a surface with flat faces.

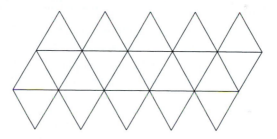

FIGURE 14.36: A pattern for the icosahedron

Observe how the pattern of five triangles around each vertex gives the surface a positive curvature and how this forces the surface to close up as you add triangles.

CP14.3. A triangular model of hyperbolic plane. Make a new supply of equilateral triangles as in the previous project. Construct a polygonal model of the hyperbolic plane in which each vertex is surrounded by seven equilateral triangles. Once again part of the challenge is to find an efficient way to do this. Make your model at least 10 triangles in diameter.

Draw a geodesic triangle on the model and measure its defect. Construct the triangle by selecting three points as far away from each other as possible. Pull on the surface so that it is as straight as possible between two of the points and place a ruler on the surface to hold it straight while you draw the side of the triangle. Use an ordinary compass to measure the angles. For this construction you will need a large ruler and a small compass. You will get the most accurate measurement of the angle if you measure it on a portion of the surface that is flat. So you should locate the vertices of your hyperbolic triangle in the interior of one of the faces of your model and try to use a compass that is small enough to measure the portion of the angle inside the flat face.

CP14.4. The spherical soccer ball. Make a supply of paper pentagons and hexagons using the templates at the end of the chapter. Construct a spherical soccer ball. The construction is a little easier if you use stiff paper, but it also works with ordinary paper. Observe how the positive curvature forces the surface to close up. How many pentagons and how many hexagons does it take to make a complete surface?

CP14.5. A hyperbolic soccer ball. Make a supply of paper hexagons and heptagons using the templates at the end of the chapter. Construct a soccer ball model of the hyperbolic plane. You should make your model large enough so that it is at least 10 polygons across. This will take quite a few polygons because the surface becomes quite floppy as it grows.

(a) Draw some lines on your model. The way to do this is to mark two points, pull the surface tight between the two points, and then place a ruler or straightedge on the surface to keep it straight between the points while you draw the line. You will need a large ruler for this.

(b) Draw parallel lines that admit a common perpendicular. To do this, draw one line ℓ as in part (a) using two points A and B that are close to the edge of your model. Then pick an external point P, about half way between ℓ and the opposite edge of your model. Drop a perpendicular from P to ℓ and call the foot F. Now explore what happens when you pull the surface tight between various points. Find two point C and D such that $\overleftrightarrow{CD}$ passes through P and is perpendicular to $\overleftrightarrow{PF}$. Observe that $\overleftrightarrow{CD}$ diverges from ℓ.

(c) Find lines through P that are not perpendicular to $\overleftrightarrow{PF}$ but still eventually start to diverge from ℓ. Use the one closest to ℓ to estimate the angle of parallelism for P and ℓ. Is your estimate an upper bound or a lower bound for the angle of parallelism?

(d) Construct some triangles on your model and calculate the defect of each. You can use an ordinary protractor to measure the angles. Make sure the vertices are situated so that enough of the angle is in a flat polygon so that you can measure the angle inside that polygon. How does the defect of the triangle relate to the size of the triangle?

CP14.6. Parallel transport. One way to measure curvature intrinsically is through a process called *parallel transport*. Here is a description of the process: Draw a triangle $\triangle ABC$ and pick a line ℓ through vertex A. Construct the line ℓ_1 such that B lies on ℓ_1 and ℓ and ℓ_1 make congruent corresponding angles with $\overleftrightarrow{AB}$. Now construct a line ℓ_2 such that C lies on ℓ_2 and ℓ_1 and ℓ_2 make congruent corresponding angles with $\overleftrightarrow{BC}$. Finally, construct a line ℓ_3 such that A lies on ℓ_3 and ℓ_2 and ℓ_3 make congruent corresponding angles with $\overleftrightarrow{AC}$. Now measure the angle between ℓ_3 and ℓ.

(a) Take a flat piece of paper and construct a Euclidean triangle on it. Take a line ℓ through vertex A and do parallel transport around $\triangle ABC$ to produce line ℓ_3. What is the angle between ℓ and ℓ_3?

(b) Do the same experiment on a sphere. You will need a large round sphere like a basketball for this experiment. How do ℓ and ℓ_3 compare in this case?

(c) Do the same experiment on the hyperbolic soccer ball that you constructed in the previous exercise. You should use a relatively large triangle for this experiment. How do ℓ and ℓ_3 compare in this case?

(d) Compare the results of (b) and (c). Did the line turn in the same direction when transported around the spherical and hyperbolic triangles?

CP14.7. Area on the hyperbolic soccer ball. Construct another hyperbolic soccer ball, this time using hexagons and heptagons that have the rectangular grid marked on them. The purpose of the squares it to allow you to estimate areas.

(a) Construct three triangles of varying sizes on the hyperbolic soccer ball. Keep this as simple as possible by using triangles that share a common angle at a vertex.

Use the squares in the grid to estimate the area of the triangles. It is not necessary to count all the small squares that lie inside the triangle. Instead you can count the number of hexagons and heptagons that are completely inside the triangle and then use formula in Exercise 14.1 to estimate the area of each of them. Both polygons are four units on a side, so the area of one of the hexagons is about 41.57 and the area of one of the heptagons is about 58.14. You can estimate the area of the remaining part of the triangle by counting the number of remaining squares. You will have to use some good judgment to estimate the fractions that lie inside the triangle, so your value for the area will be only an approximation. But you can get a fairly good estimate by this method.

For each of the three triangles, calculate the defect and the area. Make a plot of area versus defect. Do the three points you plotted lie on a straight line? Does that line pass through the origin?

(b) Construct two circles with the same center and one having twice the radius of the other. Estimate the circumference and area of each of the circles. By what multiple does the circumference grow when the radius is doubled? By what multiple does the area grow when the circumference is doubled?

In order to construct these circles you will need a large compass. If you don't have one, improvise by using a pencil on a string. Draw the circle by keeping the center fixed and pulling the surface straight in radial directions. Estimate the circumference of the circles by dividing the circle up into small pieces that are approximately straight. Estimate the area of the circles as you did the area of the triangles in the previous exercise.

14.7 TEMPLATES

On the next few pages you will find templates that you can use for the construction projects. Each diagram should be enlarged to full page size and then copied.

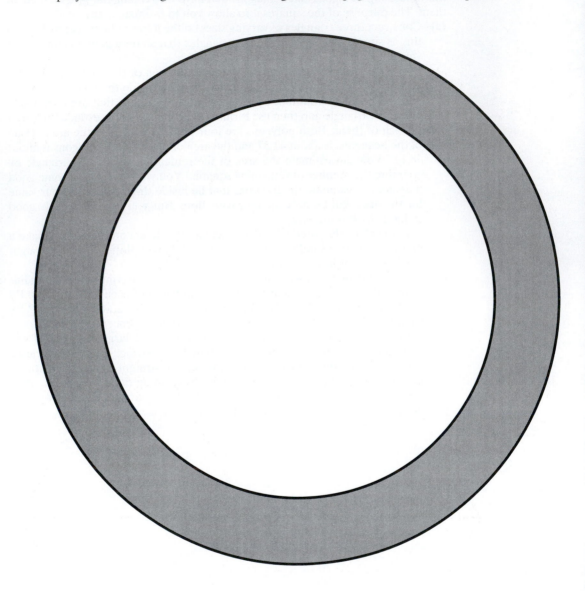

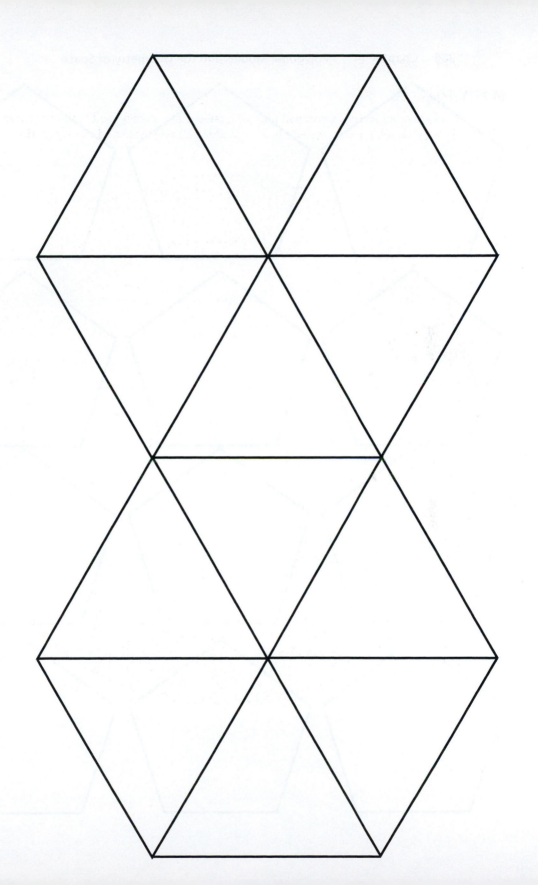

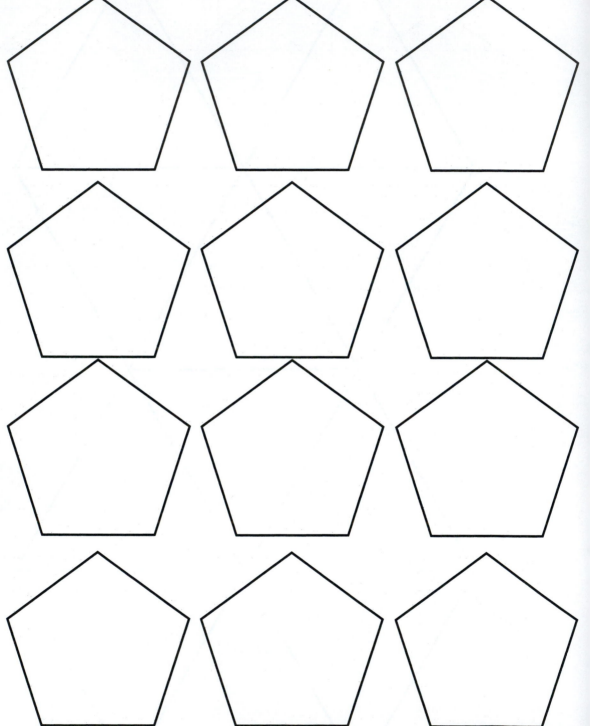

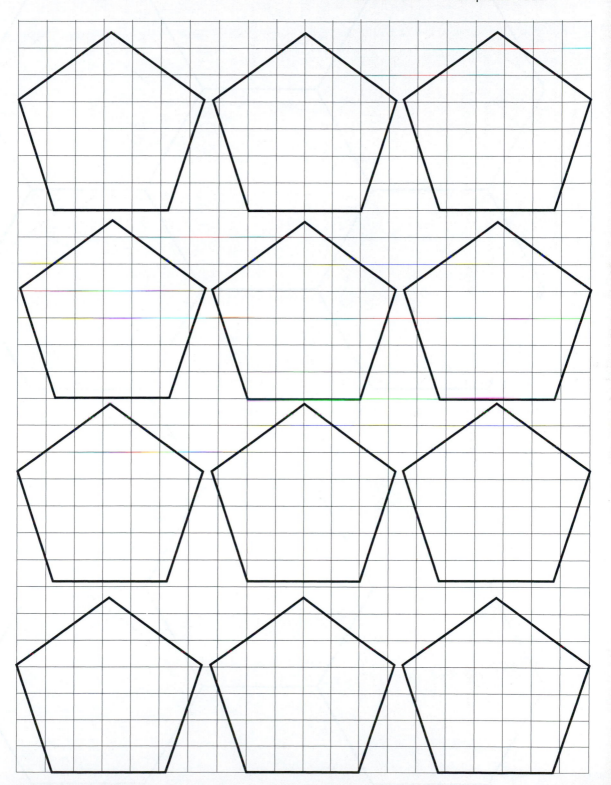

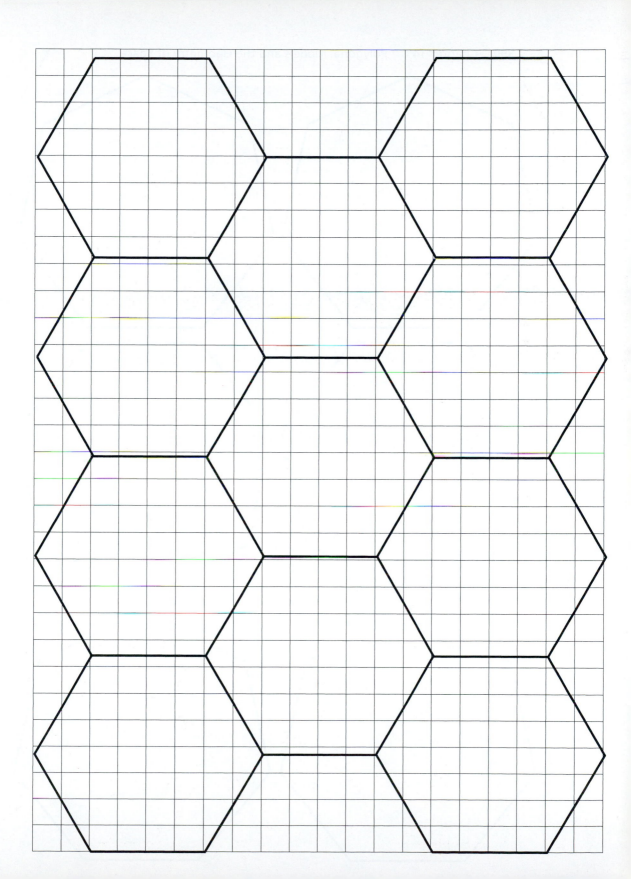

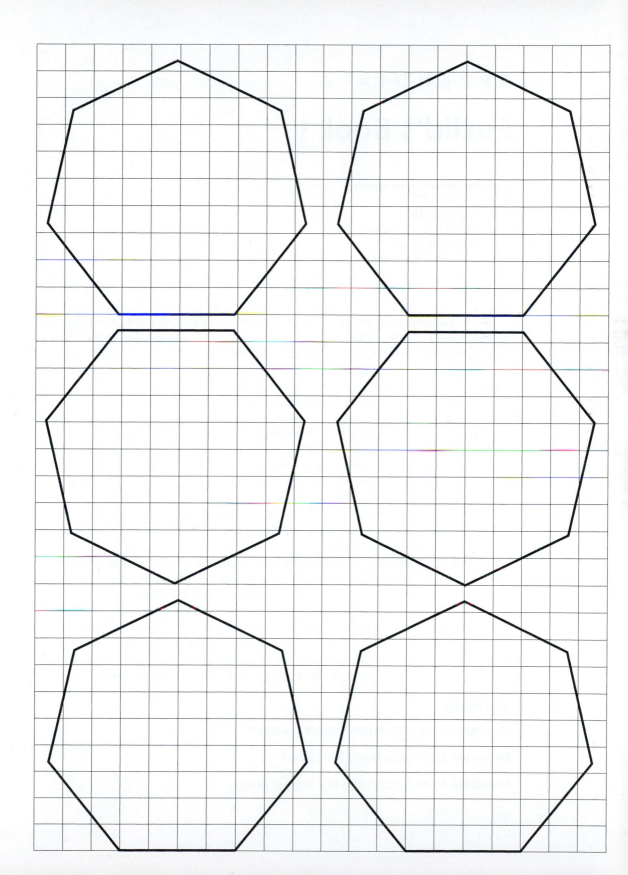

APPENDIX A

Euclid's Book I

A.1 DEFINITIONS
A.2 POSTULATES
A.3 COMMON NOTIONS
A.4 PROPOSITIONS

The statements of all the definitions and propositions from Book I of Euclid's *Elements* are reproduced below. The translation into English is due to Sir Thomas Heath and may also be found in [30], [19], or [18].

While this appendix contains enough of Euclid to allow one to read the remainder of the book, the reader is encouraged to gain a more extensive knowledge of the *Elements* by studying them more directly, if possible. The most easily accessible modern source is the three-volume Dover edition [30]. Those volumes contains a complete translation of the *Elements* into English by Sir Thomas Heath. Not only do they include Heath's translations of the *Elements* themselves, but they also contain Heath's extensive commentary on the *Elements*.

Two new editions of the *Elements* have recently appeared ([19] and [18]). These books reproduce Heath's translations in a more attractive layout. The proofs of the propositions are included in [19] but only the statements of the propositions are found in [18]. Neither includes the commentary on the proofs, so [30] is probably still preferable for serious study.

Another possibility is to find Euclid's *Elements* on the World Wide Web. Neither Euclid nor Heath holds a current copyright, so the material may be freely reproduced. A quick search of the Web will turn up numerous sources. One nice thing about some of the versions of Euclid on the Web is the fact that they include Euclid's diagrams in the form of Java applets that can be manipulated by the user.

In references to the *Elements* it is standard to use Roman numerals for the book numbers and Arabic numerals for the Proposition numbers within a book. Thus Proposition III.28 refers to Proposition 28 in Book III of the *Elements*.

A.1 DEFINITIONS

Definition 1. A *point* is that which has no part.

Definition 2. A *line* is breadthless length.

Definition 3. The extremities of a line are points.

Definition 4. A *straight line* is a line which lies evenly with the points on itself.

Definition 5. A *surface* is that which has length and breadth only.

Definition 6. The *extremities of a surface* are lines.

Definition 7. A *plane surface* is a surface which lies evenly with the straight lines on itself.

Definition 8. A *plane angle* is the inclination to one another of two lines in a plane which meet one another and do not lie in a straight line.

Definition 9. And when the lines containing the angle are straight, the angle is called *rectilinear*.

Definition 10. When a straight line standing on a straight line makes the adjacent angles equal to one another, each of the equal angles is *right*, and the straight line standing on the other is called a *perpendicular* to that on which it stands.

Definition 11. An *obtuse* angle is an angle greater than a right angle.

Definition 12. An *acute* angle is an angle less than a right angle.

Definition 13. A *boundary* is that which is an extremity of anything.

Definition 14. A *figure* is that which is contained by any boundary or boundaries.

Definition 15. A *circle* is a plane figure contained by one line such that all the straight lines falling upon it from one point among those lying within the figure equal one another.

Definition 16. And the point is called the *center* of the circle.

Definition 17. A *diameter* of the circle is any straight line drawn through the center and terminated in both directions by the circumference of the circle, and such a straight line also bisects the circle.

Definition 18. A *semicircle* is the figure contained by the diameter and the circumference cut off by it. And the *center of the semicircle* is the same as that of the circle.

Definition 19. *Rectilinear* figures are those which are contained by straight lines, *trilateral* figures being those contained by three, *quadrilateral* those contained by four, and *multilateral* those contained by more than four straight lines.

Definition 20. Of trilateral figures, an *equilateral* triangle is that which has its three sides equal, an *isosceles* triangle that which has two of its sides alone equal, and a *scalene* triangle that which has its three sides unequal.

Definition 21. Further, of trilateral figures, a *right-angled triangle* is that which has a right angle, an *obtuse-angled triangle* that which has an obtuse angle, and an *acute-angled triangle* that which has its three angles acute.

Definition 22. Of quadrilateral figures, a *square* is that which is both equilateral and right-angled; an *oblong* that which is right-angled but not equilateral; a *rhombus* that which is equilateral but not right-angled; and a *rhomboid* that which has its opposite sides and angles equal to one another but is neither equilateral nor right-angled. And let quadrilaterals other than these be called *trapezia*.

Definition 23. *Parallel* straight lines are straight lines which, being in the same plane and being produced indefinitely in both directions, do not meet one another in either direction.

A.2 POSTULATES

Let the following be postulated:

Postulate I. *To draw a straight line from any point to any point.*

Postulate II. *To produce a finite straight line continuously in a straight line.*

Postulate III. *To describe a circle with any center and radius.*

Postulate IV. *That all right angles equal one another.*

Postulate V. *That, if a straight line falling on two straight lines makes the interior angles on the same side less than two right angles, the two straight lines, if produced indefinitely, meet on that side on which are the angles less than the two right angles.*

A.3 COMMON NOTIONS

Common Notion I. *Things which are equal the same thing are also equal to one another.*

Common Notion II. *If equals be added to equals, the wholes are equal.*

Common Notion III. *If equals be subtracted from equals, the remainders are equal.*

Common Notion IV. *Things which coincide with one another are equal to one another.*

Common Notion V. *The whole is greater than the part.*

A.4 PROPOSITIONS

Proposition 1. *On a given finite straight line to construct an equilateral triangle.*

Proposition 2. *To place at a given point [as an extremity] a straight line equal to a given straight line.*

Proposition 3. *Given two unequal straight lines, to cut off from the greater a straight line equal to the less.*

Proposition 4. *If two triangles have two sides equal to two sides respectively, and have the angles contained by the equal straight lines equal, they will also have the base equal to the base, the triangle will be equal to the triangle, and the remaining angles will be equal to the remaining angles respectively, namely those which the equal sides subtend.*

Proposition 5. *In isosceles triangles the angles at the base are equal to one another, and, if the equal straight lines be produced further, the angles under the base will be equal to one another.*

Proposition 6. *If in a triangle two angles be equal to one another, the sides which subtend the equal angles will also be equal to one another.*

Proposition 7. *Given two straight lines constructed on a straight line [from its extremities] and meeting in a point, there cannot be constructed on the same straight line [from its extremities] and on the same side of it, two other straight lines meeting in another point and equal to the former two respectively, namely each to that which has the same extremity with it.*

Proposition 8. *If two triangles have the two sides equal to two sides respectively, and have also the base equal to the base, they will also have the angles equal which are contained by the equal straight lines.*

Proposition 9. *To bisect a given rectilinear angle.*

Proposition 10. *To bisect a given finite straight line.*

Proposition 11. *To draw a straight line at right angles to a given straight line from a given point on it.*

Proposition 12. *To a given infinite straight line, from a given point which is not on it, to draw a perpendicular straight line.*

Proposition 13. *If a straight line set up on a straight line makes angles, it will make either two right angles or angles equal to two right angles.*

Proposition 14. *If with any straight line, and at a point on it, two straight lines not lying on the same side make the adjacent angles equal to two right angles, the two straight lines will be in a straight line with one another.*

Proposition 15. *If two straight lines cut one another, they make the vertical angles equal to one another.*

Porism. From this it is manifest that, if two straight lines cut one another, they will make the angles at the point of section equal to four right angles.

Proposition 16. *In any triangle, if one of the sides be produced, the exterior angle is greater than either of the interior and opposite angles.*

Proposition 17. *In any triangle two angles taken together in any manner are less than two right angles.*

Proposition 18. *In any triangle the greater side subtends the greater angle.*

Proposition 19. *In any triangle the greater angle is subtended by the greater side.*

Proposition 20. *In any triangle two sides taken together in any manner are greater than the remaining one.*

Proposition 21. *If on one of the sides of a triangle, from its extremities, there be constructed two straight lines meeting within the triangle, the straight lines so constructed will be less than the remaining two sides of the triangle, but will contain a greater angle.*

Proposition 22. *Out of three straight lines, which are equal to three given straight lines, to construct a triangle: thus it is necessary that two of the straight lines taken together in any manner should be greater than the remaining one.*

Proposition 23. *On a given straight line and at a point on it to construct a rectilineal angle equal to a given rectilineal angle.*

Proposition 24. *If two triangles have the two sides equal to two sides respectively, but have the one of the angles contained by the equal straight lines greater than the other, they will also have the base greater than the base.*

Proposition 25. *If two triangles have the two sides equal to two sides respectively, but have the base greater than the base, they will also have the one of the angles contained by the equal straight lines greater than the other.*

Proposition 26. *If two triangles have two angles equal to two angles respectively, and one side equal to one side, namely, either the side adjoining the equal angles, or that subtending one of the equal angles, they will also have the remaining sides equal to the remaining sides and the remaining angle to the remaining angle.*

Proposition 27. *If a straight line falling on two straight lines makes the alternate angles equal to one another, the straight lines will be parallel to one another.*

Proposition 28. *If a straight line falling on two straight lines makes the exterior angle equal to the interior and opposite angle on the same side, or the interior angles on the same side equal to two right angles, the straight lines will be parallel to one another.*

Proposition 29. *A straight line falling on parallel straight lines makes the alternate angles equal to one another, the exterior angle equal to the interior and opposite angle, and the interior angles on the same side equal to two right angles.*

Proposition 30. *Straight lines parallel to the same straight line are also parallel to one another.*

Proposition 31. *Through a given point to draw a straight line parallel to a given straight line.*

Proposition 32. *In any triangle, if one of the sides be produced, the exterior angle is equal to the two interior and opposite angles, and the three interior angles of the triangle are equal to two right angles.*

Proposition 33. *The straight lines joining equal and parallel straight lines [at the extremities] which are in the same directions [respectively] are themselves also equal and parallel.*

Proposition 34. *In parallelogrammic areas the opposite sides and angles are equal to one another, and the diameter bisects the areas.*

Proposition 35. *Parallelograms which are on the same base and in the same parallels are equal to one another.*

Proposition 36. *Parallelograms which are on equal bases and in the same parallels are equal to one another.*

Proposition 37. *Triangles which are on the same base and in the same parallels are equal to one another.*

Proposition 38. *Triangles which are on equal bases and in the same parallels are equal to one another.*

Proposition 39. *Equal triangles which are on the same base and on the same side are also in the same parallels.*

Proposition 40. *Equal triangles which are on equal bases and on the same side are also in the same parallels.*

Proposition 41. *If a parallelogram has the same base with a triangle and be in the same parallels, the parallelogram is double of the triangle.*

Proposition 42. *To construct, in a given rectilineal angle, a parallelogram equal to a given triangle.*

Proposition 43. *In any parallelogram the complements of the parallelograms about the diameter are equal to one another.*

Proposition 44. *To a given straight line to apply, in a given rectilinear angle, a parallelogram equal to a given triangle.*

Proposition 45. *To construct, in a given rectilineal angle, a parallelogram equal to a given rectilineal figure.*

Proposition 46. *On a given straight line to describe a square.*

Proposition 47. *In right-angled triangles the square on the side subtending the right angle is equal to the squares on the sides containing the right angle.*

Proposition 48. *If in a triangle the square on one of the sides be equal to the squares on the remaining two sides of the triangle, the angle contained by the remaining two sides of the triangle is right.*

APPENDIX B

Other Systems of Axioms for Geometry

B.1 HILBERT'S AXIOMS
B.2 BIRKHOFF'S AXIOMS
B.3 MACLANE'S AXIOMS
B.4 SMSG AXIOMS
B.5 UCSMP AXIOMS

This appendix contains a description of various alternative axiom systems for geometry. It is interesting and instructive to compare and contrast the various approaches to the foundations. The systems are presented in historical order.

B.1 HILBERT'S AXIOMS

The following axioms were developed by David Hilbert.[1] Hilbert's axioms are like Euclid's in that they are completely synthetic and do not rely on any previous knowledge about the real number system. At the same time they satisfy the modern standards of rigor by completely spelling out all the assumptions that are needed in order to do Euclidean geometry.

Undefined terms

The undefined terms are *point, line, plane, lie on, between,* and *congruent.* The collections of points, lines and planes form three sets. The terms *lie on, between,* and *congruent* express relations between the elements of these sets.

Axioms of incidence

Axiom I-1. *For every two points A, B there exists a line ℓ that contains each of the points A, B.*

Axiom I-2. *For every two points A, B there exists no more than one line that contains each of the points A, B.*

Axiom I-3. *There exist at least two points on a line. There exist at least three points that do not lie on a line.*

---

[1]Reprinted from [34] by permission of the Open Court Publishing Company.

Axiom I-4. *For any three points A, B, C that do not lie on the same line there exists a plane α that contains each of the points A, B, C. For every plane there exists a point which it contains.*

Axiom I-5. *For any three points A, B, C that do not lie on one and the same line there exists no more than one plane that contains each of the three points A, B, C.*

Axiom I-6. *If two points A, B of a line ℓ lie in a plane α then every point of ℓ lies in the plane α.*

Axiom I-7. *If two planes α, β have a point A in common then they have at least one more point B in common.*

Axiom I-8. *There exist at least four points which do not lie in a plane.*

Axioms of order

Axiom II-1. *If a point B lies between a point A and a point C then the points A, B, and C are three distinct points of a line, and B then also lies between C and A.*

Axiom II-2. *For two points A and C, there always exists at least one point B on the line $\overleftrightarrow{AC}$ such that C lies between A and B.*

Axiom II-3. *Of any three points on a line there exists no more than one that lies between the other two.*

Axiom II-4. *Let A, B, C be three points that do not lie on a line and let ℓ be a line in the plane ABC which does not meet any of the points A, B, C. If the line ℓ passes through a point of the segment AB, it also passes through a point of the segment AC, or through a point of the segment BC. Expressed intuitively, if a line enters the interior of a triangle, it also leaves it.*

Axioms of congruence

Axiom III-1. *If A, B are two points on a line ℓ, and A' is a point on the same or on another line ℓ' then it is always possible to find a point B' on a given side of the line ℓ' through A' such that the segment $\overline{AB}$ is congruent or equal to the segment $\overline{A'B'}$. In symbols, $\overline{AB} \cong \overline{A'B'}$.*

Axiom III-2. *If a segment $\overline{A'B'}$ and a segment $\overline{A''B''}$, are congruent to the same segment $\overline{AB}$, then the segment $\overline{A'B'}$ is also congruent to the segment $\overline{A''B''}$, or briefly, if two segments are congruent to a third one, they are congruent to each other.*

Axiom III-3. *On the line ℓ let $\overline{AB}$ and $\overline{BC}$ be two segments which except for B have no point in common. Furthermore, on the same or on another line ℓ' let $\overline{A'B'}$ and $\overline{B'C'}$ be two segments which except for B' also have no point in common. In that case, if $\overline{AB} \cong \overline{A'B'}$ and $\overline{BC} \cong \overline{B'C'}$, then $\overline{AC} \cong \overline{A'C'}$.*

Axiom III-4. *Let $\angle BAC$ be an angle in a plane α and ℓ' a line in a plane α' and let a definite side of ℓ' in α' be given. Let $\overrightarrow{A'B'}$ be a ray on the line ℓ' that emanates from a point A'. Then there exists in the plane α' one and only one ray $\overrightarrow{A'C'}$ such that $\angle B'A'C' \cong \angle BAC$ and all interior points of $\angle B'A'C'$ lie on the given side of ℓ'.*

Axiom III-5. *If for two triangles $\triangle ABC$ and $\triangle A'B'C'$ the congruences $\overline{AB} \cong \overline{A'B'}$, $\overline{AC} \cong \overline{A'C'}$, and $\angle BAC \cong \angle B'A'C'$ hold, then the congruence $\angle ABC \cong \angle A'B'C'$ is also satisfied.*

Axiom of parallels

Axiom IV-1 (Euclid's Axiom). *Let ℓ be any line and A a point not on it. Then there is at most one line in the plane, determined by ℓ and A, that passes through A and does not intersect ℓ.*

Axioms of Continuity

Axiom V-1 (Axiom of Measure or Archimedes' Axiom). *If $\overline{AB}$ and $\overline{CD}$ are any segments then there exists a number n such that n segments $\overline{CD}$ constructed contiguously from A, along the ray from A through B, will pass beyond the point B.*

Axiom V-2 (Axiom of Line Completeness). *An extension of a set of points on a line with its order and congruence relations that would preserve the relations existing among the original elements as well as the fundamental properties of line order and congruence that follows from Axioms I–III and from V-1 is impossible.*

B.2   BIRKHOFF'S AXIOMS

The following axioms are due to George David Birkhoff. He published them in a paper in the *Annals of Mathematics* in 1932 [7]. Birkhoff was the first to build the real number system into the foundations of geometry. He gives a mathematically rigorous treatment of geometry based on these axioms in [7]. The treatment of the axioms presented in Chapter 5 owes much to that paper. Birkhoff and a colleague, Ralph Beatley, developed a high school textbook based on these axioms that is still in print today [8].

Undefined terms

Birkhoff's undefined terms are *point, line, distance*, and *angle*.

Postulates

Postulate of Line Measure. *The points $A, B, \dots$ of any line ℓ can be put into $(1, 1)$ correspondence with the real numbers x so that $|x_B - x_A| = d(A, B)$ for all points A, B.*

Point-Line Postulate. *One and only one straight line ℓ contains two given points $P, Q\ (P \neq Q)$.*

Postulate of Angle Measure. *The half-lines $\ell, m, \ldots$ through any point O can be put into $(1, 1)$ correspondence with the real numbers $a \pmod{2\pi}$ so that, if $A \neq O$ and $B \neq O$ are points on ℓ and m respectively, the difference $a_m - a_\ell \pmod{2\pi}$ is $\angle AOB$. Furthermore, if the point B on m varies continuously in a line r not containing the vertex O, the number a_m varies continuously also (Fig. B.1).*

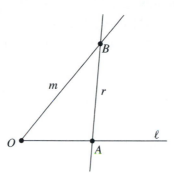

FIGURE B.1: Birkhoff's Postulate of Angle Measure

Postulate of Similarity. *If in two triangles, $\triangle ABC$, $\triangle A'B'C'$, and for some constant $k > 0$, $d(A', B') = kd(A, B)$, $d(A', C') = kd(A, C)$ and also $\angle B'A'C' = \pm \angle BAC$, then also $d(B', C') = kd(B, C)$, $\angle C'B'A' = \pm \angle CBA$, and $\angle A'C'B' = \pm \angle ACB$.*

B.3   MACLANE'S AXIOMS

In 1959 Saunders MacLane published a set of axioms for geometry in the *American Mathematical Monthly* [46]. MacLane used directed angles, which greatly simplifies the statement of the angle addition assumption, and he introduced a Continuity Axiom that replaces the Plane Separation Postulate. MacLane's Continuity Postulate contains both the Crossbar Theorem and its converse.

Undefined terms

MacLane's undefined terms are *point, distance, line,* and *angle measure.*

The axioms on distance

D1. *There are at least two points.*
D2. *If A and B are points, $d(AB)$ is a nonnegative number.*
D3. *For points A and B, $d(AB) = 0$ if and only if $A = B$.*
D4. *If A and B are points, $d(AB) = d(BA)$.*

The axioms on lines

L1. *A line is a set of points containing more than one point.*
L2. *Through two distinct points there is one and only one line.*

L3. *Three distinct points lie on a line if and only if one of them is between the other two.*

L4. *On each ray from a point O and to each positive real number b there is a point B with $d(OB) = b$.*

The axioms on angles

A1. *If r and s are rays from the same point, $\angle rs$ is a real number modulo 360.*

A2. *If r, s, and t are three rays from the same point, then $\angle rs + \angle st = \angle rt$.*

A3. *If r is a ray from O and c is a real number, then there is a ray s from O such that $\angle rs = c°$.*

A4. *If $A \neq O \neq B$, then $\angle AOB = \angle BOA \neq 0°$ if and only if $d(AB) = d(AO) + d(OB)$.*

The Similarity Axiom

If two ordered triangles $\triangle ABC$ and $\triangle A'B'C'$ have $\angle ABC = \epsilon \angle A'B'C'$, $d(AB) = kd(A'B')$, and $d(BC) = kd(B'C')$ (for $\epsilon = \pm 1$, k positive) they are similar.

The Continuity Axiom

Let $\angle AOB$ be proper (Fig. B.2). If D is between A and B, then $0 < \angle AOD < \angle AOB$. Conversely, if $0 < \angle AOC < \angle AOB$, then the ray OC meets the interval AB.

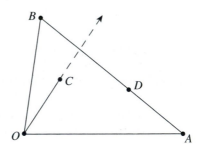

FIGURE B.2: MacLane's Continuity Axiom

B.4   SMSG AXIOMS

In the 1960s the School Mathematics Study Group (SMSG) introduced a new set of axioms.[2] These axioms follow Birkhoff's approach of building the real number system into the axioms of geometry. They differ from Birkhoff in that they do not attempt to minimize the number of axioms assumed. Instead they introduce a certain amount of redundancy in order to make the axioms simple and clear. This makes the axioms ideal for use in high school because they allow the course to

---

[2]Reprinted from [64] by permission of Yale University Press.

move quickly past many of the highly technical results that must be proved at the beginning of a rigorous course in geometry. Moise used essentially these axioms in his high school textbook [52]. He removed some of the redundancy and modified the axioms for use in his college-level textbook [51]. Some of the names used below come from [52]. The SMSG postulates continue to be influential in contemporary high school textbooks.

Undefined terms

The undefined terms are *point, line, plane, lie on, distance, angle measure, area*, and *volume*.

Postulates

Postulate 1 (The Incidence Postulate). *Given any two different points, there is exactly one line which contains both of them.*

Postulate 2 (The Distance Postulate). *To every pair of different points there corresponds a unique positive number.*

Postulate 3 (The Ruler Postulate). *The points of a line can be placed in correspondence with the real numbers in such a way that*

   1. *To every point of the line there corresponds exactly one real number,*
   2. *To every real number there corresponds exactly one point of the line, and*
   3. *The distance between two points is the absolute value of the difference of the corresponding numbers.*

Postulate 4 (The Ruler Placement Postulate). *Given two points P and Q of a line, the coordinate system can be chosen in such a way that the coordinate of P is zero and the coordinate of Q is positive.*

Postulate 5 (The Plane-Space Postulate). *(a) Every plane contains at least three non-collinear points. (b) Space contains at least four non-coplanar points.*

Postulate 6 (The Flat Plane Postulate). *If two points lie in a plane, then the line containing these points lies in the same plane.*

Postulate 7 (The Plane Postulate). *Any three points lie in at least one plane, and any three non-collinear points lie in exactly one plane. More briefly, any three points are coplanar, and any three non-collinear points determine a plane.*

Postulate 8 (Intersection of Planes Postulate). *If two different planes intersect, then their intersection is a line.*

Postulate 9 (The Plane Separation Postulate). *Given a line and a plane containing it, the points of the plane that do not lie on the line form two sets such that*

   1. *each of the sets is convex and*
   2. *if P is in one set and Q is in the other then the segment $\overline{PQ}$ intersects the line.*

Postulate 10 (The Space Separation Postulate). *The points of space that do not lie in a given plane form two sets such that*

    1. each of the sets is convex and

    2. if P is in one set and Q is in the other, then the segment $\overline{PQ}$ intersects the plane.

Postulate 11 (The Angle Measurement Postulate). *To every angle $\angle BAC$ there corresponds a real number between 0 and 180.*

Postulate 12 (The Angle Construction Postulate). *Let $\overrightarrow{AB}$ be a ray on the edge of the half-plane H. For every number r between 0 and 180 there is exactly one ray $\overrightarrow{AP}$ with P in H, such that $m\angle PAB = r$.*

Postulate 13 (The Angle Addition Postulate). *If D is a point in the interior of $\angle BAC$, then $m\angle BAC = m\angle BAD + m\angle DAC$.*

Postulate 14 (The Supplement Postulate). *If two angles form a linear pair, then they are supplementary.*

Postulate 15 (The S.A.S. Postulate). *Given a correspondence between two triangles (or between a triangle and itself). If two sides and the included angle of the first triangle are congruent to the corresponding parts of the second triangle, then the correspondence is a congruence.*

Postulate 16 (The Parallel Postulate). *Through a given external point there is at most one line parallel to a given line.*

Postulate 17 (The Area Postulate). *To every polygonal region there corresponds a unique positive number.*

Postulate 18 (The Congruence Postulate). *If two triangles are congruent, then the triangular regions have the same area.*

Postulate 19 (The Area Addition Postulate). *Suppose that the region R is the union of two regions R_1 and R_2. Suppose that R_1 and R_2 intersect at most in a finite number of segments and points. Then the area of R is the sum of the areas of R_1 and R_2.*

Postulate 20 (The Unit Postulate). *The area of a rectangle is the product of the length of its base and the length of its altitude.*

Postulate 21 (The Unit Postulate). *The volume of a rectangular parallelepiped is the product of the altitude and the area of the base.*

Postulate 22 (Cavalieri's Principle). *Given two solids and a plane. If for every plane which intersects the solids and is parallel to the given plane the two intersections have equal areas, then the two solids have the same volume.*

B.5 UCSMP AXIOMS

The last set of axioms included in this appendix is taken from the geometry textbook in the University of Chicago School Mathematics Project (UCSMP) series.[3] This is a textbook that is in wide use at the present time. The UCSMP axioms are in many ways similar to those of SMSG. The main difference is that the Reflection Postulate is used in place of the Side-Angle-Side Postulate. This alternative approach is discussed and explained in Chapter 12.

Undefined terms

The only terms that are specified as undefined are *point*, *line*, and *plane*.

Point-Line-Plane Postulate

    a. Unique Line Assumption. *Through any two points there is exactly one line.*

    b. Number Line Assumption. *Every line is a set of points that can be put in one-to-one correspondence with the real numbers, with any point corresponding to zero and any other point corresponding to 1.*

    c. Dimension Assumption. *(1) Given a line in a plane, there is at least one point in the plane that is not on the line. (2) Given a plane in space, there is at least one point in space that is not on the plane.*

    d. Flat Plane Assumption. *If two points lie in a plane, the line containing them lies in the plane.*

    e. Unique Plane Assumption. Through three noncollinear points, there is exactly one plane.

    f. Intersecting Planes Assumption. *If two different planes have a point in common, then their intersection is a line.*

Distance Postulate

    a. Uniqueness Property. *On a line, there is a unique distance between two points.*

    b. Distance Formula. *If two points on a line have coordinates x and y, the distance between them is $|x - y|$.*

    c. Additive Property. *If B is on $\overline{AC}$, then $AB + BC = AC$.*

Triangle Inequality Postulate

The sum of the lengths of two sides of a triangle is greater than the length of the third side.

Angle Measure Postulate

    a. Unique Measure Assumption. *Every angle has a unique measure from $0°$ to $180°$.*

---

[3]Reprinted from [71] by permission of Pearson Prentice Hall.

b. Unique Angle Assumption. *Given any ray $\overrightarrow{VA}$ and a real number r between 0 and 180, there is a unique angle $\angle BVA$ in each half-plane of $\overleftrightarrow{VA}$ such that $m\angle BVA = r$.*

c. Zero Angle Assumption. *If $\overrightarrow{VA}$ and $\overrightarrow{VB}$ are the same ray, then $m\angle BVA = 0$.*

d. Straight Angle Assumption. *If $\overrightarrow{VA}$ and $\overrightarrow{VB}$ are opposite rays, then $m\angle BVA = 180$.*

e. Angle Addition Property. *If $\overrightarrow{VC}$ (except for point V) is in the interior of $\angle AVB$, then $m\angle AVC + m\angle CVB = m\angle AVB$.*

Corresponding Angle Postulate

Suppose two coplanar lines are cut by a transversal.

a. *If two corresponding angles have the same measure, then the lines are parallel.*

b. *If the lines are parallel, then corresponding angles have the same measure.*

Reflection Postulate

Under a reflection:

a. *There is a 1-1 correspondence between points and their images.*

b. *Collinearity is preserved. If three points A, B, and C lie on the same line then their images A', B', and C' are collinear.*

c. *Betweenness is preserved. If B is between A and C, then the image B' is between the images A' and C'.*

d. *Distance is preserved. If $\overline{A'B'}$ is the image of $\overline{AB}$, then $A'B' = AB$.*

e. *Angle measure is preserved. If $\angle A'C'E$ is the image of $\angle ACE$, then $m\angle A'C'E' = m\angle ACE$.*

f. *Orientation is reversed. A polygon and its image, with vertices taken in corresponding order, have opposite orientations.*

Area Postulate

a. Uniqueness Property. *Given a unit region, every polygonal region has a unique area.*

b. Rectangle Formula. *The area of a rectangle with dimensions ℓ and w is ℓw.*

c. Congruence Property. *Congruent figures have the same area.*

d. Additive Property. *The area of the union of two nonoverlapping regions is the sum of the areas of the regions.*

Volume Postulate

a. Uniqueness Property. *Given a unit cube, every polygonal region has a unique volume.*

b. Box Volume Formula. *The volume V of a box with dimensions ℓ, w, and h is found by the formula $V = \ell w h$.*

c. Congruence Property. *Congruent figures have the same volume.*

d. Additive Property. *The volume of the union of two nonoverlapping solids is the sum of the volumes of the solids.*

e. Cavalieri's Principle. *Let I and II be two solids included between parallel planes. If every plane P parallel to the given planes intersects I and II in sections with the same area, then Volume(I) = Volume(II).*

APPENDIX C

The Postulates Used in this Book

C.1 THE UNDEFINED TERMS

There are six undefined terms: *point, line, distance, half-plane, angle measure*, and *area*.

C.2 THE POSTULATES OF NEUTRAL GEOMETRY

The Existence Postulate. *The collection of all points forms a nonempty set. There is more than one point in that set.*

The Incidence Postulate. *Every line is a set of points. For every pair of distinct points A and B there is exactly one line ℓ such that $A \in \ell$ and $B \in \ell$.*

The Ruler Postulate. *For every pair of points P and Q there exists a real number PQ, called the distance from P to Q. For each line ℓ there is a one-to-one correspondence from ℓ to $\mathbb{R}$ such that if P and Q are points on the line that correspond to the real numbers x and y, respectively, then $PQ = |x - y|$.*

The Plane Separation Postulate. *For every line ℓ, the points that do not lie on ℓ form two disjoint, nonempty sets H_1 and H_2, called half-planes bounded by ℓ, such that the following conditions are satisfied.*

1. *Each of H_1 and H_2 is convex.*
2. *If $P \in H_1$ and $Q \in H_2$, then $\overline{PQ}$ intersects ℓ.*

The Protractor Postulate. *For every angle $\angle BAC$ there is a real number $\mu(\angle BAC)$, called the measure of $\angle BAC$, such that the following conditions are satisfied.*

   1. $0° \leqslant \mu(\angle BAC) < 180°$ *for every angle $\angle BAC$.*

   2. $\mu(\angle BAC) = 0°$ *if and only if $\overrightarrow{AB} = \overrightarrow{AC}$.*

   3. (Angle Construction Postulate) *For each real number r, $0 < r < 180$, and for each half-plane H bounded by $\overleftrightarrow{AB}$ there exists a unique ray $\overrightarrow{AE}$ such that E is in H and $\mu(\angle BAE) = r°$.*

   4. (Angle Addition Postulate) *If the ray $\overrightarrow{AD}$ is between rays $\overrightarrow{AB}$ and $\overrightarrow{AC}$, then*

$$\mu(\angle BAD) + \mu(\angle DAC) = \mu(\angle BAC).$$

The Side-Angle-Side Postulate. *If $\triangle ABC$ and $\triangle DEF$ are two triangles such that $\overline{AB} \cong \overline{DE}, \angle ABC \cong \angle DEF$, and $\overline{BC} \cong \overline{EF}$, then $\triangle ABC \cong \triangle DEF$.*

C.3   THE PARALLEL POSTULATES

The Euclidean Parallel Postulate. *For every line ℓ and for every external point P, there is exactly one line m such that P lies on m and $m \parallel \ell$.*

The Elliptic Parallel Postulate. *For every line ℓ and for every external point P, there is no line m such that P lies on m and $m \parallel \ell$.*

The Hyperbolic Parallel Postulate. *For every line ℓ and for every external point P, there are at least two lines m and n such that P lies on both m and n and both m and n are parallel to ℓ.*

C.4   THE AREA POSTULATES

The Neutral Area Postulate. *Associated with each polygonal region R there is a nonnegative number $\alpha(R)$, called the area of R, such that the following conditions are satisfied.*

   1. (Congruence) *If two triangles are congruent, then their associated triangular regions have equal areas.*

   2. (Additivity) *If R is the union of two nonoverlapping polygonal regions R_1 and R_2, then $\alpha(R) = \alpha(R_1) + \alpha(R_2)$.*

The Euclidean Area Postulate. *If R is a rectangular region, then $\alpha(R) = length(R) \times width(R)$.*

C.5   THE REFLECTION POSTULATE

The Reflection Postulate. *For every line ℓ there exists a transformation $\rho_\ell : \mathbb{P} \to \mathbb{P}$, called the reflection in ℓ, that satisfies the following conditions.*

1. *If P lies on ℓ, then P is a fixed point for ρ_ℓ.*
2. *If P lies in one of the half-planes determined by ℓ then $\rho_\ell(P)$ lies in the opposite half-plane.*
3. *ρ_ℓ preserves collinearity.*
4. *ρ_ℓ preserves distance.*
5. *ρ_ℓ preserves angle measure.*

C.6   LOGICAL RELATIONSHIPS

In the presence of the other neutral axioms, SAS and the Reflection Postulate are equivalent (Theorem 12.6.5). Both are independent of the other axioms of neutral geometry (Example 5.8.2).

The Elliptic Parallel Postulate is inconsistent with the axioms of neutral geometry (Corollary 6.5.7). Both the Euclidean Parallel Postulate and the Hyperbolic Parallel Postulate are consistent with the axioms of neutral geometry and are independent of those axioms (Chapter 13).

It is not strictly necessary to assume everything stated in the Ruler Postulate. If the existence of a metric is assumed and it is assumed that for each ray $\overrightarrow{AB}$ and each positive real number r there is a point C on $\overrightarrow{AB}$ such that $AC = r$, then the existence of coordinate functions can be proved—see Saunders MacLane [46]. This is true even using the weak definition of metric (not including the triangle inequality) stated in the text. A distance function that satisfies the conditions listed in Chapter 5 should be called a semimetric.

The Protractor Postulate is almost entirely unnecessary. If the existence of a right angle is assumed, then the angle measurement function can be constructed using only the other neutral axioms. The existence of a right angle can be proved if the Reflection Postulate is assumed instead of SAS, so in fact the Reflection Postulate can replace both SAS and the Protractor Postulate. The construction uses standard limiting techniques of real analysis. It is worked out in [2], for example. In a course such as this it is better to assume the existence of the angle measurement function and not to get involved in the details of constructing one.

APPENDIX D

The Van Hiele Model of the Development of Geometric Thought

The following description of the "van Hiele Model of the development of geometric thought" is quoted from [17]. Sometimes the levels are numbered 1 through 5 rather than 0 through 4.

Level 0. (Basic Level) Visualization.
Students recognize figures as total entities (triangles, squares), but do not recognize properties of these figures (right angles in a square).

Level 1. Analysis.
Students analyze component parts of the figures (opposite angles of parallelograms are congruent), but interrelationships between figures and properties cannot be explained.

Level 2. Informal Deduction.
Students can establish interrelationships of properties within figures (in a quadrilateral, opposite sides being parallel necessitates opposite angles being congruent) and among figures (a square is a rectangle because it has all the properties of a rectangle). Informal proofs can be followed but students do not see how the logical order could be altered nor do they see how to construct a proof starting from different or unfamiliar premises.

Level 3. Deduction.
At this level the significance of deduction as a way of establishing geometric theory within an axiom system is understood. The interrelationship and role of undefined terms, axioms, definitions, theorems and formal proof is seen. The possibility of developing a proof in more than one way is seen.

Level 4. Rigor.
Students at this level can compare different axiom systems (non-Euclidean geometry can be studied). Geometry is seen in the abstract with a high degree of rigor, even without concrete examples.

APPENDIX E

Hints for Selected Exercises

HINTS FOR EXERCISES IN CHAPTER 1

1.9. You will probably want to use Proposition 26 (ASA) and Propositions 27–29 regarding parallel lines.

HINTS FOR EXERCISES IN CHAPTER 2

2.9. A postulate can be satisfied vacuously.

2.10. Fano's geometry satisfies this axiom. Prove that there cannot be another example with fewer points and lines.

HINTS FOR EXERCISES IN CHAPTER 4

4.5. $\pi = 3.14159265358979\ldots$.

4.14. Suggestion. Let $A = \{r \in \mathbb{Q} \mid r^2 < 2\}$. Show that A is nonempty and bounded above, but that A does not have a least upper bound (in $\mathbb{Q}$).

HINTS FOR EXERCISES IN CHAPTER 5

5.3 You must verify that both functions have the properties listed in the definition of coordinate function. Specifically, you must prove that each f is one-to-one and onto. You must also verify that the first kind of f satisfies the equation

$$\sqrt{(x_2 - x_1)^2 + (y_2 - y_1)^2} = \left| x_2\sqrt{1 + m^2} - x_1\sqrt{1 + m^2} \right|$$

for every pair of points (x_1, y_1) and (x_2, y_2) on ℓ. There is a similar equation for the second kind of f.

5.6. Start with an arbitrary coordinate function given by the Ruler Postulate. First adjust it so that P corresponds to zero and then adjust it again, if necessary, to make Q correspond to a positive number.

5.8. You must use the given condition to prove that f is one-to-one and onto.

5.15. First use Theorem 5.3.7 to prove that two edges of a triangle intersect in exactly one point.

5.16. Remember the advice back on page 44 about how to prove that two sets are equal.

5.31. Suggestion: Given $\angle BAC$, find a point C' on $\overrightarrow{AC}$ such that $AB = AC'$. Let M be the midpoint of $\overline{BC'}$. Prove that $\overrightarrow{AM}$ is the angle bisector.

5.35. Begin by giving a careful statement of the theorem in if...then form. The hypotheses are hidden in the definition of vertical angles. Apply Exercise 5.34 and be sure to understand the difference between the two exercises.

HINTS FOR EXERCISES IN CHAPTER 6

6.2. Proceed as in the second proof of the Isosceles Triangle Theorem but apply ASA instead of SAS.

6.4. First apply Axiom 5.6.2, Part 3, to construct one of the necessary angles and then apply Axiom 5.4.1. Don't forget to prove uniqueness.

6.6. Choose point B' on $\overrightarrow{CB}$ such that $\overline{CB'} \cong \overline{EF}$. Prove that $\triangle AB'C \cong \triangle DEF$. The RAA hypothesis $B' \neq B$ leads to a contradiction.

6.7. Find a point D' on $\overrightarrow{AC}$ such that $A * C * D'$ and $CD' = FD$. Prove that $\triangle D'BC \cong \triangle DEF$. Use the Isosceles Triangle Theorem to show that $\angle CD'B \cong \angle CAB$ and then apply AAS.

6.8 Find a point G, on the opposite side of $\overleftrightarrow{AB}$ from C, such that $\triangle ABG \cong \triangle DEF$. Let H be the point at which $\overline{CG}$ crosses $\overleftrightarrow{AB}$. (Be sure to prove that points G and H exist.) Use two applications of the Isosceles Triangle Theorem to prove that $\angle AGB \cong \angle ACB$ and then apply SAS. (Be careful, there are five different possibilities for the location of H relative to A and B. Three of them are shown in Fig. E.1.)

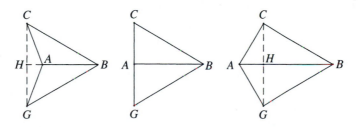

FIGURE E.1: Three of the possible locations for H in the proof of SSS

6.9 Locate a point B' on $\overrightarrow{AB}$ such that $AB' = DE$. Prove $\triangle AB'C \cong \triangle DEF$. Either $B' = B$ or $\triangle BB'C$ is isosceles.

6.10. This is a case in which a theorem can be used to prove its own converse. Start with the hypothesis $\mu(\angle ACB) > \mu(\angle BAC)$. Then apply trichotomy to the lengths AB and BC. Use the Isosceles Triangle Theorem to rule out the possibility that $AB = BC$. Use the part of the theorem that has already been proved to rule out the possibility $AB < BC$.

6.11. Find a point D such that $A * B * D$ and $\overline{BD} \cong \overline{BC}$. Prove that $\mu(\angle ACD) > \mu(\angle BCD) = \mu(\angle BDC)$ and apply Theorem 6.4.1 to conclude that $AD > AC$.

6.20. The interior angle at A and the exterior angle at A are supplements.

6.25. A diagonal splits the quadrilateral into two triangles. Be sure you understand where the fact that the quadrilateral is convex is used in the proof.

6.26. Use the definition of angle interior to show that each vertex is in the interior of the opposite angle.

6.27. Use Pasch's Axiom to prove the contrapositive of (a).

6.28. Use the previous exercise.

6.30. Apply Theorem 5.7.10 to prove that the point at which the diagonals intersect is in the interior of each of the angles. Then prove that the opposite vertex is in the interior of the angle.

6.32. (c) Apply Theorem 6.4.7.

6.36. Construct one parallel in the usual way. Use Euclid's Fifth Postulate in an RAA argument to prove that no other line could be parallel.

6.37. For EPP $\Rightarrow$ Statement: If not, then both t and ℓ are parallel to ℓ'. For Statement $\Rightarrow$ EPP: Construct one parallel in the usual way and use the statement to prove uniqueness.

6.38. For EPP $\Rightarrow$ Statement: Use the fact that EPP implies the Converse of the Alternate Interior Angles Theorem.

6.39. For EPP $\Rightarrow$ Statement: Use the first part of this theorem to prove that m intersects ℓ and that n intersects k.

6.40. For EPP $\Rightarrow$ Statement: Use the Euclidean Parallel Postulate to prove that if ℓ and n have a point in common, then they are equal.

6.41. Construct a line through C that is parallel to $\overleftrightarrow{AB}$ and apply the Converse to the Alternate Interior Angles Theorem.

6.46. For parts 1 and 2: First show that $\triangle ABC \cong \triangle BAD$, then show that $\triangle ACD \cong \triangle BDC$.

6.49. Suppose, in part 4, that $BC > AD$. Find a point C' between B and C such that $BC' = AD$. Apply Theorem 6.9.10 to $\square ABC'D$ and apply the Exterior Angle Theorem to $\triangle C'CD$ to obtain a contradiction.

6.50. Let E and F be the feet of the perpendiculars from P and Q to $\overleftrightarrow{AC}$. Let S be the foot of the perpendicular from Q to $\overleftrightarrow{PE}$. Prove that $\triangle AEP \cong \triangle QSP$ and conclude $ES = 2PE$. Use the fact that $\square EFQS$ is a Lambert quadrilateral to reach the desired conclusion.

6.51. To prove that $PE < QF$ (see Figure 6.42), proceed as follows. First prove that $\angle EPA$ is acute, so $\angle EPQ$ is obtuse. If $EP = FQ$, then $\square EFQP$ is a Saccheri quadrilateral. If $EP > FQ$, then there exists a point P' between E and P such that $\square EFQP'$ is a Saccheri quadrilateral. Both possibilities lead to contradictions. For the second part, use the Archimedean Property of Real Numbers together with the previous exercise.

HINTS FOR EXERCISES IN CHAPTER 7

7.2. Put in a diagonal and prove that the resulting triangles are congruent.

7.4. Use Theorem 6.7.9 to prove that the diagonals intersect. Prove that the small triangles you see are congruent in pairs.

7.8. If $AB = DE$, the proof is easy. Otherwise it may be assumed that $AB > DE$. (Explain.) Choose a point B' between A and B such that $AB' = DE$. Let m be the line through B' that is parallel to $\overleftrightarrow{BC}$. Show that m must intersect $\overline{AC}$ at an interior point C'. Prove that $\triangle AB'C' \sim \triangle ABC$. Then use that fact and some algebra to show that $AC' = DF$. Finally, use SAS to prove that $\triangle AB'C' \cong \triangle DEF$.

7.9. Proceed as in the previous proof, but use SSS in place of SAS in the last step.

7.11. Use the proof of the Pythagorean Theorem rather than the statement. In the notation of that proof, you are being asked to prove that $h = \sqrt{xy}$.

7.12. In the notation of the proof of the Pythagorean Theorem, you are being asked to prove that $a = \sqrt{cy}$ and $b = \sqrt{cx}$.

7.13. Construct a second triangle $\triangle DEF$ such that $DF = AC$, $EF = BC$, and $\angle EFD$ is a right angle. Apply the Pythagorean Theorem to $\triangle DEF$; then conclude that $DE = AB$ and $\angle EFD \cong \angle BCA$.

7.14. Drop a perpendicular from C to $\overleftrightarrow{AB}$ and call the foot D. Prove that $b \sin \angle A = CD = a \sin \angle B$. Be careful to make sure your proof covers the cases in which one of $\angle A$ or $\angle B$ is either right or obtuse. The other equation follows similarly.

7.15. For the case in which $\angle C$ is acute: Label vertices so that $\angle A$ is also acute. Drop a perpendicular from B to $\overleftrightarrow{AC}$ and call the foot D. Note that $BD = a \sin \angle C$ and $CD = a \cos \angle C$ so $AD = b - a \cos \angle C$. Apply the Pythagorean Theorem to the right triangle $\triangle ABD$. You must also supply proofs of the cases in which $\angle C$ is right or obtuse.

7.21. Use an RAA argument and Theorem 7.2.7 to show that two of the perpendicular bisectors intersect. Then prove that the point of intersection is also on the third perpendicular bisector.

7.24(a). $\square AFDE$ is a parallelogram; apply Theorem 7.2.10.

7.24(b). Use Exercise 7.22.

7.26. First prove that $\triangle AB'B \sim \triangle AC'C$ and then apply the SAS similarity criterion.

7.29. Assume that the medians concur and that the perpendicular bisectors of the sides concur. Construct the point H' and prove that it lies on each of the altitudes.

7.32(a) The points C, P, and N are collinear Menelaus points for $\triangle ABL$.

7.34. Use the Similar Triangles Theorem to express the ratios from Ceva's Theorems in terms of the lengths of the altitudes.

HINTS FOR EXERCISES IN CHAPTER 8

8.1. Use additivity of defect.

8.7. You must prove the following: If X is a point that is on the same side of $\overleftrightarrow{QC}$ as B' and $\overrightarrow{QX}$ is between $\overrightarrow{QC}$ and $\overrightarrow{QD'}$, then $\overrightarrow{QX}$ must intersect $\overleftrightarrow{CB}$. In order to do so, first prove that $\overrightarrow{PX}$ must intersect $\overleftrightarrow{AB}$ and then use Pasch's Axiom.

8.9. Suppose not. Then there exists a ray $\overrightarrow{AX}$ between $\overrightarrow{AB}$ and $\overrightarrow{AQ}$ such that $\overrightarrow{AX} \cap \overleftrightarrow{QD} = \emptyset$. Drop a perpendicular from P to $\overleftrightarrow{AX}$ and call the foot F. Prove that $PF < PA$ and $\kappa(PF) < \kappa(PA)$. This contradicts Theorem 8.4.6.

8.14(a). Consider Figure 8.14. Prove that A is in the interior of $\angle DPD'$ and prove that A does not lie on any crossbar for $\angle DPD'$.

8.14(b). Construct a line through P that is parallel to $\overleftrightarrow{BC}$.

8.16(c). $\square AA'B'B$ is convex by Theorem 6.7.8.

8.16(g). The angle bisectors of $\triangle DEF$ match the rays $\overrightarrow{PA}$, $\overrightarrow{PB}$, and $\overrightarrow{PC}$.

8.17. Start with an angle of measure $(180 - \delta_0)^\circ$.

8.18. Put in the diagonal $\overline{QF}$. Apply theorems from Chapter 5.

8.19. Let F be the foot of the perpendicular from R to ℓ. Put in the diagonals $\overline{SF}$ and $\overline{TF}$. Prove corresponding triangles congruent.

8.21. You must prove that f is continuous at x for every x. Given $x \in (a, b)$, find a closed interval that contains x in its interior and is contained in (a, b). Apply Lemma 5.7.26 to the function $-f$ defined on that closed interval.

HINTS FOR EXERCISES IN CHAPTER 9

9.2. A point $P \in \blacktriangle ABC$ is either on one of the edges of $\triangle ABC$ or is in the interior of $\triangle ABC$. Explain why, in each case, P must be in either $\blacktriangle ABE$ or in $\blacktriangle EBC$. This proves one-half of the set equality $\blacktriangle ABC = \blacktriangle ABE \cup \blacktriangle EBC$.

9.4. Be careful; you will probably need to use Theorem 9.1.7 in the proof.

9.5. There are three different cases to consider, one in which D is between A and B, one in which D equals either A or B, and one in which D is not in $\overline{AB}$. Again you will probably want to use Theorem 9.1.7 in the proof.

9.6. Fig. E.2 illustrates one of the equalities that must be proved. To show that $(1/2)(AC)(BD) = (1/2)(AB)(EC)$, prove that $\triangle AEC$ and $\triangle ADB$ are similar and apply the Similar Triangles Theorem.

9.7. Use the Similar Triangles Theorem to prove that both the base and height are multiplied by the common ratio.

9.10. Apply Exercise 9.9 three times and subtract.

9.14. If you run stuck on part (c), you might try working the next exercise first. Another possibility is to apply the proof of the Fundamental Theorem of Dissection Theory for Euclidean Geometry to this special case. That does give a solution, but it is considerably more complicated than the one that results

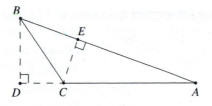

FIGURE E.2: Exercise 9.6

from applying the ideas of the Exercise 9.15.

9.15(a). Label the vertices so that $AB \geqslant BC$. If $AB > 4BC$, cut the rectangle in half and stack the pieces. Repeat as often as necessary.

9.15(b). Let $a = AB$ and $b = BC$. Choose E, F and G so that $\square BEFG$ is a square and $BE = \sqrt{ab}$. Use the Similar Triangles Theorem to prove that $\triangle AGH \cong \triangle KCE$ and then the other pieces should fall into place. Be sure you understand where you used the hypothesis $BC \leqslant AB \leqslant 4BC$. It might be helpful to note that $a \leqslant 4b$ iff $a \leqslant 2\sqrt{ab}$.

9.16. Use either Bhaskara's proof of the Pythagorean Theorem (some subdivision is necessary) or Exercise 9.13.

9.19. Use Theorem 9.3.4 to prove that $\overleftrightarrow{AB} \parallel \overleftrightarrow{MG}$.

9.20. Don't overlook the possibility that $G = E$ or $G = D$.

9.21. You must show that $\overline{BH}$ crosses $\overleftrightarrow{DE}$ at its midpoint. Use the fact that $AE = BD$.

9.22. Use the decomposition of Theorem 9.3.5 and the additivity of defect.

HINTS FOR EXERCISES IN CHAPTER 10

10.5. Suppose P does not lie on $\overleftrightarrow{O_1 O_2}$. Drop a perpendicular from P to $\overleftrightarrow{O_1 O_2}$ and call the foot F. The point Q such that $P * F * Q$ and $PF = FQ$ also lies on both circles.

10.7(a). Use Elementary Circular Continuity (Theorem 10.2.8).

10.7(b). Let $\gamma = \mathcal{C}(O, r)$. Use the first part of this exercise to find a point C such that $\angle OCP$ is a right angle and $OC = r$. Prove that $\overleftrightarrow{PC}$ is one of the tangent lines.

10.8. Let P be a point on ℓ. Use the Pythagorean Theorem to set up a quadratic equation that relates the distance $x = AP$ to the distance OP. Set $OP = r$ and solve for x. Then use the Ruler Postulate to locate the required point on γ. Don't forget to explain why the quadratic equation must have real number solutions.

10.9. For the Euclidean part use Exercise 7.23. For the hyperbolic part, consider a line ℓ and an external point P. Drop a perpendicular from P to ℓ and call the foot A. Choose a point $B \neq A$ on ℓ and a point Q on $\overrightarrow{PA}$ such that $P * A * Q$ and $PA = QA$. Locate points D and E such that $\overrightarrow{PD} \mid \overrightarrow{AB}$ and $\overrightarrow{QE} \mid \overrightarrow{AB}$. Finally, drop a perpendicular from P to a point S on $\overleftrightarrow{QE}$ and prove that $\overline{PS}$ must intersect ℓ at a point R. Prove that the three altitudes of $\triangle PQR$ are parallel.

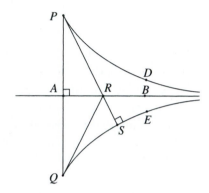

FIGURE E.3: The hyperbolic part of Exercise 10.9

10.10. Apply Theorem 6.4.6.

10.11. Let $\gamma = C(O, r)$ be a circle. Fix a positive integer $n \geq 3$ and a point P_1 on γ. For each i, $2 \leq i \leq n$, define P_i to be a point on γ such that $\mu(P_i O P_{i-1}) = (360/n)^{\circ}$, and P_i and P_{i-2} lie on opposite sides of $\overleftrightarrow{OP}_{i-1}$. Prove that $P_1 P_2 \cdots P_n$ is a regular polygon.

10.12. Choose a point C' on $\overrightarrow{MC}$ so that $MC' = AM$. By Theorem 10.4.1, $\angle AC'B$ is a right angle. Use the Exterior Angle Theorem in an RAA argument to prove that $C = C'$.

10.13. Apply Theorem 10.4.3 and then proceed as in the proof of Theorem 10.4.1.

10.16. Use Exercise 6.27 and Theorem 10.2.7.

10.17. Draw in diagonals for the quadrilateral. By the previous exercise they divide each interior angle of the quadrilateral into two inscribed angles for the circle.

10.18. First dispense with the case in which $\overleftrightarrow{BC}$ is parallel to t. In the other case, let O be the point at which $\overleftrightarrow{BC}$ intersects t. It may be assumed that $B * C * O$. Use Theorem 10.4.12 and the SAS Similarity Criterion to show that $\triangle OAC \sim \triangle OBA$.

10.19. Apply the second part of Exercise 10.6.

10.20. This lemma is neutral. Apply Theorem 6.4.1.

10.21. Use theorems in the chapter to construct the second point.

10.22. Use Circular Continuity.

HINTS FOR EXERCISES IN CHAPTER 11

11.8. Use the previous construction to copy the triangle $\triangle ABC$ and then apply SSS.

11.10 You do not have to prove that the circle has a center because the existence of a center point is part of the definition of a circle. Start with three points on the circle, and use them to locate the center. One way to do so is to identify two lines that must both pass through the center. If you do the construction that way, be sure to prove that the two lines intersect. Euclid's method is to find one line that must contain the center and then use the fact that that line must be a secant line to identify the center of the circle.

11.11 Refer to the proof of Theorem 10.3.8, if necessary.

11.14(a). You should be able to identify two lines on which the center of the circle must lie. Use the Euclidean Parallel Postulate to prove that the two lines intersect.

11.14(b). Start with two parallel lines that do not admit a common perpendicular and work part (a) backward.

11.16 Refer to the proof of Theorem 10.3.4, if necessary.

11.19. Use Exercise 10.18 to get started.

11.26. Use similar triangles to construct ab and a/b.

HINTS FOR EXERCISES IN CHAPTER 12

12.1. To prove onto, let P be a point. Choose two points A and B. If P, $f(A)$, and $f(B)$ are not collinear, then there exists two points C and C' such that $\triangle ABC \cong \triangle f(A)f(B)P$ and $\triangle ABC' \cong \triangle f(A)f(B)P$ (give a reason). Prove that either $f(C) = P$ or $f(C') = P$. If P, $f(A)$, and $f(B)$ are collinear, use a different argument.

12.7. To prove the set equality you need to prove two things. If $X \in T(\overline{AB}) = \{T(P) \mid P \in \overline{AB}\}$, then $X \in \overline{A'B'}$ and if $X \in \overline{A'B'}$ then $X \in \{T(P) \mid P \in \overline{AB}\}$.

12.10. Use the previous parts of the theorem to prove that $T(\angle BAC)$ is an angle. Use SSS to prove that it is congruent to $\angle BAC$.

12.17. Show that the two transformations agree on the three points P, O, and Q and then apply the uniqueness part of Theorem 12.2.8.

12.20 To find t, apply the first part of the proof to R_{BOA} and then apply the previous exercise.

12.22. Let C be the point used in the definition of T_{AB}. You must prove that $T_{AB}(C) = \rho_\ell \circ \rho_m(C)$ for the two lines ℓ and m shown in the figure. You could begin by verifying that $\square BACC'$ is a Saccheri quadrilateral.

12.24(b). Choose P so that $\overleftrightarrow{AP} \perp \overleftrightarrow{AB}$ and use properties of Saccheri quadrilaterals.

12.32(b) Suppose the identity transformation has been written as the composition of an odd number of reflections. Use part (a) to reduce the number of reflections to either one or three. Use properties of reflections and glide reflections to rule out these possibilities.

12.32(c) Use part (b).

12.39. Choose a point $P \in n$, define $P' = \rho_m(\rho_\ell(P))$, and let s be the perpendicular bisector of $\overline{PP'}$. Prove that this line s has the required properties.

12.47. Find a fixed point for $\rho_\ell \circ \rho_m \circ \rho_n$, where ℓ, m, and n are the perpendicular bisectors of the sides.

HINTS FOR EXERCISES IN CHAPTER 13

13.3. Check that f has limit $+\infty$ as B moves away from A in one direction and $-\infty$ in the other. Then use continuity to show that f is onto. As mentioned in the text, you must show that $d(B, C) = |f(B) - f(C)|$. Use properties of the logarithm function to do this.

13.5. Use circular continuity.

13.9. Use the two preceding exercises to show that this is true of circles centered at O. Use reflections to extend the proof to other center points.

HINTS FOR EXERCISES IN CHAPTER 14

14.4. The calculus formula for the curvature of the curve $y = c(x)$ is

$$\kappa(x) = \frac{c''(x)}{(1 + (c'(x))^2)^{3/2}}.$$

14.9(e) The assertion "nearby points on a smooth surface are the endpoints of a shortest geodesic" is quite sloppy; restate it carefully. It really means, "For every point x there exists $\epsilon > 0$ such that if $d(x, y) < \epsilon$, then ...."

14.10. The shortest path in (a) has length 6. The shortest paths in (b) and (c) have length approximately 5.657.

Bibliography

1. Edwin Abbott Abbott, *Flatland with a New Introduction by Thomas Banchoff*, Princeton University Press, Princeton, New Jersey, 1991.

2. Frederic Ancel, *2-dimensional axiomatic geometry*, Abstracts of the American Mathematical Society **25** (2004), 336.

3. _____, *Course Notes for a Course in 2-Dimensional Axiomatic Geometry*, University of Wisconsin Milwaukee, 2004.

4. W. W. Rouse Ball and H. S. M. Coxeter, *Mathematical Recreations and Essays*, 13th ed., Dover Publications, Inc., New York, 1987.

5. Arthur Baragar, *A Survey of Classical and Modern Geometries with Computer Activities*, Prentice Hall, Upper Saddle River, New Jersey, 2001.

6. Allan Berele and Jerry Goldman, *Geometry: Theorems and Constructions*, Prentice Hall, Upper Saddle River, New Jersey, 2001.

7. George David Birkhoff, *A set of postulates for plane geometry*, Annals of Mathematics **33** (1932), 329–345.

8. George David Birkhoff and Ralph Beatley, *Basic Geometry*, Third ed., AMS Chelsea Publishing, American Mathematical Society, Providence, Rhode Island, 2000 (originally published in 1940).

9. Harvey I. Blau, *Foundations of Plane Geometry*, Prentice Hall, Upper Saddle River, New Jersey, 2003.

10. Ethan D. Bloch, *A First Course in Geometric Topology and Differential Geometry*, Birkhäuser, Boston, Basel, Berlin, 1997.

11. Leonard M. Blumenthal, *A Modern View of Geometry*, Dover Publications, Inc., New York, 1980.

12. Conference Board of the Mathematical Sciences CBMS, *The Mathematical Education of Teachers*, Issues in Mathematics Education, Volume 11, The Mathematical Association of America and The American Mathematical Society, Washington, DC, 2001.

13. Judith N. Cederberg, *A Course in Modern Geometries*, second ed., Springer-Verlag, New York and Berlin, 2001.

14. Thomas Clark, *Hyperbolic Geometry and More: A Collection of Projects in non-Euclidean Geometry for High School Students*, Calvin College, Grand Rapids, Michigan, 2004.

15. David Corfield, *Towards a Philosophy of Real Mathematics*, Cambridge University Press, Cambridge, UK, 2003.

16. H. S. M. Coxeter, *Non-Euclidean Geometry*, sixth ed., Mathematical Association of America, Washington, DC, 1998.

17. Mary Crowley, *The van Hiele Model of the Development of Geometric Thought*, Learning and Teaching Geometry, K–12: the Yearbook of the NCTM, National Council of Teachers of Mathematics, Reston, Virginia, 1987, p. 1–16.

18. Dana Densmore (ed.), *The Bones*, Green Lion Press, Santa Fe, New Mexico, 2002.

19. Dana Densmore (ed.), *Euclid's Elements*, Green Lion Press, Santa Fe, New Mexico, 2002.

20. Clayton W. Dodge, *Euclidean Geometry and Transformations*, Dover Publications, Inc., Mineola, New York, 2004.

21. Underwood Dudley, *A Budget of Trisections*, Springer-Verlag, New York, Heidelberg, Berlin, 1987.

22. _____ , *Mathematical Cranks*, Mathematical Association of America, Washington, DC, 1992.

23. William Dunham, *Journey Through Genius: The Great Theorems of Mathematics*, Penguin Books, New York, 1991.

24. _____ , *Euler: The Master of Us All*, Mathematical Association of America, Washington, DC, 1999.

25. Howard Eves, *A Survey of Geometry*, revised ed., Allyn and Bacon, Boston, 1972.

26. _____ , *Modern Elementary Geometry*, Jones and Bartlett, Boston and London, 1992.

27. Marvin Jay Greenberg, *Euclidean and Non-Euclidean Geometries: Development and History*, third ed., W. H. Freeman and Company, New York, 1994.

28. Brian Greene, *The Elegant Universe: Superstrings, Hidden Dimensions, and the Quest for the Ultimate Theory*, W.W. Norton & Company, New York, 1999.

29. Robin Hartshorne, *Geometry: Euclid and Beyond*, Springer-Verlag, New York, 2000.

30. Sir Thomas L. Heath, *The Thirteen Books of Euclid's Elements with Introduction and Commentary*, Dover Publications, Inc., Mineola, New York, 1956.

31. David W. Henderson, *Experienceing Geometry in Euclidean, Spherical, and Hyperbolic Spaces*, second ed., Prentice Hall, Upper Saddle River, New Jersey, 2001.

32. _____ , *Review of "Geometry: Euclid and Beyond" by Robin Hartshorne*, Bulletin of the American Mathematical Society **39** (2002), 563–571.

33. D. Hilbert and S. Cohn-Vossen, *Geometry and the Iagination*, Chelsea Publishing Company, New York, 1952.

34. David Hilbert, *Foundations of Geometry*, Open Court Publishing Company, La Salle, Illinois, 1971.

35. Audun Holme, *Geometry: Our Cultural Heritage*, Springer-Verlag, Berlin, 2002.

36. H. Hu, *Review of "Geometry: Our Cultural Heritage" by Audun Holme*, Notices of the American Mathematical Society **51** (2004), 529–537.

37. Victor J. Katz, *A History of Mathematics*, Addison-Wesley, Reading, Massachusetts, 1998.

38. David C. Kay, *College Geometry: A Discovery Approach*, second ed., Addison Wesley Longman, Boston, 2001.

39. Clark Kimberling, *Geometry in Action*, Key College Publishing, Emeryville, CA, 2003.

40. Morris Kline, *Mathematics in Western Culture*, Oxford University Press, New York, 1953.

41. _____ , *Mathematics: The Loss of Certainty*, Oxford University Press, New York, 1982.

42. Donald Knuth, *Mathematical Writing*, Mathematical Association of America, 1990, MAA Notes, Number 14.

43. Robert P. Langlands, *Euclid's windows and our mirrors*, Notices of the American Mathematical Society **49** (2002), 554–564.

44. Elisha Scott Loomis, *The Pythagorean Proposition*, Classics in Mathematics Education, National Council of Teachers of Mathematics, Ann Arbor, Michigan, 1972.

45. Jean-Pierre Luminet, Jeffrey Weeks, Alain Riazuelo, Rolad Lehoucq, and Philippe Uzan, *Dodecahedral space topology as an explanation for weak wide-angle temperature correlations in the cosmic background*, Nature **425** (2003), 593–595.

46. Saunders MacLane, *Metric postulates for plane geometry*, American Mathematical Monthly **66** (1959), 543–555.

47. John McCleary, *Geometry from a Differentiable Viewpoint*, Cambridge University Press, Cambridge and New York, 1994.

48. James E. McClure, *Start where they are: geometry as an introduction to proof*, American Mathematical Monthly **107** (2000), 44–52.

49. Richard S. Millman and George D. Parker, *Geometry: A Metric Approach Using Models*, second ed., Springer-Verlag, New York, Berlin, Heidelberg, 1991.

50. Leonard Mlodinow, *Euclid's Window*, The Free Press, New York, 2001.

51. Edwin E. Moise, *Elementary Geometry from an Advanced Standpoint*, third ed., Addison-Wesley Publishing Company, Reading, Massachusetts, 1990.

52. Edwin E. Moise and Floyd L. Downs, *Geometry*, Addison-Wesley Publishing Company, Reading, Massachusetts, 1991.

53. José Maria Montesinos, *Classical Tessellations and Three-Manifolds*, Springer-Verlag, Berlin, 1987.

54. National Council of Teachers of Mathematics NCTM, *Principles and Standards for School Mathematics*, NCTM, 2000, available online at http://standards.nctm.org/document/index.htm.

55. M. Helena Noronha, *Euclidean and Non-Euclidean Geometries*, Prentice Hall, Upper Saddle River, New Jersey, 2002.

56. Barrett O'Neill, *Elementary Differential Geometry*, second ed., Academic Press, New York, 1997.

57. Roger Penrose, *The Geometry of the Universe*, Mathematics Today: Twelve Informal Essays, Springer-Verlag, New York, Heidelberg, Berlin, 1978, p. 83–125.

58. Henri Poincaré, *Science and Hypothesis*, Dover Publications, Inc., New York, 1952.

59. Alfred S. Posamentier, *Advanced Euclidean Geometry*, Key College Publishing, Emeryville, California, 2002.

60. Chris Pritchard (ed.), *The Changing Shape of Geometry*, Cambridge University Press, Cambridge, UK, 2003.

61. Bertrand Russell, *An Essay on the Foundations of Geometry*, Routledge, London and New York, 1996.

62. Girolamo Saccheri, *Euclides Vindicatus*, Chelsea Publishing Company, New York, 1986, translated by George Bruce Halstead.

63. James R. Smart, *Modern Geometries*, fifth ed., Brooks/Cole Publishing Company, Pacific Grove, California, 1998.

64. School Mathematics Study Group SMSG, *Geometry, Parts 1 and 2 (Student Text)*, Yale University Press, New Haven and London, 1961.

65. Michael Spivak, *A Comprehensive Introduction to Differential Geometry*, second ed., Publish or Perish, Berkeley, 1979.

66. Saul Stahl, *The Poincaré Half-Plane: A Gateway to Modern Geometry*, Jones and Bartlett, Boston and London, 1993.

67. _____ , *Geometry from Euclid to Knots*, Prentice Hall, Upper Saddle River, New Jersey, 2003.

68. Sherman Stein, *Archimedes: What Did He Do Besides Cry Eureka?*, Mathematical Association of America, Washington, DC, 1999.

69. William P. Thurston, *Three-Dimensional Geometry and Topology*, vol. 1, Princeton University Press, Princeton, New Jersey, 1997, edited by Silvio Levy.

70. Richard J. Trudeau, *The Non-Euclidean Revolution*, Birkhäuser, Boston, 2001.

71. The University of Chicago School Mathematics Project UCSMP, *Geometry, Parts I and II*, second ed., Prentice Hall, Glenview, Illinois, 2002, Teacher's Edition.

72. Edward C. Wallace and Stephen F. West, *Roads to Geometry*, second ed., Prentice Hall, Upper Saddle River, New Jersey, 1998.

73. Jeffrey R. Weeks, *The Shape of Space*, second ed., Marcel Dekker, Inc., New York and Basel, 2002.

74. _____ , *The Poincaré dodecahedral space and the mystery of the missing fluctuations*, Notices of the American Mathematical Society **51** (2004), 610–619.

Index